W9-BZJ-723

A TRANSITION TO
MATHEMATICS
WITH PROOFS

The Jones & Bartlett Learning Series in Mathematics

Geometry

Geometry with an Introduction to Cosmic Topology
Hitchman (978-0-7637-5457-0)
© 2009

Euclidean and Transformational Geometry: A Deductive Inquiry
Libeskind (978-0-7637-4366-6)
© 2008

A Gateway to Modern Geometry: The Poincaré Half-Plane, Second Edition
Stahl (978-0-7637-5381-8) © 2008

Understanding Modern Mathematics
Stahl (978-0-7637-3401-5) © 2007

Lebesgue Integration on Euclidean Space, Revised Edition
Jones (978-0-7637-1708-7) © 2001

Precalculus

Precalculus: A Functional Approach to Graphing and Problem Solving, Sixth Edition
Smith (978-1-4496-4916-6) © 2013

Precalculus with Calculus Previews, Fifth Edition
Zill/Dewar (978-1-4496-4912-8)
© 2013

Essentials of Precalculus with Calculus Previews, Fifth Edition
Zill/Dewar (978-1-4496-1497-3)
© 2012

Algebra and Trigonometry, Third Edition
Zill/Dewar (978-0-7637-5461-7)
© 2012

College Algebra, Third Edition
Zill/Dewar (978-1-4496-0602-2)
© 2012

Trigonometry, Third Edition
Zill/Dewar (978-1-4496-0604-6)
© 2012

Calculus

Multivariable Calculus
Damiano/Freije (978-0-7637-8247-4)
© 2012

Single Variable Calculus: Early Transcendentals, Fourth Edition
Zill/Wright (978-0-7637-4965-1)
© 2011

Multivariable Calculus, Fourth Edition
Zill/Wright (978-0-7637-4966-8)
© 2011

Calculus: Early Transcendentals, Fourth Edition
Zill/Wright (978-0-7637-5995-7)
© 2011

Calculus: The Language of Change
Cohen/Henle (978-0-7637-2947-9)
© 2005

Applied Calculus for Scientists and Engineers
Blume (978-0-7637-2877-9) © 2005

Calculus: Labs for Mathematica
O'Connor (978-0-7637-3425-1)
© 2005

Calculus: Labs for MATLAB®
O'Connor (978-0-7637-3426-8)
© 2005

Linear Algebra

Linear Algebra: Theory and Applications, Second Edition
Cheney/Kincaid
(978-1-4496-1352-5) © 2012

Linear Algebra with Applications, Seventh Edition
Williams (978-0-7637-8248-1)
© 2011

Linear Algebra with Applications, Alternate Seventh Edition
Williams (978-0-7637-8249-8)
© 2011

Advanced Engineering Mathematics

A Journey into Partial Differential Equations
Bray (978-0-7637-7256-7)
© 2012

Advanced Engineering Mathematics, Fourth Edition
Zill/Wright (978-0-7637-7966-5)
© 2011

An Elementary Course in Partial Differential Equations, Second Edition
Amaranath (978-0-7637-6244-5)
© 2009

Complex Analysis

Complex Analysis for Mathematics and Engineering, Sixth Edition
Mathews/Howell
(978-1-4496-0445-5) © 2012

The Jones & Bartlett Learning
Series in Mathematics (*Continued*)

*A First Course in Complex Analysis with Applications,
Second Edition*
Zill/Shanahan (978-0-7637-5772-4)
© 2009

Classical Complex Analysis
Hahn (978-0-8672-0494-0) © 1996

Real Analysis

Elements of Real Analysis
Denlinger (978-0-7637-7947-4)
© 2011

*An Introduction to Analysis,
Second Edition*
Bilodeau/Thie/Keough
(978-0-7637-7492-9) © 2010

Basic Real Analysis
Howland (978-0-7637-7318-2)
© 2010

*Closer and Closer: Introducing
Real Analysis*
Schumacher (978-0-7637-3593-7)
© 2008

*The Way of Analysis, Revised
Edition*
Strichartz (978-0-7637-1497-0)
© 2000

Topology

*Foundations of Topology,
Second Edition*
Patty (978-0-7637-4234-8)
© 2009

Discrete Mathematics and Logic

*Essentials of Discrete Mathematics,
Second Edition*
Hunter (978-1-4496-0442-4) © 2012

*Discrete Structures, Logic, and
Computability, Third Edition*
Hein (978-0-7637-7206-2) © 2010

*Logic, Sets, and Recursion,
Second Edition*
Causey (978-0-7637-3784-9) © 2006

Numerical Methods

Numerical Mathematics
Grasselli/Pelinovsky
(978-0-7637-3767-2) © 2008

*Exploring Numerical Methods:
An Introduction to Scientific
Computing Using MATLAB®*
Linz (978-0-7637-1499-4) © 2003

Advanced Mathematics

*A Transition to Mathematics
with Proofs*
Cullinane (978-1-4496-2778-2)
© 2013

Mathematical Modeling with Excel®
Albright (978-0-7637-6566-8) © 2010

*Clinical Statistics: Introducing
Clinical Trials, Survival Analysis, and
Longitudinal Data Analysis*
Korosteleva (978-0-7637-5850-9)
© 2009

*Harmonic Analysis:
A Gentle Introduction*
DeVito (978-0-7637-3893-8) © 2007

*Beginning Number Theory,
Second Edition*
Robbins (978-0-7637-3768-9)
© 2006

A Gateway to Higher Mathematics
Goodfriend (978-0-7637-2733-8)
© 2006

*For more information on this series
and its titles, please visit us online
at http://www.jblearning.com.
Qualified instructors, contact
your Publisher's Representative at
1-800-832-0034 or info@jblearning
.com to request review copies for
course consideration.*

The Jones & Bartlett Learning International Series in Mathematics

A Transition to Mathematics with Proofs
Cullinane (978-1-4496-2778-2) © 2013

Functions of Mathematics in the Liberal Arts
Johnson (978-0-7637-8116-3) © 2014

Linear Algebra: Theory and Applications, Second Edition, International Version
Cheney/Kincaid (978-1-4496-2731-7) © 2012

Multivariable Calculus
Damiano/Freije (978-0-7637-8247-4) © 2012

Complex Analysis for Mathematics and Engineering, Sixth Edition, International Version
Mathews/Howell (978-1-4496-2870-3) © 2012

A Journey into Partial Differential Equations
Bray (978-0-7637-7256-7) © 2012

Association Schemes of Matrices
Wang/Huo/Ma (978-0-7637-8505-5) © 2011

Advanced Engineering Mathematics, Fourth Edition, International Version
Zill/Wright (978-0-7637-7994-8) © 2011

Calculus: Early Transcendentals, Fourth Edition, International Version
Zill/Wright (978-0-7637-8652-6) © 2011

Real Analysis
Denlinger (979-0-7637-7947-4) © 2011

Mathematical Modeling for the Scientific Method
Pravica/Spurr (978-0-7637-7946-7) © 2011

Mathematical Modeling with Excel®
Albright (978-0-7637-6566-8) © 2010

An Introduction to Analysis, Second Edition
Bilodeau/Thie/Keough (978-0-7637-7492-9) © 2010

Basic Real Analysis
Howland (978-0-7637-7318-2) © 2010

For more information on this series and its titles, please visit us online at http://www.jblearning.com. Qualified instructors, contact your Publisher's Representative at 1-800-832-0034 or info@jblearning.com to request review copies for course consideration.

A TRANSITION TO
MATHEMATICS
WITH PROOFS

MICHAEL J. CULLINANE

Keene State College

JONES & BARTLETT
LEARNING

World Headquarters
Jones & Bartlett Learning
5 Wall Street
Burlington, MA 01803
978-443-5000
info@jblearning.com
www.jblearning.com

Jones & Bartlett Learning books and products are available through most bookstores and online booksellers. To contact Jones & Bartlett Learning directly, call 800-832-0034, fax 978-443-8000, or visit our website, www.jblearning.com.

Substantial discounts on bulk quantities of Jones & Bartlett Learning publications are available to corporations, professional associations, and other qualified organizations. For details and specific discount information, contact the special sales department at Jones & Bartlett Learning via the above contact information or send an email to specialsales@jblearning.com.

Production Credits
Publisher: Cathleen Sether
Senior Acquisitions Editor: Timothy Anderson
Managing Editor: Amy Bloom
Director of Production: Amy Rose
Senior Marketing Manager: Andrea DeFronzo
V.P., Manufacturing and Inventory Control: Therese Connell
Composition: Cenveo Publisher Services
Cover Design: Kristin E. Parker
Cover Image: © Comstock Images/age fotostock
Printing and Binding: Malloy, Inc.
Cover Printing: Malloy, Inc.

Library of Congress Cataloging-in-Publication Data
Cullinane, Michael J.
  A transition to mathematics with proofs / Michael J. Cullinane.
      p. cm.
  Includes index.
  ISBN-13: 978-1-4496-2778-2 (casebound)
  ISBN-10: 1-4496-2778-1 (casebound)
 1. Logic, Symbolic and mathematical. 2. Mathematics. I. Title.
  QA9.C84 2013
  511.3—dc23

                            2011039616

6048

Printed in the United States of America
15 14 13 12 11   10 9 8 7 6 5 4 3 2 1

Dedication

To my mother and the memory of my father.

Contents

Preface

Many colleges and universities now require a "transition" course for their mathematics majors, in which students begin to move beyond the primarily procedural methods of their calculus courses toward the more abstract and conceptual environment found in more advanced mathematics courses. These transition courses tend to emphasize mathematical rigor and are usually designed to help students learn how to develop and write mathematical proofs. This book was written for use in such courses, but with considerable attention paid to the developmental needs of students. In particular, a deliberate attempt has been made to write a text that is generally readable by students.

The text addresses key topics such as set theory, number systems, logic, relations, functions, and induction, which are usually covered in a transition course. The text also includes material on counting and graph theory so that it can be used in an introductory discrete mathematics course. The overall approach of the text is developmental. For instance, the central topic of proof is foreshadowed and motivated throughout the first three chapters, but proof strategies and structuring mechanisms are not presented until Chapter 4. Students gradually become aware of the need for rigor, proof, and precision as the formal treatment of mathematics is initiated in the early chapters through careful consideration of the roles played by mathematical definitions and axioms, logical form, and quantifiers. Generally, the text precedes the formal by the informal and the abstract by the concrete. Key mathematical ideas are motivated through examples before the consideration of general results and proofs involving the ideas. Proof techniques and strategies are thoroughly discussed so that the underlying logic behind them is made transparent.

Closely related to the developmental nature of the text are the pedagogical supports the text offers students. Each section of Chapters 2 through 6 begins with a set of *guided reading questions* intended to help students in identifying the most significant points being made in the section. Each of these sections also incorporates embedded *practice problems* designed so that students can actively work with a key idea that has just been introduced. Each chapter includes a large set of *problems* from which homework assignments can be created. The problems range in difficulty from straightforward exercises to relatively challenging proof situations.

The essential core of the book is formed by Chapters 1 through 6. Most instructors will want to cover most of the material in these chapters. Chapter 1 provides an overview of the context in which students using the book will be working. Chapter 2 initiates the study of sets and number systems, emphasizing associated vocabulary and notations, and considers the important role played by axioms in a formal mathematical study. Chapter 3 covers relevant topics in logic, such as validity of an argument, and the meaning and use of quantifiers. Chapter 4 contains fundamental material on proof development and writing. The proofs included here, together with those of the associated problem assignments, relate to students' prior work with sets and number systems in Chapter 2. Students will already be familiar with the facts about numbers being proved here, which allows them to focus on developing rigorous arguments that are structured according to the writing standards adopted by the general mathematical community. On the other hand, students' work with proofs in the less familiar area of set theory gives them a chance to learn mathematics that is new to them and to gain experience with mathematical abstractions. The text promotes set theory as a language for communicating mathematics so that students will see inherent value in learning about sets. Chapter 5 covers relations and functions, with special attention paid to equivalence relations, injections, surjections, and inverses. Chapter 6 focuses on the natural numbers and induction, and includes a section on elementary number theory and another section on basic counting techniques and combinatorial reasoning.

Chapter 7 is not part of the core. It provides introductions to three mathematical topic areas (graph theory, binary operations and groups, and set cardinality) that are sometimes addressed in transition courses. Each section of this chapter is independent of the others, so they may be taken up in any order. The pacing of the material is more deliberate than in the earlier chapters as the student is now expected to play a greater role, via strategically embedded problems, in developing the mathematics here. In fact, it is possible to use one or more of the sections in Chapter 7 as the basis for student projects and to assess the degree to which students have begun to make the transition around which courses using the text are focused.

Complete solutions to the problems found in the text and a set of PowerPoint lecture outlines are available for instructor download. Students can access answers to selected problems that do not require a proof. Please visit go.jblearning.com/Cullinane to access these resources.

I would like to thank my colleagues at Keene State College for their support during the development of this project. I am especially grateful to Professor Karen Stanish for the valuable feedback she provided based on her use of early versions of the manuscript in her introductory proof classes. Thanks are also due to the many individuals at Jones & Bartlett Learning who helped this project reach fruition. It has been a pleasure working with all of these dedicated professionals, but I would like to particularly note the contributions of Tim Anderson, Senior Acquisitions Editor; Amy Bloom, Managing Editor; and Amy Rose, Director of Production, whose guidance, suggestions, feedback, and support have greatly improved the finished product.

Chapter 1

Mathematics and Mathematical Activity

This chapter provides context and purpose for the mathematical activity you will undertake throughout the rest of the text. We address the nature of mathematics itself, along with mathematical problem solving, mathematical reasoning, and mathematical writing. A discussion of proof is initiated that will thread its way through the entire text. Suggestions for effective reading of a mathematics textbook are also offered.

1.1 What Is Mathematics?

Mathematics is sometimes referred to as *the study of quantity and space*. For instance *the study of quantity* could certainly include the notions of number and arithmetic and, at least, some aspects of algebra and calculus, and *the study of space* could encompass geometry and parts of calculus. But is every significant mathematical idea intrinsically tied to the notions of number or geometry? Is this what mathematics is about in the most universal sense? Is understanding that there is a pattern in the change of seasons over the course of a year mathematical? What about understanding that this pattern repeats itself from year to year?

A somewhat broader characterization of mathematics views the subject as *the science of patterns*. Think about your own experiences with mathematics: Solving an equation, constructing a function to model a physical situation, and performing a calculation, whether by hand or by computer, all involve patterns of some sort.

However, equating mathematics with the study of patterns probably does not adequately incorporate the impact of the subject on individuals and society. For instance, mathematics allowed for the development of the computer and the atomic bomb, two devices that have greatly altered the course of human history.

Mathematics is also used on a daily basis by people engaged in an amazingly wide variety of activities, from financial planning and product marketing to political polling and earthquake prediction. Given its utility and impact, it would be convenient to have a conception of mathematics that not only tells what mathematics is but also suggests how it is used by people and why it is important to us. One possible step in this direction is the notion that *mathematics seeks to assist in the creation of fundamental awareness and understanding.* This perspective yields an image of mathematics as the uncovering of truths, making clear what had been unclear, or creating simplicity out of complexity.

A key step that is often used in mathematics to gain greater awareness and develop better understanding is a process known as *abstraction*. In this process, some of the details or specifics of the problem situation being analyzed are stripped away so that the most fundamental concepts, those lying at the heart of the situation, may be brought into focus. Then, what can at first seem to be vastly different problems from quite distinct disciplines of study may begin to emerge as related in some ways. Often, what is revealed in the abstraction process is a mathematical foundation on which the situation being explored rests. For example, a certain abstract notion of distance studied in the branch of mathematics known as *topology* has been used to model both computer calculations and genetic mutations. By learning more about this particular mathematical concept, we may wind up gaining greater understanding of some important issues in computer science and biology (and perhaps other areas of study as well). In fact, the ability of mathematics to mesh with so many fields of study, and yet simultaneously exist as an entity unto itself, may be the subject's most distinctive feature.

In working your way through this book, you will have the opportunity to study some of the most significant and widely applied ideas in mathematics, ideas that underlie phenomena observed and studied throughout the many different fields of human endeavor. Some of these ideas are very old, whereas others were developed relatively recently. Some are concrete, others more abstract. As you interact with more of them, you can expect your own perception of mathematics to grow to accommodate the experience you will gain.

1.2 Mathematical Research and Problem Solving

Research in many academic fields, including mathematics, is vitally connected to the act of solving problems within the field. Now by *problem solving* we do not mean the mere implementation of an already existing solution or solution strategy but rather the actual development of a solution or solution strategies.

Recall for instance that any quadratic equation $ax^2 + bx + c = 0$ can be solved using the Quadratic Formula,

$$x = \frac{-b \pm \sqrt{b^2 - 4ac}}{2a}.$$

From our perspective, solving a particular quadratic equation such as $2x^2 + 7x = 4$ by using the Quadratic Formula would not be considered true problem solving because all that is required is the implementation of an existing formula. In contrast, deriving the Quadratic Formula itself, without being given a suggested strategy for doing so, would be real problem solving.

As a mathematics student, part of the process by which you learn mathematics that is new to you involves doing some research. Before you work on an assigned homework problem, you might consult your class notes or textbook for information pertaining to the problem. You might also need to make use of some prior mathematical knowledge in working out your solution. You might find yourself asking questions, conducting experiments, making informed guesses, looking for evidence, and seeking explanations. These are all attributes of a good researcher and a good problem solver.

As you progress through this book, you will learn to rely less on prepackaged procedures and more on your own problem-solving efforts. There will still be "standard approaches" for you to learn and practice, but not all problems will readily yield to them, and sometimes you will have to adapt what you are learning to a new situation.

1.3   An Example of a Mathematical Research Situation

We now provide an example of a specific mathematical research situation. Our focus will be *prime numbers*. Though you are probably already very familiar with this concept, we will review some mathematical vocabulary to make sure that we are communicating clearly. You will note that some terms in what follows appear in boldface type. Throughout this book, boldface is used to indicate that a term is being defined.

The numbers 1, 2, 3, and so on, used for counting, are called **counting numbers**. When counting numbers are multiplied together to obtain another counting number, the result is referred to as the **product** of the numbers we started with. Each of the numbers being multiplied to obtain the product is called a **factor** of the product. For example, since $4 \cdot 7 = 28$, we can say that 28 is the product of 4 and 7, and that both 4 and 7 are factors of 28.

A counting number is **prime** provided that it has exactly two counting number factors. Each of the counting numbers 2, 3, and 5 has exactly two counting number factors, the given number itself and the number 1; they are the three smallest prime numbers. The number 1 is not prime because 1 is its only counting number factor. The number 4 is not prime because it has three counting number factors; namely, 1, 2, and 4.

You might remember a general method for determining whether a given counting number is prime (or you might be able to figure out a method on your own right now). However, even if we have a method for identifying primes, you probably realize that the identification process could become time consuming (even for a computer) as the numbers being tested become larger. This leads us to the following research question:

How can we determine whether a given counting number is prime, and are there any ways to streamline the classification process?

We will not attempt to answer this question right now because our goal at the moment is simply to indicate how a research agenda might be put together, in part so that you will see your own activity in this book from Chapter 2 forward as taking place in a sort of "apprentice research" environment.

How might we develop additional research questions? One often fruitful method is to look at specific examples and see if we can discern any patterns. For instance, examining the following list of prime numbers less than 50,

$$2, 3, 5, 7, 11, 13, 17, 19, 23, 29, 31, 37, 41, 43, 47,$$

we might make the following observations and pose the following questions:

- It appears that primes other than 2 are odd. Is this really the case?
- Some primes seem to occur as a pair of consecutive odd numbers; for instance, 3 and 5, 5 and 7, and 11 and 13. Does this pattern continue to hold as we look at larger primes? If not, when does it stop? Also, except for the pair consisting of 3 and 5, the counting number in between the primes in a pair of consecutive odd primes appears to be divisible by 6. Will this pattern continue for larger pairs of consecutive odd primes?
- The counting numbers 3, 5, and 7 form a triple of consecutive odd primes. Are there any other such "prime triplets?"

Mathematical training itself can also lead us to formulate certain kinds of questions. Mathematicians tend to wonder, for instance, about *how many* things there are of a certain type. Along these lines, then, we might ask how many primes there are. If there are only finitely many, what is the largest one? If there are infinitely many, how would we show that they "go on forever?"

One thing many research questions have in common is the lack of either an immediate answer or an obvious explanation. This feature actually makes them attractive to mathematicians! For instance, it is not obvious that there are infinitely many primes, though this fact has been known since at least the time of ancient Greece. Moreover, at any given time in any given mathematical research area, there will usually be certain unresolved or "open" problems that tend to be the focus of considerable effort and activity. If such a problem remains open for a long period of time, it may gain fame within the mathematical community. One famous open problem in the study of prime numbers is the one we posed asking whether there are infinitely many pairs of consecutive odd primes (the answers to all of our other research questions have been determined).

1.4   Conjectures and Theorems

A **conjecture** or **claim** is an educated guess based on some evidence that has been collected. In mathematics, a conjecture becomes what is known as a **theorem** when a *proof* is found for the conjecture. A **proof** is a detailed argument, usually proceeding in an ordered sequence of steps, which uses logical reasoning to move from given assumptions to a new conclusion. A proof for a theorem can be viewed as providing evidence supporting the *truth* of the theorem; that is, a proof explains why the theorem is *true*. Two of the major activities in which you will take part throughout this book are the formulation of conjectures and the development of proofs that we hope will turn these conjectures into theorems.

A mathematical conjecture can be viewed as hypothesizing an answer to an open question in a particular mathematical research area. For example, the so-called *Twin Prime Conjecture* hypothesizes that there are infinitely many pairs of consecutive odd primes. Even though there is considerable numerical evidence suggesting the Twin Prime Conjecture is true, it is still possible that the conjecture is false, as no proof for it has yet been found. In other words, whether the Twin Prime Conjecture is actually a theorem is not yet known.

In contrast, because we will eventually be able to prove that all primes other than 2 are odd, this conjecture is really a theorem. Some mathematical theorems are, in fact, so famous that many people who are not directly involved in the study of mathematics are familiar with them. Probably the most famous example of a mathematical theorem is the *Pythagorean Theorem*, which can be stated as follows: *In any right triangle, the square of the hypotenuse c is equal to the sum of the squares of the legs a and b; in other words, with the given interpretations of a, b, and c, $a^2 + b^2 = c^2$.*

1.5   Methods of Reasoning

Many of us first encounter mathematical proofs in our high school geometry classes. You may recall from such an experience that a proof makes use of what is called **deductive reasoning**. Each step in a proof is either part of the initial assumptions (i.e., **premises**) on which the proof is based or else can be deduced (using principles of logic) from earlier steps in the proof or from previously proved theorems. Because no gaps are allowed in a deductive argument such as a mathematical proof, the ultimate conclusion drawn may be viewed as inevitable under the assumption that the initial premises are all true.

To illustrate, suppose that it is known that a and b are numbers and $a = b$. Assume it has also been proved that "equals added to equals yield equals." Then we may deduce that $a + 2 = b + 2$. Note that we are not saying that $a + 2 = b + 2$ in isolation from any other facts. What we are saying is that under the assumption that $a = b$ is true, we may use what we are citing as a previously proved theorem, "equals added to equals yields equals," to logically deduce the conclusion $a + 2 = b + 2$. Whether the assumption $a = b$ is true is open to question, but if $a = b$ really is true and addition really preserves equality, then $a + 2 = b + 2$ must also be true.

It is instructive to contrast the deductive (i.e., proof-oriented) foundation for truth in mathematics with the approach used by natural scientists in building scientific theories. Charles Darwin's Theory of Natural Selection, which provides the basis for human evolution from other life forms, is one example of a scientific theory. Another is Albert Einstein's Theory of General Relativity. Scientists base their conclusions on the so-called *scientific method*. In this method experiments are conducted, and the results of the experiments are then used as evidence to promote or refute a given conjecture. By replicating the results of an experiment, a scientist becomes more convinced that the conjecture being tested should be accepted as part of a scientific theory. There is always, however, the possibility that new evidence will force the revision, or even the rejection, of a scientific theory. For instance, various aspects of Isaac Newton's theories concerning gravity and motion have been revised in light of Einstein's relativity theory. Moreover, a revised scientific theory is itself subject to revision and, perhaps, rejection. In addition, a scientific theory is *not* subject to proof. For example, it is not possible, nor will it ever be possible, to *prove* the Theory of Natural Selection, though there is considerable scientific evidence for this theory within the field of biology.

But in mathematics, once we have a theorem, that is, a "proven conjecture," the theorem is not capable of being rejected. For this reason, many mathematical results from long ago, for example the Pythagorean Theorem (which goes back at least to the time of ancient Greece), are still important today.

Often, a mathematical conjecture emerges when we notice a pattern after examining specific examples or cases. The form of reasoning in which we generalize from particular examples is called **inductive reasoning** and is crucial to the process of trying to discover what might be true. For instance, once we observe enough examples involving the addition of even counting numbers 2, 4, 6, 8, and so forth, we might formulate the following conjecture:

The sum of two even counting numbers is always even.

In doing so, we have used inductive reasoning to make a generalization based on what we observed with specific examples. However, to establish this conjecture as a theorem, we would need to develop a proof for it. Because there are infinitely many even counting numbers, it would not be enough simply to cite a few, or even many, examples in which two even counting numbers are being added and the resulting sum turns out to be even. We would *never* be able to consider all possible examples of the addition of even counting numbers. Whereas inductive reasoning has enabled us to discover something that might be true, we would need to use deductive reasoning to argue that our conjecture really is true.

Inductive reasoning is the basis for the scientific method. Because mathematics makes extensive use of this sort of reasoning, it is possible to view mathematics as a science. But we want to emphasize that mathematics distinguishes itself from the other sciences by its ability to draw upon proof. Mathematics does not confirm its results via inductive evidence, even if that evidence is considerable. Inductive analysis is often what leads mathematicians to their conjectures, but nothing less than a proof can establish a conjecture as a theorem.

1.6   Why Do We Need Proofs?

One reason for requiring a proof for a mathematical conjecture is fairly obvious. Unless we have a proof, we do not *really* know that our conjecture is true. No matter what kind of a hunch we might have about the truth of a conjecture, no matter what our prior experience with the mathematical ideas bound up in a seemingly true conjecture, it is possible that the conjecture is false. Once we have a proof, though, we have eliminated the possibility of the conjecture being false.

Students sometimes object to reading and writing proofs for mathematical results they already know to be true or for which proofs already exist. But proofs are also important because they can provide insights that help us to better understand the mathematics we are studying. Though the Pythagorean Theorem is a well-known mathematical result, its truth is not readily apparent. With no prior knowledge of the content of this theorem, one is unlikely to look at a drawing of a right triangle and exclaim, "Of course! $a^2 + b^2 = c^2$." However, exploring a proof of the Pythagorean Theorem can help us to better understand why and how this famous equation "works," how it generalizes to the Law of Cosines for an arbitrary triangle, under what circumstances it might be usefully applied, and so on.

Thus, proof plays a dual role in mathematics:

- Proof is used to establish the truth of mathematical conjectures (i.e., proof in the role of helping us to be certain of our mathematical claims).

- Proof is used to gain insight into mathematical concepts (i.e., proof in the role of helping us better to understand mathematics).

1.7 Mathematical Writing

To communicate mathematics and our understanding of mathematics, we speak, demonstrate with physical models, draw pictures, and create computer-generated images. The great variety of forms in which mathematical communication can occur allows us to approach a mathematical concept from multiple perspectives. Ultimately, though, any mathematical idea we wish to study and communicate about must be expressible in written form. We write in mathematics in order to

- state mathematical assumptions, definitions, conjectures, and theorems;
- discuss problem-solving strategies and problem solutions;
- describe the methodology and results of mathematical investigations; and
- convey mathematical arguments, usually in the form of proofs.

Because these are our writing objectives, the mathematical community values writing that is clear, precise, concise, and logically organized.

Mathematics admits considerable use of symbolic expressions, so we also agree to allow ourselves to integrate appropriate symbolism into our writing. This does not, however, give us free reign to write in an unorganized fashion or to abuse mathematical notation. Even when we incorporate symbols into our writing, we should still use English sentence structures and organize our sentences into paragraphs. In fact, mathematical writing, with the exception of a few discipline-specific conventions such as the frequent inclusion of symbols, tends to adhere to the traditional expectations of quality academic writing.

For instance, if we wish to communicate the fact that the equation $x^2 = 4$ has 2 and −2 as its only solutions, we should write a sentence such as

The only solutions to the equation $x^2 = 4$ are 2 and −2

rather than isolated symbolic expressions like

$$x^2 = 4$$
$$x = 2, -2$$

on our paper. This is not to say that when we are writing informally, say in developing ideas for a problem solution or a proof on "scratch paper," we must avoid such shorthand. Our "scratch work" can take whatever form we find convenient and may be very "bare bones." But the written work we submit for others to read should adhere to the professional standards we have outlined.

1.8   Reading a Mathematics Textbook

You are about to begin what we hope will be a rewarding, yet challenging, mathematical journey. One of the challenges for many people is actually reading a mathematics textbook with the goal of full comprehension of what has been written. So we offer some advice for you to keep in mind as you read. Further suggestions will be provided at appropriate points in later chapters.

First, keep in mind that most mathematics texts used for courses beyond the calculus level incorporate a more formal approach than you are likely to have experienced before. You cannot expect to learn from such a text by simply skimming some of the examples in the hope of finding one similar to a homework exercise you are trying to do. *You must carefully read the text and process what you read.* In this book some ideas will be totally new, but even ones with which you are familiar will be discussed from a more sophisticated perspective. This is intentional! We are attempting to provide a transition from relatively informal and intuitive ways of thinking and communicating to modes of reasoning and writing that will help you gain a much deeper understanding of mathematical concepts and their applications.

To understand and fully process what you read, you need to be an *active* reader. Primarily, this means you should read the text slowly and carefully, with pencil and paper at hand to work out details the text omits, take additional notes, and record any questions you are not able to resolve. It may be necessary to re-read a passage, a paragraph, or even an individual sentence or example, perhaps several times, to achieve full understanding of what is being conveyed. As you read, pause regularly to think about the ideas being discussed and to assess your understanding of them.

This book includes two features specifically designed to help you become an active reader of mathematics:

- *Questions to Guide Your Reading*
- *Practice Problems*

Beginning in the next chapter, each section of the text opens with a set of questions intended to guide your reading. These questions will help you identify the most important points being discussed. We strongly suggest you carefully read through the questions before reading a section of the book, then look for the answers to the questions as you read. Doing so is likely to make it easier for you to understand what you are reading and will also better prepare you to work on the problems your instructor assigns for homework.

Also commencing with Chapter 2, a series of practice problems is embedded in the narrative. These practice problems provide you with an opportunity to actively work with an idea or technique immediately after it is introduced. You should try each practice problem before reading any further in the text.

Again, doing so is highly likely to make what you are reading clearer and more understandable, though you may have to do some re-reading to complete a practice problem. The answers to practice problems appearing in a particular section can be found at the end of that section.

As you read, you will find that certain items such as examples, theorems, proofs, and some definitions are usually set apart from the main narrative of the text and given labels to indicate the categories to which they belong. This allows such items to be located relatively easily, provides a degree of formal organization to the writing, and usually makes it easier to study what has been written. In this book, we number such items chapter by chapter so that, for instance, *2.5* refers to the fifth formally designated section or item appearing in Chapter 2.

It is also common practice in mathematics textbooks to sometimes place explanatory remarks and informal discussion *after* a theorem, definition, or proof. If you do not understand something you have read, it is therefore a good idea to continue reading a bit further as the immediately following sentences or paragraphs may provide the necessary clarification or explanation.

Up to this point, you may have paid little attention to the formal definitions appearing in mathematics textbooks. But because most mathematical arguments depend on the definitions of the terms appearing within them, it will now be of the greatest importance that you study and learn definitions as they arise. A term's definition provides the basis for any further use of the term, whether in the statement of a theorem or another definition or within a proof. For example, suppose you read the following sentence in a geometry textbook:

> A **circle** is the set of all points in a plane equidistant
> from a given point referred to as the circle's **center**.

In this one sentence two terms, *circle* and *center of a circle*, are defined. The other mathematical words used in the sentence, *points*, *plane*, and *equidistant*, must have been introduced earlier in the book so that the newly defined words make sense. Any future reference in the book to the terms *circle* and *center of a circle* can be traced back to the definitions created for these terms. For instance, if the book later states

> A **radius** of a circle is a segment joining a point
> on the circle to the center of the circle

we should realize that each occurrence of the word *circle* in this new definition for *radius* is referring to a specific "set of points in a plane equidistant from a given point." If the book goes on to state and prove a theorem such as

> *All radii of a circle are equal in length*

the meaning assigned to the terms *radii* and *circle* in this statement come from their respective definitions.

   In much of your prior mathematics coursework, the focus was likely on procedures and technique; for instance, how to solve a certain type of equation or how to differentiate a certain kind of function. As you continue to move into more advanced mathematics, however, you will notice a greater emphasis on concepts over procedures, *why* over *how*. The examples in advanced textbooks also reflect this shift in priority. Rather than illustrating each type of problem appearing in the end-of-chapter exercises, these examples are instead designed to provide you with concrete instances of the often abstract "big picture" ideas being developed in the text. They also tend to link new ideas to your earlier mathematical experiences. Whenever you find yourself having trouble coming to terms with a new mathematical concept, we recommend you carefully work through any associated examples the text provides. They often supply the necessary "fertile soil" for deeper understanding to take root and grow.

   One last bit of advice: Make sure you can justify each sentence in an example or proof. This means asking "Why?" *Why* does this example illustrate that concept? *Why* does this statement in a proof follow from earlier statements? *Why* are we applying a certain technique or a certain theorem here? The key to *learning* mathematics is *understanding* mathematics. And understanding generally follows if we can say *why* something works the way it does.

Chapter 1 Problems

1A. Abstraction

Try to find some part of mathematics that you have studied that links all of the following:

- ocean tides
- the motion of the moon around the earth
- sound
- sales of skis at a local sporting goods store.

1B. Formulating and Gathering Evidence for a Conjecture

Consider the polynomial $n^2 + n + 17$.

   1. Evaluate this polynomial for each counting number $n = 1$ through $n = 9$. Then formulate a conjecture about the kind of number that results when this polynomial is evaluated for counting numbers in general. Test your conjecture with other counting numbers. If you believe your conjecture is true, try to explain why without just citing numerical examples. If you think your conjecture will not turn out to be true, try to find a counting number for which your conjecture is false. Write a paragraph in which you report on your conclusion and how you reached it.

2. If your conjecture turns out to be true and we are able to construct a proof for it, what can we then call the conjecture?

3. What sort of reasoning, *inductive* or *deductive*, are you using when you evaluate the polynomial for some specific counting numbers and then form a conjecture about the kind of number that results in general?

4. What sort of reasoning, *inductive* or *deductive*, does a proof utilize?

1C. The Pythagorean Theorem and Fermat's Last Theorem

1. Consider again the Pythagorean Theorem, which is stated earlier in this chapter.

   (a) This theorem gives us information about what kind of mathematical objects?

   (b) According to the Pythagorean Theorem, can a right triangle have sides of lengths 3, 4, and 5 units?

   (c) According to the Pythagorean Theorem, can a right triangle have sides of lengths 4, 5, and 6 units?

   (d) Find counting numbers x, y, and z, different from those mentioned previously in this problem, for which $x^2 + y^2 = z^2$.

2. *Fermat's Last Theorem*, originally stated by Pierre de Fermat in a letter believed to have been written in 1637 but not proved until 1995 by Andrew Wiles, states that for any counting number n that is greater than 2, the equation $a^n + b^n = c^n$ has no solutions where each of a, b, and c is a counting number.

   (a) According to Fermat's Last Theorem, and without actually performing any calculations, determine whether it is possible for $102^6 + 273^6 = 285^6$.

   (b) Observe that $1^3 + 2^3 = (\sqrt[3]{9})^3$. Explain why this does not contradict Fermat's Last Theorem.

   (c) Observe that $2^3 + 2^3 = 2^4$. Explain why this does not contradict Fermat's Last Theorem.

1D. Mathematical Research: Prime Numbers

In this problem, you will continue the research situation concerning prime numbers that we initiated in this chapter.

1. We listed all the prime numbers less than 50. Extend the list to include all primes less than 100.

2. Use the list you created in (1) to find any pairs of consecutive odd primes lying between 50 and 100.

3. *Goldbach's Conjecture*, another open problem in the study of prime numbers, states that every even counting number larger than 2 can be written as the sum of two primes. Gather evidence for Goldbach's Conjecture by writing each even counting number from 4 through 20 as the sum of two primes.

4. Find an even counting number that can be written as the sum of two primes in at least two different ways.

5. Which primes less than 100 exhibit the form $n^2 + 1$ where n is a counting number? (It has been conjectured that there are infinitely many primes possessing this form.)

6. Explain why the following argument is not convincing:

   The counting numbers go on forever. So the prime numbers must also go on forever.

1E. Mathematical Research: Operations with Even and Odd Numbers

In this chapter, we conjectured that the sum of even counting numbers is also even. Try to develop additional conjectures concerning sums and products of combinations of even and odd counting numbers (e.g., what do you think you can say about the sum of two odd counting numbers? what about the sum of an even counting number and an odd counting number? what about the product of two even counting numbers? etc.). Can you find any deductive evidence to support your conjectures?

1F. Mathematical Research: Sums of Consecutive Odd Numbers

Consider the following sums:

$$1$$
$$1 + 3$$
$$1 + 3 + 5$$
$$1 + 3 + 5 + 7.$$

1. Calculate these sums. Do you notice a pattern? What is it?

2. Evaluate each of the following sums:

$1 + 3 + 5 + 7 + 9$

$1 + 3 + 5 + 7 + 9 + 11.$

Does the pattern you observed in (1) still apply?

3. Make a conjecture based on the research you have conducted in (1) and (2). Write your conjecture as a sentence.

4. What is the mathematical term that describes the kind of numbers the sums above produce?

5. Your answer to (4) should also make you think of a specific geometric shape. Try to connect the sums considered above to the area of this geometric shape. Can you create a pictorial argument supporting the conjecture you made in (3)?

1G. Mathematical Research: Inequalities

1. When we add 3 to both sides of the true inequality $1 < 2$, we get the inequality $4 < 5$, which is also true. Formulate a conjecture about what happens in general when the same number is added to both sides of a true inequality. Test your conjecture with five different numerical examples, being sure to consider the effects of various kinds of numbers (e.g., fractions, positive numbers, negative numbers, 0, etc.). If necessary, revise your original conjecture based on your numerical findings. Then try to explain why you think your (original or revised) conjecture might be true.

2. When we multiply both sides of the true inequality $1 < 2$ by 3, we get the inequality $3 < 6$, which is also true. Formulate a conjecture about what happens in general when both sides of a true inequality are multiplied by the same number. Test your conjecture with five different numerical examples, being sure to consider the effects of various kinds of numbers (e.g., fractions, positive numbers, negative numbers, 0, etc.). If necessary, revise your original conjecture based on your numerical findings. Then try to explain why you think your (original or revised) conjecture might be true.

1H. Mathematical Research: Perfect Numbers

A counting number is called **perfect** if it is the sum of all of its positive factors other than itself. For example, 6 is perfect because the positive factors of 6, other than 6 itself, are 1, 2, and 3, and $1 + 2 + 3 = 6$. Find all perfect numbers less than 30.

Chapter 2

Sets, Numbers, and Axioms

Set theory is the language within which mathematics is communicated. In this chapter, we introduce this language, including standard set-theoretic terminology and notation, and apply it while exploring various kinds of *numbers*. We also take our first look at the *axiomatic* approach to mathematics by establishing some foundational assumptions that reflect our intuition about sets and numbers.

2.1 Sets and Numbers from an Intuitive Perspective

Questions to guide your reading of this section:

1. How can listing be used to describe a set? What is the significance of the *ellipsis notation* (...)?

2. How can we symbolically indicate that a certain object is a member of a particular set?

3. Which numbers are *real numbers*? Which numbers are *natural numbers*? Which numbers are *integers*? Which numbers are *rational numbers*? What symbols are used to represent these sets of numbers?

4. What is *set-builder notation* and how is it used to describe a set?

The idea of a *set* arises naturally in all areas of mathematics. We speak of, for instance, the set of solutions to an equation, the set of points lying inside a sphere, and the set of functions satisfying some particular property; for example, the set of functions whose graphs are straight lines. From an intuitive point of view, a *set* is obtained by collecting together some objects into a single entity. The collected objects are referred to as the *members* or *elements* or *points* of the set.

If a set has relatively few members, we can describe the set by listing its members in between the symbols { and }, which are called **set braces** or **curly brackets**, separating the members from one another by commas.

Example 2.1   To describe the set whose members are precisely the numbers 2, 5, and 9, we may write {2, 5, 9}.

To make it easier to refer to a set, it is customary to name it using an upper-case letter.

Example 2.1 (continued)   It can be both cumbersome and tedious to keep writing {2, 5, 9} when we want to refer to this set. By letting $A = \{2, 5, 9\}$, we are giving the set the name A. When we subsequently refer to the set A, our reader knows we are speaking of the set whose only members are the numbers 2, 5, and 9. Of course once we finish working in our current situation and want to move on to another one, we are free to use the letter A to name a different set.

It does not matter in what order or how many times the members of a set are listed, as our only goal in forming a set is to collect together certain objects.

Example 2.1 (continued)   Thus, {2, 9, 5} and {2, 5, 9, 5} each describe the same set as {2, 5, 9}.

Providing a complete list of the members of a set may not be efficient or feasible once the number of members becomes large or infinite. When a set has too many members to conveniently list within set braces, but there is a relatively simple and unambiguous pattern relating the members, we simply list enough of the members to make the pattern clear and use the so-called **ellipsis notation** consisting of three consecutive dots (...) to indicate there is a gap in our listing.

Example 2.2   The set whose members are the first 100 counting numbers can be expressed as {1, 2, 3, 4,..., 100}. The set of all counting numbers can be described by writing {1, 2, 3, 4,...}. The set {10, 12, 14,..., 20} also includes 16 and 18 among its members.

Practice Problem 1   List all members of the set {24, 12, 6, 3, ..., 0.375}.

A set having exactly one member is called a **singleton**. Also, given objects x and y, the set $\{x, y\}$ having x and y as its only members is called a **doubleton**. Thus, the set whose only member is the number 8 is the singleton {8}, and the set whose only members are the numbers 0 and −1/2 is the doubleton {0, −1/2}.

Note that the doubleton $\{x, y\}$ has the potential for two distinct members, but it is possible that x and y could be the same. See Problem 2C.3.

Practice Problem 2   Is the set $\{3, 3\}$ a singleton?

Indicating Membership in a Set

To indicate that an object x is a member of a set A, we write $x \in A$, and to indicate that x is *not* a member of A, we write $x \notin A$.

Example 2.3   Observe that $6 \in \{2, 6, 8, 13\}$, $4 \notin \{2, 6, 8, 13\}$, and $10 \in \{2, 4, 6, \ldots, 44\}$.

The Greek letter $\in$ has been chosen as our set membership symbol because it corresponds with the letter e. You should think e for *element* and e for $\in$. Thus,

$$x \in A$$

may be read as

x is an element of A.

Practice Problem 3   Let L be the set of counting numbers larger than 9. Write a symbolic expression that makes the statement that 12 is a member of L. Then write another symbolic expression that makes the statement that 4 is not an element of L.

The Real Numbers

In your study of algebra and calculus, you likely learned that all those numbers that can be written in decimal form or plotted on a number line are referred to as *real numbers*.

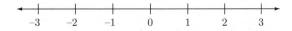

A number line provides us with a useful visual representation of the set of real numbers. For instance, the number line shown above specifies a location for the origin (where the number 0 has been plotted) and marks off a scale showing the counting numbers radiating off to the right and the negatives of the counting numbers radiating off to the left. All of these numbers are real numbers, but the scale also helps us locate other real numbers that have not been specifically marked on the line; for instance, we know that the real number 1.5 corresponds with the point midway between the points corresponding with the numbers 1 and 2 on the line.

Example 2.4 Each of the numbers 2, 0, –5, and $\frac{11}{4}$ is a real number because each can be expressed in decimal form and plotted on a number line. For example, in decimal form $\frac{11}{4}$ is 2.75, and on the number line $\frac{11}{4}$ lies three quarters of a unit to the right of the point corresponding with 2.

The numbers $\sqrt{3}$ and π are also real numbers. You probably recall that each of these numbers has a decimal representation that never ends and does not repeat. For instance, 3.1415926 is an approximation of the decimal representation for π to seven decimal places, but the actual decimal representation for π goes on forever without repeating the same sequence of digits.

Providing a formal mathematical description for the real numbers that goes beyond the informal interpretations we have discussed here was one of the greatest achievements of 19th century mathematics. But as developing an understanding of such a formal description first requires one to have mastered the material in a text such as the one you are currently reading, we will not pursue such formalism here. Instead, we will assume that the real numbers are familiar enough to you, from your prior work with them and the examples we have provided, that we may use them as members of a set.

The set of all real numbers is denoted by **R**, the set of all positive real numbers by **R**⁺, and the set of all negative real numbers by **R**⁻. Because 0 is considered neither positive nor negative, it is a member of neither **R**⁺ nor **R**⁻.

Example 2.5 Because 2.75 is a real number, we may write $2.75 \in \mathbf{R}$. Furthermore, because 2.75 is positive, we may write $2.75 \in \mathbf{R}^+$.

Note that the symbol **R** used to represent the set of all real numbers is uppercase and bold. Mathematicians distinguish among uppercase, lowercase, plain, italic, bold, and script letters. Thus, neither R nor r represents the set of all real numbers. Moreover, uppercase R and lowercase r should not be used interchangeably. And $\mathcal{R}$ ("script" R) is considered to be different from each of R, r, and **R**.

The Natural Numbers and the Integers

Various other sets of numbers, besides the set **R** of all real numbers, are also designated by special symbols. The set {1, 2, 3, ...} of all **natural numbers** is denoted by **N** and the set { ..., –3, –2, –1, 0, 1, 2, 3, ...} of all **integers** by **Z**.

We also use $\mathbf{Z}^+$ to denote the set of all positive integers and $\mathbf{Z}^-$ to denote the set of all negative integers.

Example 2.6 Because $\mathbf{N}$ includes only those numbers used for counting, $8 \in \mathbf{N}$, $-8 \notin \mathbf{N}$, and $0 \notin \mathbf{N}$. Because $\mathbf{Z}^+$ also represents the set of counting numbers, we also have $8 \in \mathbf{Z}^+$, $-8 \notin \mathbf{Z}^+$, and $0 \notin \mathbf{Z}^+$. In contrast, the set $\mathbf{Z}$ of all integers includes not only the counting numbers but their negatives and 0 as well, so $8 \in \mathbf{Z}$, $-8 \in \mathbf{Z}$, and $0 \in \mathbf{Z}$. And as $\mathbf{Z}^-$ includes only the negatives of the counting numbers, we have $8 \notin \mathbf{Z}^-$, $-8 \in \mathbf{Z}^-$, and $0 \notin \mathbf{Z}^-$. Finally, because 2/3 is a real number, but is neither a natural number nor an integer, we have $2/3 \in \mathbf{R}$, $2/3 \notin \mathbf{N}$, and $2/3 \notin \mathbf{Z}$.

Rational and Irrational Numbers

Real numbers that can be obtained by dividing integers are called *rational numbers* because such numbers represent *ratios* of integers. Real numbers that cannot be obtained in this way are called *irrational numbers*. Keep in mind that a "fraction bar" is often used to represent the operation of division and that division by 0 is undefined.

Definition 2.7 Any real number that can be written in the form a/b, where a is an integer and b is a nonzero integer, is a **rational number**.

The set of all rational numbers is denoted by $\mathbf{Q}$ (as rational numbers are *quotients* of integers). We also use $\mathbf{Q}^+$ to denote the set of positive rational numbers and $\mathbf{Q}^-$ to denote the set of negative rational numbers. Any real number that is not rational is called an **irrational number**.

Example 2.8 The number 24.51 is rational because $24.51 = 2451/100$, the quotient of the integer 2451 by the nonzero integer 100. Thus, we may write $24.51 \in \mathbf{Q}$. We will eventually prove that there are no integers a and b for which $\sqrt{2} = a/b$. Thus, $\sqrt{2}$ is irrational and we may write $\sqrt{2} \notin \mathbf{Q}$.

Practice Problem 4 Determine which of the following statements are true:

(a) $-\frac{5}{2} \in \mathbf{Z}$.   (b) $-\frac{5}{2} \in \mathbf{Q}^-$.   (c) $|-10| \in \mathbf{N}$.   (d) $0 \in \mathbf{R}^+$.

The Complex Numbers

There is no real number solution to the equation $x^2 = -1$ because whenever a real number is squared, the resulting number is nonnegative. To provide this equation with a solution, mathematicians introduced a new number that is not a real number. Specifically, the **imaginary unit** i is defined to be a number, often designated by $\sqrt{-1}$, whose square is -1. That is, i has the property that $i^2 = -1$. The imaginary unit is *not* a real number as it cannot be plotted on a number line or expressed in decimal form; thus, $i \notin \mathbf{R}$. Any number that can be written in the form $a + bi$, where both a and b are real numbers, is called a **complex number**. The set of all complex numbers is denoted by $\mathbf{C}$.

Example 2.9   The number $\sqrt{2}i$ is a complex number because $\sqrt{2}i = 0 + \sqrt{2}i$, where both 0 and $\sqrt{2}$ are real numbers. Hence, we may write $\sqrt{2}i \in \mathbf{C}$.

Set-Builder Notation

Mathematicians commonly use so-called *set-builder notation* to describe sets, especially sets that are too large to allow for their elements to be listed.

Notation 2.10   The set of all members of a given set A possessing a certain property P can be expressed as

$$\{x \in A \mid x \text{ has property } P\},$$

which can be read as *the set of all members x of the set A such that x has property P*. This type of notation for describing a set is referred to as **set-builder notation**.

Example 2.11   The set of all real numbers between 2 and π includes infinitely many members (remember, this set includes all numbers corresponding with points on the number line lying between 2 and π, and there are infinitely many such points). It turns out that there is no possible way to use listing/ellipsis to describe this set. This set can, however, be expressed using set-builder notation as

$$\{x \in \mathbf{R} \mid 2 < x < \pi\},$$

which can be read as *the set of all real numbers x such that x is between 2 and π*.

In learning how to use set-builder notation, we must be sure to understand the individual components of the notation as well as how these components work together to help us determine exactly what is and what is not a member of the set being formed.

Example 2.11 (continued)   In the expression $\{x \in \mathbf{R} \mid 2 < x < \pi\}$:

- The statement $x \in \mathbf{R}$ to the left of the vertical bar tells us that each member x of the set being formed is a real number.

- The property $2 < x < \pi$ to the right of the vertical bar must be satisfied by each member x of the set.

- The vertical bar itself can be read as any of the following:

such that   *for which*   *with the property that.*

Note that, for instance,

$$2.4 \in \{x \in \mathbf{R} \mid 2 < x < \pi\}$$

because 2.4 is a real number and, when 2.4 is assigned as the value of x, the property $2 < x < \pi$ is satisfied because $2 < 2.4 < \pi$ is true. In contrast,

$$2 \notin \{x \in \mathbf{R} \mid 2 < x < \pi\}$$

because even though 2 is a real number, when 2 is assigned as the value of x, the property $2 < x < \pi$ is *not* satisfied as $2 < 2 < \pi$ is *not* true because a number cannot be less than itself. Also, if we again take i to be the imaginary unit $\sqrt{-1}$, it follows that

$$i \notin \{x \in \mathbf{R} \mid 2 < x < \pi\}$$

because i is not even a real number.

Practice Problem 5   Use set-builder notation to describe the set of all natural numbers larger than 99.

In our general set-builder notation template

$$\{x \in A \mid x \text{ has property } P\},$$

the letter x is being used as a placeholder and is called a **dummy variable**. But there is no reason that a different letter or symbol could not be used in place of x.

Example 2.11 (continued)   The set $\{x \in \mathbf{R} \mid 2 < x < \pi\}$ can also be described by writing, for instance, $\{b \in \mathbf{R} \mid 2 < b < \pi\}$ or $\{* \in \mathbf{R} \mid 2 < * < \pi\}$. However, writing $\{x \in \mathbf{R} \mid 2 < b < \pi\}$ would be incorrect because the property $2 < b < \pi$ has been expressed in terms of a placeholder b rather than the placeholder x that appears to the left of the vertical bar.

This is similar to standard function notation that you have no doubt used in your study of calculus. To define f as the function that squares a number, we could write $f(x) = x^2$ or $f(b) = b^2$, but it would not make sense to write $f(x) = b^2$. Whether with function notation or set-builder notation, the same placeholder must appear throughout the expression.

If it concerns you that different letters or symbols may be used as dummy variables to express the same set, bear in mind that the dummy variable is needed only for the *symbolic* expression. It is always possible to suppress any explicit mention of the dummy variable when reading an expression involving set-builder notation or writing out the set in words.

Example 2.11 (continued)   What we may choose to write as

$$\{x \in \mathbf{R} \mid 2 < x < \pi\}$$

may be read aloud as

the set of all real numbers between 2 and π

with no mention of the placeholder x.

Practice Problem 6   Describe the set $\{y \in \mathbf{Q} \mid y^2 < 5\}$ in words and without mentioning a dummy variable.

Of course, by its very nature, set-builder notation is particularly useful when we want to form the set of all members of a given set that possess a certain property.

Example 2.12   The set of all prime numbers is really the set of all natural numbers having the property that they are prime. This set can be described using set-builder notation as $\{k \in \mathbf{N} \mid k \text{ is prime}\}$.

Practice Problem 7   Use set-builder notation to describe the set of all negative integers that are odd.

Many of the sets of numbers we introduced earlier in this section can be expressed using set-builder notation. The set $\mathbf{R}^+$ of all positive real numbers is $\{x \in \mathbf{R} \mid x > 0\}$, the set $\mathbf{R}^-$ of all negative real numbers is $\{x \in \mathbf{R} \mid x < 0\}$, the set $\mathbf{Q}$ of all rational numbers is $\{a/b \mid a,b \in \mathbf{Z}, \ b \neq 0\}$, the set $\mathbf{C}$ of all complex numbers is $\{a + bi \mid a,b \in \mathbf{R}\}$, and so forth. Note the custom of writing, for instance,

$$a, \ b \in \mathbf{Z}$$

in place of

$$a \in \mathbf{Z} \text{ and } b \in \mathbf{Z}.$$

We will do our best to alert you to such conventions as they arise.

Answers to Practice Problems

1. Note that we can divide by 2 to get from 24 to 12 and from 12 to 6. Continuing this pattern, we would conclude that the members of the set are 24, 12, 6, 3, 1.5, 0.75, and 0.375.

2. Yes, this set is a singleton as it has exactly one member; namely, 3.

3. $12 \in L$ and $4 \notin L$.

4. Only (b) and (c) are true.

5. $\{x \in \mathbf{N} \mid x > 99\}$.

6. This set is the set of all rational numbers whose squares are less than 5.

7. Two possibilities are $\{x \in \mathbf{Z}^- \mid x \text{ is odd}\}$ and $\{x \in \mathbf{Z} \mid x \text{ is negative and odd}\}$.

2.2 Set Equality and Set Inclusion

Questions to guide your reading of this section:

1. What is meant by a *defined term* in mathematics? Is it possible to define all terms used in mathematics?

2. How should we translate the logical phrase *if and only if*? How is this phrase often abbreviated?

3. What different purposes are served by informal and formal versions of a mathematical definition?

4. How is the idea of *set equality* expressed informally? formally?

5. How can we recognize when one set is a *subset* of another set?

6. What is the difference in meaning and use among the symbols $\in$, $\subseteq$, and $\subset$?

7. What members does an *empty* set have?

8. How do we form the *power set* of a set?

In mathematics, some new words or phrases are given meaning through the use of previously introduced words or phrases. Words or phrases established in this way are called **defined terms**.

Example 2.13   On an intuitive level, the only thing that distinguishes one set as being different from another set is the existence of an object that is a member of one of the sets but not the other. To be able to talk more easily about the idea of two sets being the same or different, we introduce a new defined term called *set equality*. Our intuition leads us to the following definition:

> *To say that two sets are* **equal** *means that they have exactly the same members.*

Practice Problem 1   Explain why the sets {5, 10, 15} and $\{5x \mid x \in \mathbf{N}\}$ are not equal.

A sentence used to define a term is referred to as a **definition**. Generally speaking, a mathematical definition is used to capture an idea or property related to the previously existing words or phrases appearing in the definition. The new term that is being defined is really just a form of shorthand for the idea or property.

Example 2.13 (continued)   Our definition of *set equality* expresses a property two sets may possess, that of *having exactly the same members*, and introduces a new, more compact way of expressing this property, "*A* equals *B*" instead of "*A* and *B* have exactly the same members."

Some Notions Will Necessarily Remain Undefined

Besides the new term being established, a mathematical definition can make use of only words that have already been introduced. But what about the *first* word or words introduced? How can they be defined if we do not yet have any "official" terminology? Simple: They can't! At the very beginning of a particular mathematical study, there is actually no way to avoid introducing some notions that will have to remain *undefined*.

Example 2.13 (continued)   Our definition for *set equality* makes use of two previously introduced notions, *set* and *membership*. But looking back, we never defined these notions, nor will we. We take them as *undefined*.

Informally, this should cause you no trouble. You already have an *intuitive* sense of what sets are and what it means for an object to be a member of a set. If you are worried about using terms that have not (and will not) be defined, it turns out that definitions are not the only way to assign meaning to mathematical notions. Later in this chapter, we will show how to describe the undefined terms *set* and *membership* without resorting to definitions.

For now, though, it is important for you to understand that we could not begin our study of set theory by announcing that there will be no undefined terms and then just *define* a set to be a *collection of elements*. Confronted with this alleged "definition," one could reasonably ask, "What is meant by *collection* and what is meant by *element*?" *Collection* and *element* are really no better than *set* and *member*, so such an attempt at a definition for *set* does not get us any closer to truly understanding what a set is.

Mathematical Definitions Cannot Be "Circular"

Unlike mathematics, everyday English admits no undefined words. A suitably extensive dictionary of the English language will provide a definition for any word we want to look up. But there are only a finite number of words in any language. This means that dictionary definitions, if traced far enough, tend to become "circular," with a word ultimately being defined by itself.

Example 2.13 (continued) A dictionary might include the following "definitions":

A set is a collection.
A collection is an aggregate.
An aggregate is a set.

Putting these together, the dictionary is really "circling" around to tell us that *a set is a set*. Though true, this hardly seems worth bringing to our attention. Mathematical definitions must not be "circular" in the sense illustrated here.

Definitions and Mathematical Notation

When a mathematical definition is being formulated, it is often accompanied by the introduction of some related mathematical notation. It is customary to choose mathematical terminology and notation so that they reflect the inherent idea being communicated.

Example 2.13 (continued) We can formulate the definition of *set equality* as follows:

*To say that two sets A and B are **equal**, denoted $A = B$, means that A and B have exactly the same members.*

In developing terminology and notation for the notion of *two sets having exactly the same members*, the term *equal* and the notation "=" have been chosen to appeal to our prior experience with the notion of *equality* in the context of numbers. Intuitively, *equality*, in whatever setting, conveys the notion of *being the same as*. Our choice of terminology and notation reflect our belief that two sets should be regarded as being the same provided that they have the same members.

Practice Problem 2   Which of the following statements are true?

(a) $\{2, 7, -4, 0\} = \{-4, 0, 2, 7\}$.   (b) $\mathbf{N} = \mathbf{Z}^+$.

(c) $\{x \in \mathbf{R} \mid 2 < x < 4\} = \{3\}$.

If and Only If

The phrase **if and only if**, which is often abbreviated to **iff**, is used as a synonym for *means the same thing as* in the context of two mathematical statements or two statements about mathematics. Any mathematical definition can be written so as to incorporate this phrase.

Example 2.13 (continued)   Our definition for *set equality* can be expressed as follows:

*Two sets A and B are **equal**, denoted A = B, iff A and B have exactly the same members.*

Do not confuse the phrase *if and only if* with the idea of *equality*. *If and only if* serves as a linking device between two *sentences* that express the same idea, whereas *equality* is used in a particular mathematical setting to suggest that two *objects* in that setting are actually identical to one another.

Example 2.14   Two sets may be equal and two numbers may be equal, but two sentences cannot be equal because the notion of *equality* does not apply to sentences. Similarly, one should not write "5 iff 2 + 3" because neither 5 nor 2 + 3 is a sentence; rather, we should simply write "5 = 2 + 3."

Practice Problem 3   Which of the following statements are correctly formed based on proper usage of "=" and *iff*?

(a) 3^2 iff 9.   (b) $x + 4 = 9$ iff $x = 5$.   (c) n is even $= n + 1$ is odd.

Set Equality and the Development of a Formal Version of a Definition

A relatively informal statement of a mathematical definition is often best for providing conceptual insight into the idea the definition is putting across. But when one is formulating an argument that involves a given term, a very careful and precise *formal* version of the term's definition facilitates clear thinking and unambiguous communication. Hence, we will need formal versions of most of the mathematical definitions we introduce.

What do we mean by a *formal* version of a definition? Earlier we stated that, except for the new terminology being introduced, a definition may only incorporate already existing words, phrases, and notation. A formal version of a definition adheres more noticeably to this requirement by using only undefined terms, previously defined terms, and allowable logical phrases such as *if and only if* (we will fully inventory and explain such logical phrases in Chapter 3).

Example 2.13 (continued)   Note that the statement

The sets A and B have exactly the same members

tells us that

The objects that meet the condition for membership in the set A are exactly the objects that meet the condition for membership in the set B

or, in other words, for any object x,

$$x \in A \text{ iff } x \in B.$$

Definition 2.15   Two sets A and B are **equal**, denoted $A = B$, iff for every x, $x \in A$ iff $x \in B$.

The advantage of this formal definition is that it is so precise and carefully constructed that there can be no disagreement over what it says among those who understand the $\in$ notation for membership in a set and the logical phrases *for every* and *if and only if*. It expresses our intuitive feeling for the equality of sets in formal mathematical language. When we begin to write proofs for theorems about sets, we will want to make use of this formal version of the definition of set equality.

Note also that the choice of the letter x used in the formal statement of this definition is really arbitrary. Another letter could be used in place of x because the letter is really just a placeholder for an arbitrary object (this is another instance of the "dummy variable" idea we saw earlier when discussing set-builder notation).

Example 2.16   The equality of sets A and B could also be expressed by writing $A = B$ iff for every t, $t \in A$ iff $t \in B$.

Set Inclusion

Sometimes all of the members of one set are also members of another set.

Definition 2.17   A set A is a **subset** of a set B, denoted $A \subseteq B$, iff for every x, if $x \in A$, then $x \in B$.

Example 2.18   If E is the set of all even integers, then as each of 2, 4, and 6 is an even integer, every member of $\{2, 4, 6\}$ is a member of E. That is, if $x \in \{2, 4, 6\}$, it follows that $x \in E$. Thus, $\{2, 4, 6\} \subseteq E$, which can be read as "$\{2, 4, 6\}$ is a subset of E."

Practice Problem 4   Explain why $\{-6, 0, 2, 3\} \subseteq \{k \in \mathbf{Z} \mid k < 5\}$.

The symbolic statement $A \subseteq B$ can also be read in any of the following ways: A is **contained in** B; A is **included in** B; B is a **superset** of A; B **contains** A.

Example 2.19   Note that $\mathbf{N} \subseteq \mathbf{Z} \subseteq \mathbf{Q} \subseteq \mathbf{R} \subseteq \mathbf{C}$. Thus, we may say that $\mathbf{N}$ is contained in $\mathbf{Z}$, $\mathbf{Z}$ is included in $\mathbf{Q}$, $\mathbf{R}$ is a superset of $\mathbf{Q}$, and $\mathbf{C}$ contains $\mathbf{R}$.

Intuitively, a mathematical *relation* somehow "relates" or "connects" mathematical objects to one another. Note that $\subseteq$ is a relation, referred to as **set inclusion** or **set-theoretic containment**, because it connects certain sets to each other. Specifically, we may think of a set A as being related to a set B via the relation $\subseteq$ exactly when A is a subset of B.

If a set A is not a subset of a set B, we write $A \not\subseteq B$. By definition, for $A \subseteq B$, *every* member of A must be a member of B. Thus, $A \not\subseteq B$ if there is even a *single* member of A that is not a member of B.

Example 2.20   Because $6 \notin \{1, 2, 3, 4\}$, it follows that $\{2, 4, 6\} \not\subseteq \{1, 2, 3, 4\}$.

The Symbol $\in$ Versus the Symbol $\subseteq$

It is important to understand the distinction between the use of the symbol $\in$ to indicate membership in a set and the use of the symbol $\subseteq$ to indicate one set is a subset of another. Remember that:

- Sets should appear on both the left and right sides of the symbol $\subseteq$.

- Whereas a set should appear on the right side of $\in$, a member of that set should appear on the left side of $\in$.

Example 2.21 We may write $1 \in \mathbf{N}$ because 1 is a member of the set $\mathbf{N}$ of natural numbers. We may also write $\{1\} \subseteq \mathbf{N}$ because each member of the set $\{1\}$ is a member of the set $\mathbf{N}$.

However, it is incorrect to write $\{1\} \in \mathbf{N}$ because $\{1\}$ is not a member of the set $\mathbf{N}$ (note that $\{1\}$ is not 1; $\{1\}$ is the set whose only member is 1). It is also incorrect to write $1 \subseteq \mathbf{N}$ because 1 is a number, not a set.

Practice Problem 5 Fill in the blanks with $\in$ and $\subseteq$ to make a true statement: 0 _____ $\mathbf{Z}$ _____ $\mathbf{R}$.

Proper and Improper Subsets

A set A is a **proper subset** of a set B iff $A \subseteq B$ and $A \neq B$. To indicate that A is a proper subset of B, we write $A \subset B$. The **improper subset** of a set is the set itself; that is, the improper subset of a given set A is A.

Example 2.18 (continued) If E again represents the set of all even integers, we may write $\{2, 4, 6\} \subset E$ because $\{2, 4, 6\} \subseteq E$ and $\{2, 4, 6\} \neq E$. In other words, $\{2, 4, 6\}$ is a proper subset of E. The improper subset of E is E and the improper subset of $\{2, 4, 6\}$ is $\{2, 4, 6\}$.

Practice Problem 6 Which of the following statements are true?

  (a) $\{1, -1\} \subset \{1, -1\}$.   **(b)** $\{1, -1\} \subseteq \{1, -1\}$.

  (c) $\{-1\} \subset \{1, -1\}$.   **(d)** $\{-1\} \subseteq \{1, -1\}$.

The Empty Set

A set having no members is called *empty*. That is, a set A is **empty** iff for every x, $x \notin A$. Later we shall prove that any two empty sets must be equal. For now, we mention that the one and only empty set is denoted by $\varnothing$ and that it is also referred to as the **null set**. Thus, as there are no positive real numbers less than 0, we may write $\{x \in \mathbf{R}^+ \mid x < 0\} = \varnothing$.

Observe that $\varnothing$ is a subset of every set A because as $\varnothing$ has no members in the first place, it follows that every member of $\varnothing$ is a member of A.[1]

Practice Problem 7 Explain why $\{0\} \neq \varnothing$.

---

1. If you find this difficult to believe, hang on until Chapter 3 where we investigate the logical basis for this conclusion.

The Power Set of a Set

There is nothing to prevent one set from being a member of another set. In particular, we may form the set of all subsets of a given set A, which is called the **power set** of A and is denoted $\mathcal{P}(A)$.

Example 2.22 The power set of {1, 2, 3} is

$$\mathcal{P}(\{1, 2, 3\}) = \{\varnothing, \{1\}, \{2\}, \{3\}, \{1, 2\}, \{1, 3\}, \{2, 3\}, \{1, 2, 3\}\},$$

the set of all subsets of {1, 2, 3}.

Example 2.23 The set $V = \{3, \{1, 3\}\}$ has two members, the number 3 and the set {1, 3}. Observe that 1 is *not* a member of V, though 1 is a member of a member of V. There are four subsets of V, namely, $\varnothing$, {3}, {{1, 3}}, and V itself; they are the members of $\mathcal{P}(V)$, the power set of V.

A set whose elements all happen to be other sets is often referred to as a **collection of sets**. Thus, the power set of a set is an example of a collection of sets.

Practice Problem 8   List the members of the power set of {−2, 2}.

Answers to Practice Problems

1. The set $\{5x \mid x \in \mathbf{N}\}$ includes many other members besides just 5, 10, and 15. For instance, $20 \in \{5x \mid x \in \mathbf{N}\}$, but $20 \notin \{5, 10, 15\}$. So, as the sets {5, 10, 15} and $\{5x \mid x \in \mathbf{N}\}$ do not have exactly the same members, they are not equal.

2. The only true statements are (a) and (b). Remember that even though 3 is the only natural number between 2 and 4, there are infinitely many real numbers between 2 and 4.

3. Only statement (b) is correctly formed. The correct way to express statement (a) is

$$3^2 = 9,$$

   and the correct way to express statement (c) is

   n is even iff n + 1 is odd.

4. The reason that $\{-6, 0, 2, 3\}$ is a subset of $\{k \in \mathbf{Z} \mid k < 5\}$ is that, as each of -6, 0, 2, and 3 is an integer less than 5, all of the members of $\{-6, 0, 2, 3\}$ are members of $\{k \in \mathbf{Z} \mid k < 5\}$.

5. The correct statement is $0 \in \mathbf{Z} \subseteq \mathbf{R}$.

6. Only the statement (a) is not true.

7. Because $0 \in \{0\}$, but $0 \notin \varnothing$, it follows that $\{0\} \neq \varnothing$.

8. The members of $\mathcal{P}(\{-2, 2\})$ are $\varnothing$, $\{-2\}$, $\{2\}$, and $\{-2, 2\}$.

2.3   Venn Diagrams and Set Operations

Questions to guide your reading of this section:

1. What are *Venn diagrams* and how are they used to visualize sets?

2. How do we form the *intersection, difference,* and *union* of two sets?

3. What are the two *distributive laws* for sets?

4. What are the two *DeMorgan's laws* for sets?

It can sometimes be helpful to use pictures to represent sets. In the figure that follows, the set A is depicted as a circular region. Intuitively, whatever members A has lie within the circle, whereas nonmembers of A lie outside the circle.

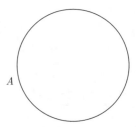

Thus, if $2 \in A$ and $5 \notin A$, we might make a diagram such as the following:

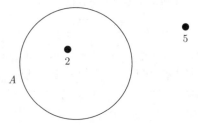

These sorts of pictorial representations of sets are called **Venn diagrams**. In such a diagram, a set is represented as the region lying within some closed curve. Because we are only using the diagram to indicate membership, the shape of the curve is irrelevant. If points representing specific objects are placed in the diagram, the only issue of significance concerning these points is the specific regions in which they are located.

The two regions into which the generic Venn diagram for a single set A separates the plane exactly correspond with the two possible membership scenarios for an object x relative to this set. Either

- $x \in A$ (i.e., x lies within the region representing A)

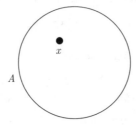

- or else $x \notin A$ (i.e., x lies outside the region representing A).

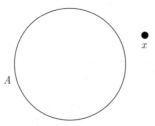

It is this correspondence between regions and membership possibilities that makes Venn diagrams a useful way to depict sets.

Practice Problem 1   Suppose it is known that $4 \in S$ and $6 \notin S$. Draw a Venn diagram so that it depicts this information.

Now consider the situation in which we are working with two sets A and B. The Venn diagram that follows can help us to explore the notion of *membership* in the context of these two sets.

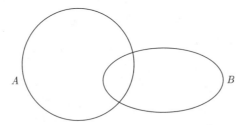

In this diagram, the set A is represented by the circular region on the left and the set B by the elliptical region on the right. The reason that the circle and the ellipse have been drawn so that they overlap is that it is possible the sets A and B have one or more members in common. Of course, this is just a possibility. Each region in the diagram has the potential for members, though it is also possible that a region could be empty.

Example 2.24   Suppose $A = \{2, 4, 6\}$ and $B = \{2, 3\}$. The following Venn diagram depicts these sets:

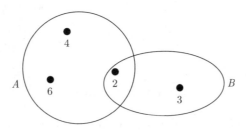

However, if $A = \{2, 4, 6\}$ and $B = \{2\}$, the Venn diagram would be as follows:

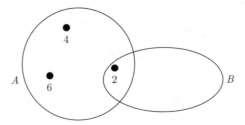

If we still have $A = \{2, 4, 6\}$ and $B = \{2\}$, and we want to emphasize the fact that 7 is a member of neither A nor B, we could create the following Venn diagram:

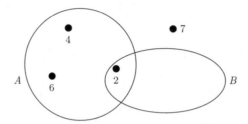

Practice Problem 2   Make a Venn diagram displaying the sets $G = \{1, 2, 3\}$ and $H = \{2, 8\}$.

Let us now go back to the generic Venn diagram

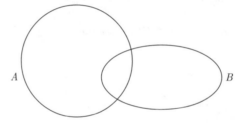

depicting two arbitrary sets A and B. Note that this diagram partitions the plane into four distinct non-overlapping regions:

- the overlap of the circular region and the elliptical region;
- the part of the circular region outside the elliptical region;
- the part of the elliptical region outside the circular region;
- the region outside both the circular region and the elliptical region.

Each of these regions may or may not have members. That is, there is a *potential* for membership within each region, although it is possible that one or more of the regions has no members, depending, of course, on exactly what members the sets A and B have.

Now let us think about membership conditions for an object x relative to the sets A and B. We know that either $x \in A$ or $x \notin A$, and we also know that either $x \in B$ or $x \notin B$. Thus, there are four distinct *logical* possibilities regarding membership relative to the sets A and B, only one of which can actually hold for a specific object x. Moreover, these membership possibilities correspond with the regions in the Venn diagram:

- The membership condition

$$x \in A \quad \text{and} \quad x \in B$$

  puts x in the overlap of the circular region and the elliptical region:

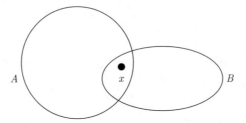

- The membership condition

$$x \in A \quad \text{and} \quad x \notin B$$

  puts x in the part of the circular region lying outside the elliptical region:

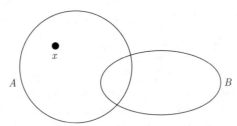

- The membership condition

$$x \notin A \quad \text{and} \quad x \in B$$

  puts x in the part of the elliptical region lying outside the circular region:

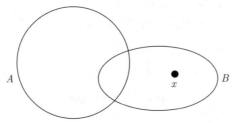

- The membership condition

$$x \notin A \quad \text{and} \quad x \notin B$$

puts x outside both the circular and elliptical regions:

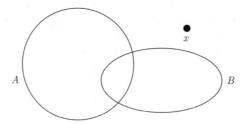

The Intersection of Sets

In the Venn diagram

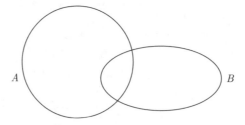

we have agreed that the region in which the circular and the elliptical regions overlap encompasses precisely the members of A that are also members of B. Just as the place where two roads meet is referred to as an *intersection*, we refer to the set of elements A and B have in common as the *intersection* of A and B.

Definition 2.25   The **intersection** $A \cap B$ of sets A and B is the set $\{x \mid x \in A \text{ and } x \in B\}$.

Example 2.26   Because $\{2, 4, 6\}$ and $\{1, 2, 3, 4\}$ have only the numbers 2 and 4 in common, $\{2, 4, 6\} \cap \{1, 2, 3, 4\} = \{2, 4\}$. Also, because it is precisely the counting numbers that are both positive real numbers and integers, $\mathbf{R}^+ \cap \mathbf{Z} = \mathbf{N}$.

Practice Problem 3   Find the intersection of the sets $\{2, 4, 6, 8, 10\}$ and $\{n \in \mathbf{N} \mid n < 7\}$.

Using Venn Diagrams To Depict Sets

When a set is formed out of some given sets, we can use a Venn diagram to represent the set by shading, checking off, or otherwise marking those regions within the diagram that correspond with membership in the set. When we use a Venn diagram to depict a set in this text, we will shade those regions within the diagram that correspond with membership in the set.

Example 2.27   In the context of two given sets A and B, the Venn diagram

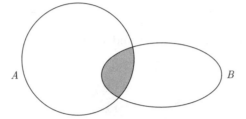

depicts the set $A \cap B$, and the Venn diagram

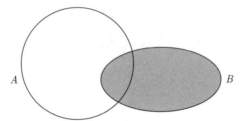

depicts the set B.

The Difference of Sets

In the Venn diagram

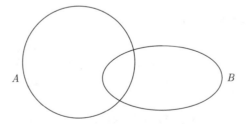

the region lying inside the circle but outside the ellipse corresponds with the set of members of A that are not members of B. Intuitively, we can obtain this region by *taking B away from A*. We will use the notation $A - B$, which can be read A *minus* B, to represent the set whose members are exactly the members of A that are not members of B.

Definition 2.28   The **difference** $A - B$ of sets A and B is the set $\{x \mid x \in A$ and $x \notin B\}$. The set $A - B$ is also referred to as the **relative complement of B in A** and as the **complement of B relative to A**.

Example 2.29   Because 6 is the only member of $\{2, 4, 6\}$ that is not a member of $\{1, 2, 3, 4\}$, we have $\{2, 4, 6\} - \{1, 2, 3, 4\} = \{6\}$.

Practice Problem 4   Determine $\mathbf{R} - \mathbf{R}^-$.

In the context of two given sets A and B, note that the Venn diagram

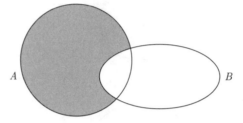

represents the set $A - B$, whereas the Venn diagram

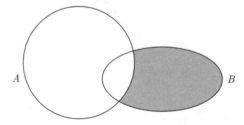

represents the set $B - A$.

The Union of Two Sets

We can also put the members of a set A together with the members of a set B, creating a new set that is depicted by the Venn diagram

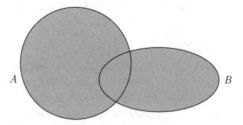

as the region obtained by combining the regions corresponding with $A - B$, $A \cap B$, and $B - A$ into one larger region. This set is called the *union* of A and B because it brings the members of the two sets together into one set.

Definition 2.30 The **union** $A \cup B$ of sets A and B is the set of all objects that are members of at least one of A or B; that is, $A \cup B = \{x \mid x \in A \text{ or } x \in B\}$.

Thus, to be a member of $A \cup B$, an object must be a member of A or a member of B. Here, as is the case throughout mathematics, the word *or* is used in its *inclusive* sense: Any object x that is a member of both A and B satisfies the membership condition

$$x \in A \quad \text{or} \quad x \in B,$$

as does any object x that is a member of exactly one of the sets A and B. Hence, the members of $A \cup B$ should be viewed as those objects that are members of *at least one* of the sets A or B.

Example 2.31 As it is precisely 1, 2, 3, 4, and 6 that are members of at least one of the sets $\{2, 4, 6\}$ or $\{1, 2, 3, 4\}$, we have $\{2, 4, 6\} \cup \{1, 2, 3, 4\} = \{1, 2, 3, 4, 6\}$.

Practice Problem 5 Find $\mathbf{N} \cup \mathbf{Z}^-$.

Venn Diagrams in the Context of Three Sets

Suppose now that we are working with three sets A, B, and C. The following figure depicts a generic Venn diagram for these three sets in which the set A is represented by the circular region, the set B by the elliptical region, and the set C by the triangular region (there is nothing special about the choice of shapes):

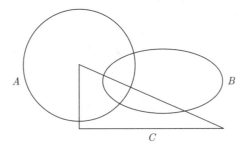

The diagram indicates that the three given sets create eight non-overlapping regions (don't forget the region outside all three sets), each corresponding with a distinct membership possibility relative to the sets A, B, and C. Keep in mind that the set corresponding with a particular region may or may not have members; as before, we are only attempting to indicate the potential for members.

Example 2.32   For instance, the membership condition

$$x \in A \quad \text{and} \quad x \in B \quad \text{and} \quad x \notin C$$

puts x inside both the circular and elliptical regions but outside the triangular region, as depicted in the following diagram:

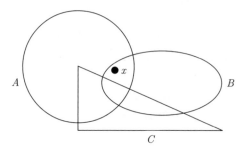

The Algebra of Sets

When we use the set operations union, intersection, and difference to construct "new" sets out of given sets, we say that we are constructing these new sets **algebraically**.

Example 2.33   One algebraic representation of the set depicted in the Venn diagram

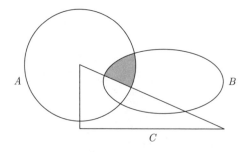

is $(A \cap B) - C$, and there are others; for example, $(A - C) \cap B$ and $(A - C) \cap (B - C)$. Thus, we also see that it may be possible for a given set to be expressed algebraically in multiple ways.

Venn diagrams can also help us formulate conjectures about possible algebraic relationships between sets that we might later try to prove.

Example 2.34   Drawn in the context of three arbitrary sets A, B, and C, a Venn diagram for the set $A \cap B$

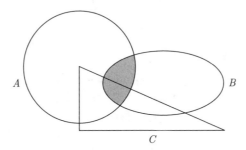

seems to indicate that

$$A \cap B = [(A \cap B) \cap C] \cup [(A \cap B) - C]$$

for any sets A, B, and C. This is because the diagram shows $A \cap B$ as being composed of two regions, one that overlaps C and another that does not overlap C.

There are certain algebraic relationships among the set operations of union, intersection, and difference that are so important, because they are used so often, that we want to be sure to point them out. Two such relationships are the *distributive laws* stated in the following theorem. The first one tells us that union distributes over intersection and the second one that intersection also distributes over union.

Theorem 2.35   (*Distributive Laws for Union and Intersection*) For all sets A, B, and C,

$$A \cup (B \cap C) = (A \cup B) \cap (A \cup C)$$

and

$$A \cap (B \cup C) = (A \cap B) \cup (A \cap C).$$

Example 2.36   We will verify that the distributive law

$$A \cup (B \cap C) = (A \cup B) \cap (A \cup C)$$

holds true for the sets $A = \{1, 2, 3\}$, $B = \{2, 3, 4, 5\}$, and $C = \{2, 4, 6\}$. Observe that

$$
\begin{aligned}
A \cup (B \cap C) &= \{1, 2, 3\} \cup (\{2, 3, 4, 5\} \cap \{2, 4, 6\}) \\
&= \{1, 2, 3\} \cup \{2, 4\} \\
&= \{1, 2, 3, 4\},
\end{aligned}
$$

and

$$
\begin{aligned}
(A \cup B) \cap (A \cup C) &= (\{1, 2, 3\} \cup \{2, 3, 4, 5\}) \cap (\{1, 2, 3\} \cup \{2, 4, 6\}) \\
&= \{1, 2, 3, 4, 5\} \cap \{1, 2, 3, 4, 6\} \\
&= \{1, 2, 3, 4\}.
\end{aligned}
$$

It is important to recognize that this verification with these specific sets does not generally establish the property, however; it simply illustrates the property with a particular example. In Problem 2Q.1, you are asked to examine both distributive laws via Venn diagrams and to attempt to explain why they hold using only the definitions of union, intersection, and set equality.

Two other often-used algebraic properties of sets are known as *DeMorgan's Laws* and are stated in the next theorem.

Theorem 2.37 (*DeMorgan's Laws for Sets*)   For all sets A, B, and C,

$$C - (A \cup B) = (C - A) \cap (C - B)$$

and

$$C - (A \cap B) = (C - A) \cup (C - B).$$

In words, DeMorgan's Laws say that *the complement of the union is the intersection of the complements*, and *the complement of the intersection is the union of the complements.*

Practice Problem 6   Verify that DeMorgan's Law $C - (A \cup B) = (C - A) \cap (C - B)$ holds true for $A = \{1, 2, 3\}$, $B = \{2, 3, 4, 5\}$, and $C = \{2, 4, 6\}$.

The verification requested in Practice Problem 6 does not establish that the property $C - (A \cup B) = (C - A) \cap (C - B)$ holds in general; it simply provides an example illustrating the property with specific sets. You will investigate both of DeMorgan's Laws further in Problem 2Q.2(a,b). Complete verification of DeMorgan's Laws, the distributive laws, and other algebraic properties of sets will have to wait until Chapter 4 when we begin to develop and write mathematical proofs.

Variables Appearing in the Statement of a Definition

It is important to remember that the letters being used as variables in the statement of a definition serve only as placeholders and should not be regarded as the names of particular objects.

For example, in Definition 2.17 we stated the definition of subset as

$$A \subseteq B \text{ iff for every } x, \quad \text{if} \quad x \in A, \quad \text{then } x \in B \tag{1}^2$$

using the placeholder variables A and B. To use this definition to understand what is meant by

$$A - B \subseteq A \cup B, \tag{2}$$

where the letters A and B in (2) are being used as the names of particular sets, we replace each occurrence of the variable A in (1) with the set $A - B$ and each occurrence of the variable B in (1) with the set $A \cup B$, giving us

$$A - B \subseteq A \cup B \text{ iff for every } x, \quad \text{if} \quad x \in A - B, \quad \text{then } x \in A \cup B. \tag{3}$$

We can do this because the variables A and B appearing in (1) are placeholders and can be replaced by *any* sets, in particular the sets $A - B$ and $A \cup B$ mentioned in (2). It does not matter that the same letters, A and B, are being used in two different ways, as variables in (1) and as the names of particular sets in (2), because letters used as variables can always be replaced by other letters. For instance, the definition of subset could be stated using placeholder variables G and H rather than A and B, yielding

$$G \subseteq H \text{ iff for every } x, \quad \text{if} \quad x \in G, \quad \text{then } x \in H. \tag{4}$$

Note that replacing G with $A - B$ and H with $A \cup B$ in (4) produces the same meaning (3) for (2).

---

2. Rather than rewriting a lengthy mathematical expression or statement every time we want to refer to it, we can use a number as a label and make references to the expression or statement via the number. As we move to each new section of the book, we will restart our numbering of such expressions with (1).

Practice Problem 7   The definition of union states that

$$x \in A \cup B \text{ if and only if } x \in A \text{ or } x \in B.$$

By definition of union, then, what does it mean for $x \in (A - B) \cup (B - A)$?

Answers to Practice Problems

1.

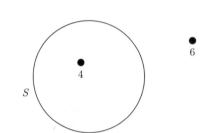

2.

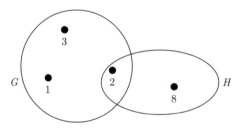

3. $\{2, 4, 6\}$.

4. The real numbers that are not negative form the set $\{t \in \mathbf{R} \mid t \geq 0\}$.

5. We have $\mathbf{N} \cup \mathbf{Z}^- = \{m \in \mathbf{Z} \mid m \neq 0\}$ because bringing all of the natural numbers and all of the negative integers into one set creates the set of all nonzero integers.

6. Observe that

$$C - (A \cup B) = \{2, 4, 6\} - (\{1, 2, 3\} \cup \{2, 3, 4, 5\})$$
$$= \{2, 4, 6\} - \{1, 2, 3, 4, 5\} = \{6\},$$

and

$$(C - A) \cap (C - B) = (\{2, 4, 6\} - \{1, 2, 3\}) \cap (\{2, 4, 6\} - \{2, 3, 4, 5\})$$
$$= \{4, 6\} \cap \{6\} = \{6\}.$$

7. $x \in A - B$ or $x \in B - A$.

2.4   Undefined Notions and Axioms of Set Theory

Questions to guide your reading of this section:

1. What are *undefined notions* and *axioms*?
2. What role do axioms serve relative to undefined notions?
3. What are the two undefined notions of set theory?
4. What axioms are we assuming in our study of sets? What does each of these axioms tell us?

An approach to mathematics that relies exclusively on pictures, intuition, and informal experience ultimately puts at risk the goals of *objective certainty* and *precise communication*. At some point, we want to be *certain* that the conclusions we think we have reached in our investigations are correct. And we want to be able to *communicate* to others our findings and our reasoning in a precise and clear manner.

Launching a Formal Study

At the very beginning of any *formal* mathematical study, we have no "official" recorded facts, no terminology, and no notation. All we have are one or more ideas that appear worthy of study, together with some intuitive feeling for these ideas and how they might relate to one another, and perhaps a certain amount of experience working with the ideas informally. These very first ideas that are formally introduced, without definition, are known as **undefined** (or **fundamental** or **primitive**) **notions**.

Example 2.38   In your study of Euclidean geometry in high school, you probably started with the following *undefined* notions:

- some undefined objects, for example, *point, line,* and *plane*;
- some undefined relationships that may or may not exist among these objects, for instance, *the notion of a point lying on a line or in a plane* and *the notion of a line lying in a plane.*

These undefined or fundamental notions represent at least part of what we believe is worthy of study in the field of geometry.

Once the fundamental notions we wish to study have been identified, we work toward creating an inventory of assumptions that, to some degree, describe the fundamental notions and how they relate to each other. These "initial facts," which we assume without proof, are called **axioms**.

Example 2.38 (continued)   In Euclidean geometry, each of the following is usually taken as an axiom:

- *Given two distinct points, there is exactly one line on which both points lie.*

- *Given three points for which there does not exist a line on which all of them lie, there is exactly one plane in which all three points lie.*

As axioms, neither of these statements is capable of being proved. They are assumptions that help to describe how the undefined geometric objects *point*, *line*, and *plane* interact with one another, and they reflect some of our intuition and experience working informally with these objects.

By making our assumptions explicit, we create a base of facts (the *axioms*) from which other facts, which we call *theorems*, can be determined through the use of deductive reasoning (proof). This framework for formal mathematical activity is called the **axiomatic method** and is the characteristic methodology of pure mathematics.

Formal Assumptions About Sets and Set Membership

We intend to take the notion of *set* as fundamental; that is, undefined. This means we will not be creating a definition for *set*. However, to begin a study of sets that is more careful and formal than what we have conducted thus far, we do need to think about what gives sets their value and significance. Why do we care about sets? How are they helpful to us? Given a set, what do we want to know about it?

Our informal work with sets has all along been concerned with the question of *membership*. Given a set, we want to know "who" its *members* are. That is, naturally intertwined with the undefined object *set* is another undefined notion, that of *membership in a set*. Even though these ideas are, and will remain, undefined, we already have a good intuitive feeling for them. Informally, we think of a set as *something that is made up of members*. It is this relationship that we must capture in the formal assumptions we make.

Axiom 2.39 (*Axioms of Set Theory*).

   1.   *Fundamental Assumption About Set Membership:* Given any set A and any object x, either $x \in A$ or $x \notin A$, but not both.

   2.   *Forming Unions of Sets:* Given any sets, there is a set, called the **union** of the given sets, whose members are precisely those objects that are members of at least one of the given sets.

   3.   *Forming Singletons:* Given any object x, we can form a set $\{x\}$ having x as its only member.

4. *Forming Sets Using Set-Builder Notation:* Given a set A and a sentence $P(x)$ that is unambiguously true or unambiguously false depending on the particular member x of A, there exists a set $\{x \in A \mid P(x)\}$ whose members are precisely those members x of A for which $P(x)$ is true.

5. *Existence of an Empty Set:* There is a set having no members.

The remainder of this section is devoted to discussing the meaning of each of the assumptions stated in Axiom 2.39 as well as illustrating the application of these assumptions.

Our Fundamental Assumption About Set Membership

On an intuitive level, we expect any object we might encounter to be either a member or a nonmember of a given set. Furthermore, we would not expect an object to be simultaneously both a member and a nonmember of a particular set. Axiom 2.39(1), our Fundamental Assumption About Set Membership, formalizes this expectation. For instance, given any set S, the axiom allows us to conclude that either $-1/2 \in S$ or $-1/2 \notin S$, but not both; that is, the number $-1/2$ is either a member or a nonmember of S, but it is impossible for $-1/2$ to be *both* a member *and* a nonmember of S.

> **Practice Problem 1**   According to our Fundamental Assumption About Set Membership, what can we conclude about the college or university you attend relative to any given set?

Unions and Singletons

Axiom 2.39(2) extends the notion of union so that it applies to any number of sets and guarantees that anytime we form the union of sets, the result is itself a set.

Example 2.40   For each counting number n, define $U_n = \{x \in \mathbf{R} \mid x > 1/n\}$. For instance, $U_3 = \{x \in \mathbf{R} \mid x > 1/3\}$, the set of all real numbers larger than 1/3. Axiom 2.39(2) allows us to form

$$U_1 \cup U_2 \cup U_3 \cup \dots$$

and guarantees that the result is actually a set. Problem 2T.2 asks you to explain why this union turns out to be the set $\mathbf{R}^+$.

Now intuitively you have probably always believed you could take some given objects and form a set having these objects as its members. Axiom 2.39(3) makes explicit our ability to do this on an extremely limited scale. For example, this axiom allows us to form the singleton set {2} having the number 2 as its only member.

Of course, we want to be able to construct sets having more than just one member. One way to do this is to use Axiom 2.39(3) to "place" each object that is to be a member of the set under construction into its own singleton set, then "union up" all of these singletons by applying Axiom 2.39(2). We shall refer to this procedure as the **union of singletons set construction process (USSCP)**.

Example 2.41   The set {1, 2, 3} can be constructed as {1} ∪ {2} ∪ {3}. Note also how the USSCP would allow us to construct the set of all copies of this book, the set of all persons living in your hometown at this very moment, and the set of all movies released before the year 2010.

Example 2.42   Consider the numbers we refer to as real numbers, the numbers that can be plotted on a number line, which we are regarding as "given" objects studied in calculus and other mathematics courses you have taken. To form the set **R** of all real numbers, for each real number r we first form the singleton set {r}. After doing this, we take the union of all of these singletons, the result being the set **R**.

There are limitations in the formation of sets using the USSCP. The process can only be applied to clearly identified and inventoried objects whose existence has already been established *before* we try to put them together into a set.

Example 2.42 (continued)   The real numbers are very specific mathematical objects you have worked with on numerous occasions. They are clearly identifiable, for instance as those numbers that can be written in decimal form or as coordinates of points on a number line. Even though there are infinitely many of them, the real numbers are completely "known" to us. There is no possibility that a "new" real number will someday be discovered. As well-defined objects that exist *before* we attempt to collect them into a set, we may use the USSCP to form the set **R** of all real numbers.

Example 2.43   We cannot apply the USSCP to create a "set of all sets." One problem is that the set of all sets would have to include itself as a member if it is to be complete. But if we are trying to construct the set of all sets, we are admitting that we do not already have it available to "place" into a singleton set, which would be necessary to apply the USSCP. Because we cannot completely inventory all of the objects (in this case *all sets*) we are attempting to collect together before beginning the USSCP, the process cannot be used.

Similarly, we cannot create a set of *all* lines. The geometric object known as a line is so omnipresent that it is not possible to imagine cataloging all of them. How would we ever know that we have found all possible lines? Is it not always possible that there are other lines awaiting our discovery?

In contrast, we can conceive of all lines lying in a particular plane, so it would be possible to form the set of all lines lying in this plane.

> **Practice Problem 2**   Explain how the USSCP could be used to create the set of all cars that park on your school's campus today.

Set-Builder Notation Revisited

Quite often, the objects we wish to collect together into a set are identified only by a property they have in common. Axiom 2.39(4) makes official our ability to create a subset of a set consisting of all members of the set that possess a certain property. It is the formal basis for the construction of sets via set-builder notation.

Example 2.44   Consider the set $\mathbf{Q}$ of rational numbers and the property of being less than 3. For any member x of $\mathbf{Q}$, this property can be expressed by the symbolic sentence $x < 3$, which is playing the role of $P(x)$ in Axiom 2.39(4). The axiom tells us that there really is a set whose members are precisely those rational numbers less than 3. In set-builder notation, this set can be denoted $\{x \in \mathbf{Q} \mid x < 3\}$.

> **Practice Problem 3**   Use set-builder notation to describe the set of all natural numbers that are multiples of 5.

Example 2.45   We can use Axiom 2.39(4) to explain why the intersection $A \cap B$ of two sets A and B is also a set. Note that we may view *being a member of the set B* as a property that a member of the set A may or may not possess [unambiguously, according to our Fundamental Assumption About Set Membership, Axiom 2.39(1)]. Axiom 2.39(4) allows us to form the set of all members of A possessing this property; that is, the set $\{x \in A \mid x \in B\}$, which is really $A \cap B$. In Problem 2T.4, you will show how Axiom 2.39(4) allows us to conclude that *the difference $A - B$ of sets A and B is also a set.*

We want to mention that care must be taken in attempting to define a set via a property using set-builder notation. The potential pitfalls of too casual an approach are illustrated in Problems 2T.6 and 2T.7.

Empty Sets Revisited

Recall that a set is **empty** provided it has no members. Empty sets arise naturally in mathematics. For instance, the set of real number solutions to the equation $x = x + 1$ is empty. Because none of the other assumptions made in Axiom 2.39 guarantees the existence of an empty set, we explicitly assert the existence of such a set in 2.39(5). As mentioned earlier, we will eventually prove that all empty sets are equal to one another, and we use the symbol $\varnothing$ to denote the empty set.

The Importance of Set Theory to Mathematics

It is actually possible to trace all mathematical ideas back to the study of sets. Because of this, the terminology, notations, and conventions of set theory often make it easier to communicate mathematics than the use of English alone. Much of our work in this chapter is intended to help you gain experience in using set-theoretic notation and terminology; that is, in using the language of set theory. As you proceed further in the book, you will see how this language can be used in making mathematical ideas—even those that do not appear to be set-theoretic in nature—clear and precise. In addition, by formulating axioms that make explicit our assumptions about sets, we have provided ourselves with a foundation for our later work in Chapter 4 where we learn how to develop and write proofs for some of our set-theoretic conjectures.

Answers to Practice Problems

1. For any given set, the college or university you attend is either a member of this given set or it is not a member of this given set, but it is impossible for the college or university you attend to be both a member and a nonmember of this set.

2. First, form a singleton set corresponding with each car that parks on your school's campus today. Then take the union of all of these singletons. The result is the set of all cars that park on your school's campus today.

3. $\{x \in \mathbf{N} \mid x \text{ is a multiple of } 5\}$.

2.5   Axioms for the Real Numbers

Questions to guide your reading of this section:

1. What notions are we taking as undefined in our study of the real numbers?

2. What does it mean to say that a certain set of numbers is *closed* under addition or multiplication?

3. What do the *commutative* and *associative properties* of addition and multiplication allow us to do? What does the *distributive property* allow us to do?

4. What number serves as the *additive identity* in the set of real numbers? What number serves as the *multiplicative identity*?

5. How do we obtain the *additive inverse* of a given real number? How do we obtain the *multiplicative inverse* of a given nonzero real number?

6. What does the *Law of Trichotomy* tell us about a given real number?

Recall that the real numbers may be informally viewed as those numbers that can be expressed as decimals or that can be plotted on the number line. We now initiate a more formal study of the real numbers, including some of their mathematical properties and certain operations that can be applied to them. We take the notions of *real number, addition, multiplication*, and *positive* as undefined. The most essential properties that we believe govern how these undefined notions relate to one another are taken as our axioms for the real numbers and are listed in Axiom 2.46. The rest of the section further clarifies these assumptions, describes how they can be applied, and discusses some of the terminology associated with them.

Axiom 2.46 (*Axioms for the Real Numbers*)   There is a set **R** whose members are called *real numbers*. There are two operations, *addition* and *multiplication*, that can be applied to any two real numbers a and b to produce, respectively, a **sum** $a + b$ and a **product** $a \cdot b$, or ab. Some real numbers possess the property of being *positive*, and the set of all positive real numbers is designated by $\mathbf{R}^+$. We also make the following assumptions:

1. *Closure Properties:* The following subsets of the set of real numbers are **closed** under both addition and multiplication, meaning that the sum and product of members of the subset always results in a member of that same subset: **R**, **Q**, **Z**, **N**, and $\mathbf{R}^+$.

2. *Commutative Properties:* For all real numbers a and b, $a + b = b + a$ and $ab = ba$.

3. *Associative Properties:* For all real numbers a, b, and c,

$$(a + b) + c = a + (b + c) \text{ and } (ab)c = a(bc).$$

4. *Distributive Property:* For all real numbers a, b, and c, $a(b + c) = ab + ac$.

5. *Existence of an Additive Identity:* There is a real number 0, called the **additive identity**, with the property that for every real number a, $0 + a = a$.

6. *Existence of a Multiplicative Identity:* There is a real number 1, called the **multiplicative identity**, which is distinct from 0 and with the property that for every real number a, $1a = a$.

7. *Existence and Fundamental Property of Additive Inverses:* Every real number a has an **opposite** or **additive inverse** $-a$ that is also a real number and that has the property that $a + (-a) = 0$.

8. *Existence and Fundamental Property of Multiplicative Inverses:* Every nonzero real number a has a **reciprocal** or **multiplicative inverse** $1/a$ that is also a real number and that has the property that $a \cdot (1/a) = 1$.

9. *Law of Trichotomy:* For any real number a, exactly one of the following is true: a is positive, a is negative, or $a = 0$, where, by **negative**, we mean that $-a$ is positive.

Closure

Axiom 2.46(1) makes explicit that when certain types of numbers are combined under either addition or multiplication, the computed result is again a number of the same type as that with which we started.

Among other things, the axiom states that the set **Z** of all integers is *closed* under both addition and multiplication. This means that whenever we add two integers, the resulting number is also an integer, and whenever we multiply two integers, the resulting number is also an integer. This should seem reasonable to you. For instance, adding the integers 3 and −5 yields another integer, namely −2, whereas multiplying the integers 3 and −5 yields the integer −15. The closure of **Z** under the operations of addition and multiplication simply tells us that there is never a case where adding or multiplying integers could produce a non-integer result.

Moreover, the axiom tells us that each of the sets **R**, **R**⁺, **Q**, and **N** is also closed under both addition and multiplication. So whenever we add or multiply two real numbers the result is a real number, whenever we add or multiply two positive real numbers the result is a positive real number, whenever we add or multiply two rational numbers the result is a rational number, and whenever we add or multiply two natural numbers the result is a natural number. Thus, even though we may not be able to give a precise accounting of exactly which real number results when $\sqrt{2}$ and π are added, we can be sure, as $\sqrt{2}$ and π are both real numbers, that the result $\sqrt{2} + \pi$ is a real number as well. Moreover, because $\sqrt{2}$ and π are both *positive* real numbers, their sum $\sqrt{2} + \pi$ is also positive.

It is not always the case, though, that a set is closed under a specified operation. For example, because both $\sqrt{2}$ and $-\sqrt{2}$ are irrational, but $\sqrt{2} + (-\sqrt{2}) = 0$, and 0, being rational, is not irrational, it follows that the set **R** − **Q** of all irrational numbers is *not closed* under addition.

Practice Problem 1   Is the set of all negative real numbers closed under addition? Is it closed under multiplication? Is it closed under subtraction?

Commutativity and Associativity

Imagine that we want to compute the sum $7 + 4 + 3$. We know from experience that we may switch the order in which the numbers are being added to $4 + 7 + 3$ without affecting the value of the sum. This ability to switch the order of computation is known as *commutativity*. The commutativity of both the addition and multiplication of real numbers is an essential property that is assumed in Axiom 2.46(2). Note, however, that not all operations are commutative. For instance, subtraction is not because, for example, $11 - 6 \neq 6 - 11$.

Now, in calculating $4 + 7 + 3$, we may certainly proceed directly from left to right as

$$4 + 7 + 3 = (4 + 7) + 3 = 11 + 3 = 14,$$

but some of us may believe that by computing $7 + 3$ first, the computation is a bit easier to handle:

$$4 + 7 + 3 = 4 + (7 + 3) = 4 + 10 = 14.$$

Of course, either approach is legitimate, because in Axiom 2.46(3) we are assuming the associativity of both the addition and multiplication of real numbers. In other words, when adding or multiplying three (or more) real numbers, it does not matter where in the calculation we begin computing. The specific application of the associativity of addition that we used above is $(4 + 7) + 3 = 4 + (7 + 3)$.

Not all operations are associative, however. For example, division is not associative because, for instance, $(12 \div 6) \div 2 = 1$, whereas $12 \div (6 \div 2) = 4$.

Practice Problem 2   Consider the calculation

$$5 \cdot 7 \cdot 2 = 7 \cdot 5 \cdot 2 = 7 \cdot 10 = 70.$$

Which property, the commutative property of multiplication or the associative property of multiplication, justifies the first equality? Which of these two properties justifies the second equality?

The Distributive Property

Because multiplication distributes over addition, we may rewrite $2(x + 5y)$ as $2x + 10y$. This *distributive property*, Axiom 2.46(4), is the only one of our assumptions about the real numbers that involves both addition and multiplication. Thus, it provides essential information about how these operations relate to each other.

For example, the distributive property encodes the notion that multiplication of natural numbers may be viewed as repeated addition. For instance,

$$3 \cdot 4 = (1 + 1 + 1) \cdot 4 = 1 \cdot 4 + 1 \cdot 4 + 1 \cdot 4 = 4 + 4 + 4,$$

indicating that multiplying 3 by 4 is the same as adding three 4's.

Practice Problem 3   Indicate how the distributive property can be used to justify the fact that $(a + b)(c + d) = ac + ad + bc + bd$.

Additive and Multiplicative Identities

Whenever we add 0 to a real number, the result is the same number we started with. Similarly, whenever we multiply a real number by 1, we obtain the original number back again. These fundamental properties, along with the very existence of the real numbers 0 and 1 possessing them, are assumed in Axioms 2.46(5) and 2.46(6).

The reason for using the terminology **additive identity** for 0 and **multiplicative identity** for 1 is that the process of adding 0 or multiplying by 1 causes a number to remain unchanged, thus retaining its original identity.

Additive and Multiplicative Inverses

The use of the terms *additive inverse* and *multiplicative inverse* has to do with the way in which addition and multiplication can be "undone" or "inverted."

Example 2.47   To undo addition by 3.6, we can add the number −3.6, the *opposite* or *additive inverse* of 3.6. This principle is used to solve the equation $x + 3.6 = 4.7$ as we know $3.6 + (-3.6) = 0$:

$$x + 3.6 = 4.7 \implies x + 3.6 + (-3.6) = 4.7 + (-3.6) \implies x + 0 = 1.1 \implies x = 1.1.$$

Similarly, to undo multiplication by 3.6, we can multiply by $\frac{1}{3.6}$, the *reciprocal* or *multiplicative inverse* of 3.6. This principle is used to solve the equation $3.6x = 10.8$ as we know $\frac{1}{3.6} \cdot 3.6 = 1$:

$$3.6x = 10.8 \implies \tfrac{1}{3.6} \cdot 3.6x = \tfrac{1}{3.6} \cdot 10.8 \implies 1 \cdot x = 3 \implies x = 3.$$

Note that the real number 0 does not have a multiplicative inverse because there is no real number r for which $0 \cdot r = 1$ (we will formally establish this when we prove in Chapter 4 that $0 \cdot r = 0$ for any real number r).

Because we want to be able to solve equations of the types illustrated in Example 2.47, we assumed in Axiom 2.46(7) that each real number has an additive inverse and in Axiom 2.46(8) that each real number other than 0 has a multiplicative inverse.

Practice Problem 4   Solve the equation $5x + 8 = 23$ for x, pointing out where you make use of the notions of additive inverse and multiplicative inverse.

The Law of Trichotomy

When we draw a number line to model the set of real numbers, we indicate a point intended to represent the number 0. All numbers corresponding with points on the line lying to one side of 0 are referred to as *positive*, whereas all numbers corresponding with points on the line lying on the other side of 0 are referred to as *negative*. Thus, we think of real numbers as being of three types: those that are positive, those that are negative, and the number 0 that separates the positives from the negatives. Axiom 2.46(9) formally expresses this assumption.

Note particularly that whereas the notion of *positive* is undefined, we defined **negative** real numbers to be those real numbers whose opposites are positive. For example, the real number -7 is negative because its opposite $-(-7)$ is actually the positive number 7.

Practice Problem 5   Suppose the real number y is neither positive nor negative. What can you conclude based on the Law of Trichotomy?

When we begin to write proofs in Chapter 4, we will see how other facts about the real numbers, many of which you are familiar with already, follow from the formal assumptions we have made here.

Answers to Practice Problems

1. The set of all negative real numbers is closed under addition (as the sum of two negative numbers is also negative) but is not closed under multiplication (as the product of two negative numbers is positive) or subtraction (e.g., $-3 - (-5) = 2$, which is not negative).

2. The commutative property justifies the first and the associative property the second.

(continues)

Answers to Practice Problems (continued)

3. First, the distributive property allows us to distribute the number $a + b$ to each of the terms c and d of the expression $c + d$, giving us

$$(a + b)(c + d) = (a + b)c + (a + b)d.$$

Then, applying the distributive property two times on the right side of the resulting equation gives us

$$(a + b)(c + d) = ac + bc + ad + bd.$$

Using the commutativity of addition to reorder the terms on the right side of the equation just obtained gives us our desired conclusion that

$$(a + b)(c + d) = ac + ad + bc + bd.$$

4. To solve $5x + 8 = 23$, we first add the additive inverse -8 of 8 to both sides of the equation to get $5x = 15$. Then, we multiply both sides of this resulting equation by the multiplicative inverse 1/5 of 5 to get $x = 3$.

5. We may conclude that $y = 0$.

Chapter 2 Problems

2A. Forming Sets by Listing

1. Use set braces and listing to describe the set whose only members are 4, 10, and 0.

2. Use the ellipsis notation, listing, and set braces to express the following sets:

    (a) the set of all counting numbers from 112 through 256;

    (b) the set of all multiples of 3 that lie between 5 and 100;

    (c) the set of all negative multiples of 3.

2B. Set Membership

1. Write a mathematical statement using the symbol $\in$ to communicate the fact that 4 is a member of the set E of even integers. How could we symbolically communicate how 5 is related to the set E?

2. Consider the set $M = \{1, 2, \{3, 4\}, \{5, \{6, 7\}\}\}$.

    (a) Is 1 a member of M?

    (b) Is 3 a member of M?

    (c) List all members of M.

2C. Singletons and Doubletons

1. Is there a distinction between $\{5\}$ and 5? If so, what is it?

2. Is there a distinction between $\{5\}$ and $\{5, 5\}$? If so, what is it?

3. Let $\{r, s\}$ be the doubleton set whose members r and s satisfy $r + s = 10$ and $4r - s = 15$. Find the actual values of r and s. Then express the set $\{r, s\}$ in as simple a form as possible.

2D. Set-Builder Notation

1. Wherever possible, list all the members of the given set. If this is not possible, write a sentence that precisely describes the set's members.

    (a) $\{x \in \mathbf{N} \mid 3 < x \leq 10\}$.   (b) $\{x \in \mathbf{R} \mid x = x^2\}$.   (c) $\{x \in \mathbf{R} \mid |x| \neq x\}$.

    (d) $\{x \in \mathbf{R} \mid x = x + 1\}$.   (e) $\{k \in \mathbf{N} \mid k^3 = k\}$.   (f) $\{k \in \mathbf{Z} \mid k^3 = k\}$.

    (g) $\{q \in \mathbf{Q} \mid (q^2 - 1)(q^2 - 0.25)(q^2 - 5) = 0\}$.

    (h) $\{q \in \mathbf{R}^+ \mid (q^2 - 1)(q^2 - 0.25)(q^2 - 5) = 0\}$.

    (i) $\{z \in \mathbf{C} \mid z \in \mathbf{R}\}$.   (j) $\{3n \mid n \in \mathbf{Z}^-\}$.

2. Express the given set using set-builder notation. Use symbols only, no words.

    (a) The set of all real number solutions to the equation $x^2 + x - 1 = 0$.

    (b) The set of integers greater than 1000.

    (c) The set of all real numbers at which the sine and cosine functions are equal.

3. Translate the symbolic descriptions of the sets in 1(a, b, c) into coherent English that suppresses the use of dummy variables.

4. Suppose we want to use set-builder notation to describe the set of all negative real numbers and we write $\{y \in \mathbf{R} \mid t < 0\}$. Tell what is wrong with what we have written and how it can be fixed.

5. Can we express the set of positive rational number solutions to the equation $81x^4 - 16 = 0$ as $\{r \in \mathbf{Q}^+ \mid 81r^4 - 16 = 0\}$?

2E. Rational Numbers

Recall that, by definition, the set of rational numbers is

$$\mathbf{Q} = \{a/b \mid a, b \in \mathbf{Z}, b \neq 0\}.$$

1. Why is it necessary to impose the condition $b \neq 0$ in this description of $\mathbf{Q}$?
2. Explain why, according to the given definition, 5, –5, and 0 are rational numbers. Is every integer a rational number?
3. Explain why, according to the given definition, 0.461 is a rational number. Does every terminating decimal represent a rational number?
4. Explain why, according to the given definition, 0.3333... is a rational number. Does every repeating decimal represent a rational number?

2F. Complex Numbers

Recall that, by definition, the set of complex numbers is

$$\mathbf{C} = \{a + bi \mid a, b \in \mathbf{R}\}, \text{ where } i = \sqrt{-1}.$$

1. Explain why each of the following is a complex number.
    (a) $\frac{1}{2}i - \sqrt{6}$. (b) i. (c) 0. (d) π. (e) $i^2 - i$.
2. Are there any integers that are not complex numbers? Are there any real numbers that are not complex numbers?

2G. Mathematical Definitions

1. Is it possible to define all notions in a formal mathematical study? Why or why not?
2. Why do we often have both intuitive and formal versions of the definition of a given term? Of what particular importance is the intuitive version? Of what particular importance is the formal version?

2H. Equality of Sets

1. Explain why {2, 4, 6} = {4, 2, 6}.
2. Explain why the following is *not* a correct definition for set equality:

    Two sets A and B are equal iff for every x, x ∈ A and x ∈ B.

2I. Subsets and Supersets

1. Determine all subset relationships among the sets $\{1, 2, 3, 4\}$, $\{2, 4, 6\}$, $\{2, 4\}$, $\{x \in \mathbf{N} \mid x < 6\}$, and $\varnothing$.

2. Write a sentence that expresses the notion that M is a *superset* of N in terms of membership in the sets M and N. Make use of the $\in$ notation for membership.

3. Give two examples of proper subsets of the set $\mathbf{Q}$. What is the improper subset of $\mathbf{Q}$? Can a proper subset of $\mathbf{Q}$ be infinite?

4. Correctly fill in each blank with the appropriate statement chosen from among $x \in A$, $x \notin A$, $x \in B$, and $x \notin B$ (some may be used more than once, some not at all).

    $A \subset B$ if and only if both of the following are true:

    (a) for every x, if _____, then _____;

    and

    (b) there exists x such that _____ and _____.

2J. Membership Versus Subset

1. Students often loosely translate each of the symbols $\in$, $\subseteq$, and $\subset$ as "is a part of." Although this is somewhat correct on a very informal level, each symbol really has a different meaning from the others. Create as many true statements as you can using these three symbols along with 3, $\{3\}$, $\{3, \{3\}\}$, and $\mathbf{N}$.

2. Which of the following statements are true?

    (a) $\varnothing \in \varnothing$.   (b) $\varnothing \in \{\varnothing\}$.   (c) $\varnothing \subseteq \{\varnothing\}$.   (d) $\varnothing \subset \{\varnothing\}$.

    (e) $\{\varnothing\} \subseteq \{\varnothing\}$. (f) $\{\varnothing\} \subset \{\varnothing\}$. (g) $\{\varnothing\} \in \{\varnothing\}$. (h) $\mathbf{R} \subseteq \mathbf{R}$.

    (i) $\mathbf{R} \in \mathbf{R}$.   (j) $\mathbf{R}^- \subset \mathbf{R}$.   (k) $\mathbf{R}^- \subseteq \mathbf{R}$.   (l) $\mathbf{R}^- \in \mathbf{R}$.

2K. The Power Set of a Set

1. Find the power set of each set.

    (a) $\varnothing$.   (b) $\{1\}$.   (c) $\{1, 2\}$.   (d) $\{1, 2, 3, 4\}$.

2. Fill in the blank with the correct symbol, $\in$ or $\subseteq$:

    For any set S, $x \in \mathcal{P}(S)$ if and only if x _____ S.

2L. The Algebra of Sets

1. Suppose that $A = \{1, 2, 3, 4, 5\}$, $B = \{2, 3, 5, 7\}$, and $C = \{2, 4\}$. Find:

   (a) $A \cup B$. (b) $A \cup C$. (c) $A \cap B$. (d) $A \cap C$. (e) $A - B$.

   (f) $B - A$. (g) $(B \cap C) \cup A$. (h) $(B - C) \cap A$.

2. Suppose $X = \{x \in \mathbf{N} \mid x < 10\}$ and $Y = \{x \in \mathbf{N} \mid x \geq 4\}$. Find:

   (a) $X \cap Y$. (b) $X \cup Y$. (c) $X - Y$. (d) $Y - X$.

3. Consider an arbitrary set A (meaning we know A is a set, but not which particular set). Provide an intuitive explanation for each of the following based on our definitions of union, intersection, and difference of sets and our understanding that two sets are equal if and only if they have exactly the same members.

   (a) $A \cup A = A$. (b) $A \cap A = A$. (c) $A - A = \varnothing$.

4. Find a way to form the doubleton set $\{3, 5\}$ from the singleton sets $\{3\}$ and $\{5\}$ using one of the set operations we have introduced.

5. Which of the following are *members* of the set $\mathbf{R} \cup \mathcal{P}(\mathbf{R})$? Which are *subsets* of $\mathbf{R} \cup \mathcal{P}(\mathbf{R})$?

   (a) π. (b) $\{\pi\}$. (c) $\{x \in \mathbf{R} \mid x > \pi\}$. (d) $\{\pi, \{1, 2\}\}$.

   (e) $\mathbf{R}$. (f) $\varnothing$.

2M. The Empty Set

1. Is there any difference between $\varnothing$ and $\{\varnothing\}$? If so, what is it?

2. Let A be an arbitrary set. Make conjectures about what you think each of the following should simplify to.

   (a) $A - \varnothing$. (b) $A \cap \varnothing$. (c) $A \cup \varnothing$. (d) $\varnothing - A$.

2N. Venn Diagrams, Membership Conditions, and Set Algebra

1. Let A and B be two generic sets. For each given set constructed from A and B:

   - Draw a Venn diagram and mark regions within the diagram so as to depict the set.

   - Find the simplest way to express the notion that x is a member of the set using only the statements $x \in A$, $x \notin A$, $x \in B$, and $x \notin B$, along with the logical connectives *and* and *or*.

- Use this expression to determine if there is a simpler way (one that uses fewer symbols) to express the set algebraically using only the symbols A, B, $\cup$, $\cap$, and $-$.

   (a) $(A \cap B) \cup B$.   **(b)** $(A \cup B) \cap B$.   **(c)** $A \cup (B - A)$.

   (d) $(A \cup B) - A$.   **(e)** $A - (A \cap B)$.   **(f)** $(A - B) \cup (A \cap B)$.

   (g) $(A - B) \cup (B - A)$.   **(h)** $[(A - B) \cup (B - A)] - A$.

2. The Venn diagram

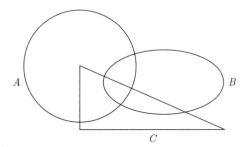

   separates the plane into eight non-overlapping regions, each corresponding with a potentially different set. For each of these eight sets, find the simplest way to express the notion that x is a member of the set using only the statements $x \in A$, $x \notin A$, $x \in B$, $x \notin B$, $x \in C$, and $x \notin C$, along with the logical connective *and*.

3. Let A, B, and C be three generic sets. For each given set constructed from A, B, and C:

   - Draw a Venn diagram and mark regions within the diagram so as to depict the set.

   - Find the simplest way to express the notion that x is a member of the set using only the statements $x \in A$, $x \notin A$, $x \in B$, $x \notin B$, $x \in C$, and $x \notin C$, along with the logical connectives *and* and *or*.

   - Use this expression to determine if there is a simpler way (one that uses fewer symbols) to express the set algebraically using only the symbols A, B, C, $\cup$, $\cap$, $-$, (, and).

   (a) $(A - B) \cap C$.   **(b)** $C - (A \cap B)$.   **(c)** $(A \cap B) - (B \cap C)$.

   (d) $(C - A) \cap (C - B)$.   **(e)** $(C - A) \cup (C - B)$.

   (f) $(C - A) - (C - B)$.   **(g)** $(C - A) - (B - A)$.

4. For the set represented by the given Venn diagram:

   - Find the simplest way to express the notion that x is a member of the set using only the statements $x \in A$, $x \notin A$, $x \in B$, $x \notin B$, $x \in C$, and $x \notin C$, along with the logical connectives *and* and *or*.

- Use this expression to represent the set algebraically using only the symbols A, B, C, $\cup$, $\cap$, $-$, $($, and $)$.

- If possible, simplify the algebraic representation you just obtained so that it uses as few symbols as possible.

(a)

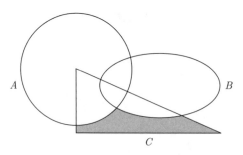

(b)

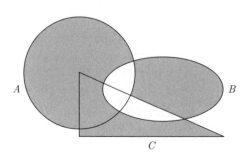

(c)

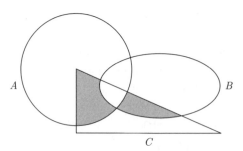

5. Consider the sets $C - (A \cap B)$ and $(C - A) - (C - B)$.

   (a) Use the Venn diagrams you created for these sets in 3(b) and 3(f) earlier to determine which one of them could possibly include a member of $A - B$.

(b) For the one of the sets, $C - (A \cap B)$ or $(C - A) - (C - B)$, that you believe could possibly include a member of $A - B$, explain why using the definitions of difference and/or intersection.

(c) For the one of the sets, $C - (A \cap B)$ or $(C - A) - (C - B)$, that you believe could not include a member of $A - B$, explain why using the definitions of difference and/or intersection.

2O. Sorting Out the Language

Suppose A and B are sets.

1. Determine which of the four connectives

$$\cup \quad \cap \quad \text{and} \quad \text{or}$$

can be placed in the blank to make a well-formed mathematical statement. There may be more than one correct answer.

(a) $x \in A$ _____ B.   (b) $x \in A$ _____ $x \in B$.

2. Write a sentence that explains the difference between $A \cap B$ and $x \in A \cap B$.

2P. Symmetric Difference

Once we have in hand some set operations such as $\cup$, $\cap$, and $-$, we can use them to define other set operations. For example, the **symmetric difference** of two sets A and B, which we shall denote by $A \triangle B$, is defined by the following equation:

$$A \triangle B = (A - B) \cup (B - A).$$

1. Draw a Venn diagram and mark regions to represent $A \triangle B$.

2. Using only the statements $x \in A$, $x \notin A$, $x \in B$, and $x \notin B$, along with the logical connectives *and* and *either... or*, complete the following statement that characterizes symmetric difference in terms of the notion of membership:

$x \in A \triangle B$ *if and only if* _____.

3. Give intuitive arguments supporting each of the following:

(a) For any sets A and B, $A \triangle B = B \triangle A$.

(b) For any sets A and B, $A \triangle B = (A \cup B) - (A \cap B)$.

2Q. Set-Theoretic Research

1. Let A, B, and C be arbitrary sets. Try to determine which of the follow-ing "distributive properties" are true in two ways:

   - First, make a Venn diagram for each of the two sets on either side of the equals sign (mark regions appropriately).

   - Then, use the definitions

     $x \in G \cap H$ iff $x \in G$ and $x \in H$

     $x \in G \cup H$ iff $x \in G$ or $x \in H$

     $x \in G - H$ iff $x \in G$ and $x \notin H$

     to translate the notion that x is a member of each of the two sets on either side of the equals sign and determine whether the result-ing membership conditions are equivalent, explaining how you reach your conclusion. For example, in (a) you would have to translate $x \in A \cap (B \cup C)$ and compare what you come up with to what $x \in (A \cap B) \cup (A \cap C)$ translates as to decide if they really mean the same thing or are fundamentally different requirements.

   - (a) $A \cap (B \cup C) = (A \cap B) \cup (A \cap C)$.
   - (b) $A \cup (B \cap C) = (A \cup B) \cap (A \cup C)$.
   - (c) $A \cap (B - C) = (A \cap B) - (A \cap C)$.
   - (d) $A \cup (B - C) = (A \cup B) - (A \cup C)$.
   - (e) $C - (A \cup B) = (C - A) \cup (C - B)$.
   - (f) $C - (A \cap B) = (C - A) \cap (C - B)$.

2. Let A, B, and C be arbitrary sets. Follow the same directions as in (1) above:

   - (a) $C - (A \cup B) = (C - A) \cap (C - B)$.
   - (b) $C - (A \cap B) = (C - A) \cup (C - B)$.
   - (c) $(A \cap B) \cup C = A \cap (B \cup C)$.

2R. Disjoint Sets

Two sets A and B are said to be **disjoint** iff none of the members of A are members of B.

1. Give a formal version of the statement of the definition of disjoint sets that makes use of the $\in$ notation for membership and the logical phrase *if . . . then.*

2. Create an alternative definition for the notion of disjoint sets using the set operation $\cap$.

3. Explain why, for any sets G and H, the sets $G \cap H$ and $G - H$ must be disjoint.

2S. Axioms in Mathematics Generally

1. What general purpose do axioms serve in mathematics?

2. What role do axioms play relative to the undefined notions in a given mathematical study?

3. Can an axiom be proved?

2T. Axioms of Set Theory

1. According to our Fundamental Assumption About Set Membership, given a set M, what can we conclude about the number 0 relative to M?

2. Recall that in Example 2.40 we defined, for each counting number n,

$$U_n = \{x \in \mathbf{R} \mid x > 1/n\}.$$

    Explain why $U_1 \cup U_2 \cup U_3 \cup \ldots = \mathbf{R}^+$.

3. Use the union of singletons set construction process to construct the given set.

    (a) $\{0, \sqrt{2}, -5\}$.   **(b)** $\mathbf{N}$.   **(c)** $P(A)$ for a given set A.

4. In Example 2.45, we used Axiom 2.39(4) to explain why the intersection of two sets is also a set. Use a similar argument to explain why the difference $A - B$ of sets A and B must be a set.

5. Recall that *point* is an undefined term in Euclidean geometry. Is it possible to use any of our set-theoretic axioms to form the set of *all* points? What about the set of all points lying on some specified line or in some specified plane? If one or more of these can be formed via the axioms, suggest how.

6. Sometimes a property we might formulate is not of the type acceptable for applying Axiom 2.39(4). The axiom requires that the property used be unambiguously true or unambiguously false for each object to which it is being applied. We now illustrate how this can fail to happen.

    Suppose the barber in a certain town is a man. Also suppose the barber shaves exactly those men in the town who do not shave themselves, and every man in the town shaves or is shaved. If we let T be the set of all men in the town and think of *not shaving oneself* as a property a member of T may possess, it seems that Axiom 2.39(4) allows for the construction of

    $$\{x \in T \mid x \text{ does not shave himself}\},$$

    the set of all men in the town who do not shave themselves. Now consider the sentence

    The barber does not shave himself.   (1)

(a) Explain why the sentence (1) cannot be true.

(b) Explain why the sentence (1) cannot be false.

Because we cannot mark (1) as unambiguously true or unambiguously false, we must conclude that, in the presence of what we have been told, *not shaving oneself* is a property for which Axiom 2.39(4) cannot be applied. Thus, it is not possible to use this axiom to construct a set of all men in the town who do not shave themselves.

The situation discussed in this problem is known as *The Barber's Paradox*. In general, a **paradox** is a statement or situation that is self-contradictory or otherwise logically unresolvable.

(c) How could we change the assumptions in The Barber's Paradox so that it no longer forces us into a contradiction?

7. It is important to understand that Axiom 2.39(4) does not allow us to form the set of all objects possessing a particular property, only the set of all objects *that are members of some already existing set* having the property. This is because allowing the formation of a set via a property without restriction can sometimes lead to a contradiction.

For example, suppose we think of *being a set* as a property, and we try to form the set Ω of all objects having this property; in other words, the "set of all sets."[3] Observe that the sets we have worked with in this book so far all have the property that they are not members of themselves. For instance, $\mathbf{N}$ is not a member of $\mathbf{N}$ because the set of all counting numbers is not itself a counting number. Now let Γ be the set of *all* sets that are not members of themselves;[4] that is, let

$$\Gamma = \{S \in \Omega \mid S \notin S\}.$$

According to Axiom 2.39(4), Γ should be a set because it consists of those members of the set Ω with the property that they *are not members of themselves*. Now, if we regard Γ as both a set and an object (this is possible because sets are, after all, objects), our Fundamental Assumption About Set Membership tells us that either $\Gamma \in \Gamma$ or $\Gamma \notin \Gamma$, but not both.

(a) Suppose $\Gamma \in \Gamma$. Explain how this allows us to conclude that $\Gamma \notin \Gamma$.

(b) Suppose $\Gamma \notin \Gamma$. Explain how this allows us to conclude that $\Gamma \in \Gamma$.

Thus, if one of the two sentences $\Gamma \in \Gamma$ or $\Gamma \notin \Gamma$ is true, so is the other. They are either both true or both false, a direct contradiction to our Fundamental Assumption About Set Membership. What has led

---

3. Ω is the uppercase Greek letter "omega."
4. Γ is the uppercase Greek letter "gamma."

to these contradictions is the attempt to create Ω, a set of all sets. The property used in the attempt to construct Ω, that of *being a set*, was applied without restriction rather than just to the objects in some given set. Axiom 2.39(4) does not allow this, and the situation presented here illustrates why.[5]

2U. Axioms and Properties of the Real Numbers

1. Intuitively, when a set is *closed* under a certain operation, the set is "closed off" from the "world" outside itself in the sense that applying the operation to members of the set will never produce anything outside that set. Determine whether the given set is closed under the given operation.

   (a) The set of natural numbers under the operation of subtraction.

   (b) The set of nonzero rational numbers under the operation of multiplication.

   (c) The set of nonzero real numbers under the operation of addition.

2. Which of the following numbers are real numbers?

   (a) 1000. (b) $-\dfrac{1}{2}$. (c) 0. (d) $\dfrac{1-\sqrt{5}}{2}$. (e) e. (f) i.

3. Correctly fill in the blanks to justify the conclusions being made.

   (a) Suppose it is known that both m and n are integers. Then we may conclude that $m + n$ is an integer because _____.

   (b) Suppose it is known that x is a rational number. Then we may conclude that $5x$ is rational because _____.

   (c) Assume that we know t is positive. It follows that $t^2 + t$ is positive because _____.

   (d) We may conclude that $x^2 + 0 = x^2$ because 0 is the _____.

   (e) We may conclude that $\frac{1}{2}y + \left(-\frac{1}{2}y\right) = 0$ because $-\frac{1}{2}y$ is the _____ of _____.

   (f) We may conclude that $1 \cdot (x + y) = x + y$ because 1 is the _____.

   (g) We may conclude that $\dfrac{3}{x^2 + 1} \cdot \dfrac{x^2 + 1}{3} = 1$ because _____ is the _____ of _____.

   (h) Because addition is _____, we may conclude that $5 + (x + y) + x^2 = 5 + x + (y + x^2)$.

---

5. The paradox we have run into here is known as *Russell's Paradox*, named after the great philosopher Bertrand Russell who first discovered it. The Barber's Paradox of Problem 2T.6 is actually an adaptation of Russell's Paradox to a less abstract setting.

(i) Because multiplication is _____, we may conclude that $2xy = 2yx$.

(j) According to the _____, we know that $4(x + 2y) = 4 \cdot x + 4 \cdot 2y$.

(k) If we know that the real number r is neither positive nor 0, the _____ allows us to conclude that r is _____.

Chapter 3

Elementary Logic

We take up the study of logic with two primary objectives in mind. First, we need to reach agreement as to the meaning of commonly used logical terms such as *if...then* and *for all*. Unless we do so, our ability to communicate will be seriously impaired. Second, we realize that constructing a proof requires that we reason *deductively* from known information. Therefore, we want to analyze the logical basis for the deductive process in order to establish what constitutes a valid argument.

3.1 Statements and Truth

Questions to guide your reading of this section:

1. What are the two undefined notions in the study of logic?

2. What are the two possible *truth values* a statement may possess?

3. What is meant by an *open statement*? What is meant by the *domain of a variable* in an open statement?

4. What is the only way that $\sim P$ can be true? What is the only way that $P \wedge Q$ can be true? What is the only way that $P \vee Q$ can be false? What is the only way that $P \Rightarrow Q$ can be false? How can $P \Leftrightarrow Q$ be true? How can $P \Leftrightarrow Q$ be false?

5. How can we distinguish between the *hypothesis* and the *conclusion* of a *conditional statement*? How do we form the *converse* and *contrapositive* of a conditional statement?

Part of our need for precise communication is our desire to state the mathematical facts we discover. Intuitively, a fact is a sentence that is *true*. Thus, our concern for communicating facts leads us to a study of the notion of *truth*. In addition, we want to write proofs for those mathematical facts (theorems) that follow from our assumptions (axioms). So we are also aware of the need for a deeper understanding of the process known as *deductive reasoning*.

The study of truth and reasoning is called **logic**. In logic, the fundamental object of interest is that of a *statement*, and the fundamental notion associated with a statement is *truth*. Although we shall take the notions of *statement* and *truth* as undefined, our prior experience with them leads us to formulate the following axiom.

Axiom 3.1 (*Fundamental Assumption About Truth*)   A given statement is either true or false but cannot be both true and false.

True and *false* are the possible **truth values** a statement may possess. Our Fundamental Assumption About Truth tells us that any statement must possess one of these truth values but cannot possess both of them at the same time.

Example 3.2   Both of the sentences

The integer 7 is prime

and

The integer 7 is not prime

are statements. The first one is true whereas the second one is false.

A sentence for which it is not known whether it is true or false will still be a statement as long as it is clear that the sentence must be either true or false and cannot be both at the same time.

Example 3.3   Back in Chapter 1, we noted that it is not known whether the Twin Prime Conjecture

There are infinitely many pairs of primes that differ by 2

is true. However, because it must be either true or false, and it cannot be both true and false, this sentence is a statement.

Not all sentences are statements. For instance, the notion of truth does not apply to questions or exclamations. Thus, the sentences

Is the enrollment of our class greater than 10?

and

Go ahead, make my day!

are not statements. Also, the sentence

The sentence you are currently reading is false

though declarative, cannot be considered a statement because it violates our Fundamental Assumption About Truth. If the sentence is true, its content allows us to conclude that it is also false. But if it is false, then we may note that it "speaks the truth," and so we would also regard it as being true.

Variables and Open Statements

A **variable** is a symbol, often a letter, that serves as a placeholder for an object of some particular type. A sentence that incorporates one or more variables and that becomes a statement when the variables are given specific values is called an **open statement**.

Example 3.4   The sentences

$$x = x^2$$

and

$$A \subseteq B$$

are both open statements because they become true or false once, respectively, a specific number is assigned as the value of x, and specific sets are assigned as the values of A and B. The terminology *open statement* reflects the idea that until values have been substituted for the variables, the question of truth remains unresolved or "open." For instance, the sentence $x = x^2$ may be true or it may be false; it depends on what number replaces x. Replacing x with 1 yields the true statement $1 = 1^2$, but replacing x with $1/2$ produces the false statement $1/2 = (1/2)^2$. Similarly, whether $A \subseteq B$ is true depends on exactly what sets play the roles of A and B. Replacing A with $\mathbf{N}$ and B with $\mathbf{R}$ yields the true statement $\mathbf{N} \subseteq \mathbf{R}$, whereas replacing A with $\mathbf{Z}$ and B with $\mathbf{N}$ gives the false statement $\mathbf{Z} \subseteq \mathbf{N}$.

The **domain** of a variable specifies the type of object that can replace the variable. We usually have in mind some *domain* in which the values of a variable appearing in an open statement should be taken.

Example 3.4 (continued)   It is reasonable to assume that the variable x in the open statement $x = x^2$ could be replaced by any real number because the operation of *squaring* applies to such numbers. Hence, we could think of the set $\mathbf{R}$ as the domain of x.

For the open statement $A \subseteq B$, the variables A and B must be replaced by sets if the resulting statement is to make any sense, so the domain for each of A and B consists of sets.

From now on, we will join the majority of the mathematical community and generally use the term *statement* to refer to both statements and open statements, but this should not cause any confusion in what follows.

Logical Connectives

New statements can be created by combining given statements through the use of so-called *logical connectives*.

Definition 3.5 Given statements P and Q, we define the following **compound statements** using the **logical connectives** ~ (**not**), ∧ (**and**), ∨ (**or**), ⇒ (**implies**), and ⇔ (**if and only if**):

- ~P, read "not P," is the **negation** of P.
- $P \wedge Q$, read "P and Q," is the **conjunction** of P and Q.
- $P \vee Q$, read "P or Q," is the **disjunction** of P and Q.
- $P \Rightarrow Q$, read either "if P, then Q" or "P implies Q," is the **conditional** or **implication** with **hypothesis** P and **conclusion** Q.
- $P \Leftrightarrow Q$, read "P if and only if Q," is the **biconditional** of P and Q.

Example 3.6 The statement

I have not taken a physics course[1]

is the negation of the statement

I have taken a physics course.

Example 3.7 The statement

Tomorrow I will go to the beach and I will go to the movies

is the conjunction of the statements

Tomorrow I will go to the beach

and

Tomorrow I will go to the movies.

---

1. Whenever a pronoun such as *I* appears in a statement, it must be clear to whom it refers. In this book, think of the appearance of such a pronoun as referring to the author, yourself, or some other fixed individual. Alternatively, such pronouns may be regarded as placeholder variables so that statements containing them are really open statements.

Example 3.8 The statement

$$x \in A \text{ or } x \notin B$$

is the disjunction of the statements

$$x \in A$$

and

$$x \notin B.$$

Example 3.9 The statement

$$\text{If } x = 3, \text{ then } x^2 = 9$$

is the conditional statement with hypothesis

$$x = 3$$

and conclusion

$$x^2 = 9.$$

It is extremely important to be able to distinguish the *hypothesis* of a conditional statement from its *conclusion*. Remember that the **hypothesis** is the statement following the word *if* and the **conclusion** is the statement following the word *then*. Interchanging the hypothesis and conclusion can have an effect on a conditional statement's truth.

Example 3.9 (continued) You probably believe the conditional statement

$$\textit{If } x = 3, \textit{ then } x^2 = 9$$

to be true. But interchanging its hypothesis and conclusion produces a new implication

$$\textit{If } x^2 = 9, \textit{ then } x = 3,$$

which you probably believe to be false. We shall soon discuss a process for determining when a conditional statement should be considered true.

Example 3.10 The statement

$$x \in A \textit{ if and only if } x \in B$$

is the biconditional of the statements

$$x \in A$$

and

$$x \in B.$$

A statement that is not capable of being created from other statements through the use of logical connectives such as *and, or, not, if . . . then,* and *if and only if* is called a **simple statement**. Each of the following nine sentences is a simple statement: *I have taken a physics course. The integer 7 is prime. Multiplication of real numbers is commutative. The set of integers is closed under addition.* $x = 3$. $x \in A$. $A = B$. $A \subseteq B$. $A \cap B = A$.

Practice Problem 1   Let P be the statement *I am a math major* and let Q be the statement *I like doing math.* Form each of the following compound statements:

(a) $\sim P$.   (b) $P \wedge Q$.   (c) $P \vee Q$.   (d) $P \Rightarrow Q$.   (e) $Q \Rightarrow P$.

(f) $P \Leftrightarrow Q$.   (g) $P \vee (\sim Q)$.

It is usually assumed that negation takes precedence over the other logical connectives unless there are parentheses to indicate otherwise. Thus, rather than write $(\sim P) \wedge Q$, we may write $\sim P \wedge Q$. If what we really intend, though, is $\sim(P \wedge Q)$, the parentheses must be included.

Generally, when more than one logical connective appears in a statement, we must use parentheses or wording to sort out the intended grouping. For example, the symbolic expression $P \wedge Q \vee R$ is ambiguous; parentheses should be inserted to make it clear whether the statement under consideration is $(P \wedge Q) \vee R$ or $P \wedge (Q \vee R)$. Similarly, writing

The integer x is odd and less than 100 or a perfect square

is ambiguous. We could write either

The integer x is both odd and less than 100 or else is a perfect square

or

The integer x is odd and is also either less than 100 or a perfect square

depending on what we intend to communicate.

When stating a mathematical definition, *iff* acts as a grouping device in the sense that a statement such as

$$\textit{Sets A and B are } \textbf{equal} \textit{ iff for every x, } x \in A \textit{ iff } x \in B$$

should be interpreted as

$$\textit{Sets A and B are } \textbf{equal} \textit{ iff (for every x, } x \in A \textit{ iff } x \in B)$$

because the entire statement

$$(\textit{for every x, } x \in A \textit{ iff } x \in B)$$

following

$$\textbf{equal} \textit{ iff}$$

is being used to define what it means for sets A and B to be equal.

The Truth of Statements Involving Logical Connectives

We must reach agreement as to when statements involving the logical connectives *not, and, or, if...then,* and *if and only if* should be considered true. The next few examples are designed to help guide your thinking on this matter.

Example 3.11 (*Exploring Our Intuition Concerning Negations*)   Consider the statement

$$\textit{I will go to the beach tomorrow}$$

and its negation

$$\textit{I will not go to the beach tomorrow.}$$

Note that, intuitively, the negation is true precisely when the original statement is false. That is, our prior experience with the use of the word *not* leads us to believe that a statement and its negation must have *opposite* truth values.

Example 3.12 (*Exploring Our Intuition Concerning Conjunctions*)   Now think about the statements

$$\textit{Tomorrow I will go to the beach}$$

and

$$\textit{Tomorrow I will go to the movies.}$$

When would we consider the conjunction

Tomorrow I will go to the beach and I will go to the movies

of these two statements to be true? Our prior experience with use of the word *and* suggests that the only way for a conjunction to be true is if *both* of its constituent statements are true, so this conjunction will be true only if tomorrow I really do go to both the beach and the movies.

Example 3.13 (*Exploring Our Intuition Concerning Disjunctions*) In everyday language, the word *or* should sometimes be interpreted *inclusively* and at other times *exclusively*. When used *inclusively*, the intention is to *include* the possibility that both statements joined by *or* may be true. For example, a college graduation requirement that states

A student must take physics or chemistry

would usually be interpreted to mean that a student must take *at least one* of the two courses but would be allowed to take both if desired. The logical connective *or* is being interpreted in its inclusive sense here. In contrast, the statement

You may vote for the Republican or the Democratic candidate for president

should be interpreted *exclusively* because voting regulations *exclude* the possibility of one person casting votes for both candidates.

The contexts in which these two statements are made allow us to choose the correct interpretation, inclusive or exclusive. But in mathematics, we cannot always expect context to help us because we may not have yet learned enough in a given situation to be able to choose one interpretation over the other. Instead, we agreed back in Section 2.3, when we introduced the notion of the union of two sets, to interpret *or* inclusively in all of our mathematical work.

Thus, we agree that the only way for the disjunction

$$x = 4 \text{ or } x^2 = 16$$

to be false is if both of its component simple statements

$$x = 4$$

and

$$x^2 = 16$$

are false. As long as one or both components are true, the disjunction will be true. So if the actual value of x is 4, the disjunction is true as both components

$x = 4$ and $x^2 = 16$ are true. And if the actual value of x is -4, the disjunction is also true as, although the first component $x = 4$ is false, the second component $x^2 = 16$ is true.

Example 3.14 (*Exploring Our Intuition Concerning Conditionals*)   If I tell you that

If it is sunny out tomorrow, then I will go to the beach

under what conditions could you call me a liar? That is, when would you consider this statement to be false? Certainly, it is false if it is sunny out tomorrow and I do not go to the beach. But note that the statement itself makes no claim about whether or not I will go to the beach if it is *not* sunny out tomorrow. The only way I can be called a liar is if I do not go to the beach when it really is sunny out. Our intuition leads us to believe that a conditional statement is true unless the hypothesis is true while at the same time the conclusion is false.

As further evidence, consider the implications

If $x = 3$, then $x^2 = 9$

and

If $x = 4$, then $x^2 = 9$.

Intuitively, the first of these statements appears to be true whereas the second appears to be false. Note that when the hypothesis $x = 4$ of the second statement is true, the conclusion $x^2 = 9$ is false. It seems, once again, that a conditional statement allowing for a false conclusion while the hypothesis is true should be considered false. In contrast, observe that there is no way for the conclusion $x^2 = 9$ of the first implication to be false when the hypothesis $x = 3$ is true. Our intuition tells us that the first conditional statement should be considered true, and it seems to be because we do not, in this case, have the possibility of a true hypothesis leading to a false conclusion. Note that if the hypothesis of the first implication is false, the conclusion may be either true or false, but this seems to have no effect on our regarding the implication itself as being true.

Example 3.15 (*Exploring Our Intuition Concerning Biconditionals*)   Under what circumstances would we consider the biconditional

I will go to the beach tomorrow if and only if it is sunny out

to be true? Recall that we may think of *if and only if* as a way of expressing the notion *means the same thing as*. It then seems reasonable to say that there are exactly two ways for this biconditional to be true. One way is if both of the constituent statements are true; that is, I actually do go to the beach tomorrow and it really is sunny out. The other is if both constituent statements are false; in other words, I do not go to the beach and it is not sunny out. Thus, we are led to believe that a biconditional is true exactly when both of its constituent statements have the same truth value, both true or both false.

We now present an axiom that formally expresses what we reflected upon in Examples 3.11 through 3.15. This axiom also makes clear that sentences constructed through the appropriate application of logical connectives to given statements are themselves statements.

Axiom 3.16 (*Determining the Truth of a Compound Statement*)   For any statements P and Q,

- $\sim P$ is a statement that is true exactly when P is false.

- $P \wedge Q$ is a statement that is true exactly when both P and Q are true.

- $P \vee Q$ is a statement that is false exactly when both P and Q are false.

- $P \Rightarrow Q$ is a statement that is false exactly when P is true and Q is false.

- $P \Leftrightarrow Q$ is a statement that is true exactly when P and Q have the same truth value, both true or both false.

Practice Problem 2   What truth values would the statements P, Q, and R have to possess in order for the compound statement $\sim[P \Rightarrow (Q \vee R)]$ to be true?

The Converse and Contrapositive of a Conditional Statement

If we interchange the hypothesis and conclusion of a conditional statement $P \Rightarrow Q$, we obtain another conditional statement $Q \Rightarrow P$ that is called the **converse** of $P \Rightarrow Q$. For instance, the converse of

$$\text{If } x = 3, \text{ then } x^2 = 9 \tag{1}$$

is

$$\text{If } x^2 = 9, \text{ then } x = 3. \tag{2}$$

Note that just because a conditional statement is true does not mean its converse has to be true. For example, whereas (1) is true, if $x = -3$ its converse (2) is false.

When we interchange the hypothesis and the conclusion of a conditional statement $P \Rightarrow Q$ and then negate both of them, we obtain the so-called **contrapositive** $\sim Q \Rightarrow \sim P$ of $P \Rightarrow Q$. Thus, the contrapositive of

$$\textit{If } x = 3, \textit{ then } x^2 = 9$$

is

$$\textit{If } x^2 \neq 9, \textit{ then } x \neq 3.$$

The contrapositive of a conditional statement is important because it turns out, as we shall see in Section 3.3, that a conditional statement and its contrapositive say the same thing logically.

Practice Problem 3   Form the converse and the contrapositive of the conditional statement *If it is sunny out, then it is not raining.*

Converse Assumed in Stating a Mathematical Definition

It is customary to state a mathematical definition using *if* rather than *if and only if*. For instance, we would usually write

$$\textit{Sets A and B are } \textbf{equal} \textit{ IF for every } x, \ x \in A \textit{ iff } x \in B \qquad (3)$$

rather than

$$\textit{Sets A and B are } \textbf{equal} \textit{ IF AND ONLY IF for every } x, \ x \in A \textit{ iff } x \in B \qquad (4)$$

even though in writing (3) we really mean (4). It is implicitly assumed in (3), as the boldface indicates a definition is being communicated, that the converse

$$\textit{If sets A and B are } \textbf{equal}, \textit{ then for every } x, \ x \in A \textit{ iff } x \in B$$

is also true. We emphasize, however, that this convention can only be used when stating a definition and that we are *not* saying that *if . . . then* and *if and only if* are synonymous generally. For example, in (3) we could not replace the occurrence of *iff* between $x \in A$ and $x \in B$ with *if* because neither of the statements $x \in A$ or $x \in B$ is being used to define the other.

Answers to Practice Problems

1. (a) I am not a math major. (b) I am a math major and I like doing math. (c) I am a math major or I like doing math. (d) If I am a math major, then I like doing math. (e) If I like doing math, then I am a math major. (f) I am a math major if and only if I like doing math. (g) I am a math major or I do not like doing math.

2. The statement P would have to be true and both of the statements Q and R would have to be false.

3. The converse is *If it is not raining, then it is sunny out.* The contrapositive is *If it is raining, then it is not sunny out.*

3.2   Truth Tables and Statement Forms

Questions to guide your reading of this section:

1. What are *statement variables* and *statement forms*?

2. How do we set up and create a *truth table* for a statement form that includes two statement variables? three statement variables?

3. How can we use a truth table to recognize a *tautology*? a *contradiction*?

It is possible to represent our assumptions about truth using what are called *truth tables*. Shown below is the truth table for $\sim P$.

P	$\sim P$
T	F
F	T

The first column displays the possible truth values for the statement P, T for *true* and F for *false*, and the second column gives the *resulting* truth values for the statement $\sim P$ according to Axiom 3.16. That is, the row consisting of a T followed by an F indicates that if the statement P is true, then the statement $\sim P$ must be false. And the row consisting of an F followed by a T indicates that if P is false, $\sim P$ must be true.

A letter, such as P, that is used to represent an arbitrary statement is called a **statement variable**. An expression, such as $\sim P$, formed using statement variables and logical connectives and that itself becomes a statement when the variables are replaced with statements is called a **statement form**. A **truth table** for a statement form displays all of the possible combinations of truth values that can be assigned to the statement variables appearing in the form along with the resulting truth assignments to the form itself. Beginning in the next chapter, we shall see that when we are attempting to prove a statement we believe to be true, the *form* of the statement will usually be helpful in deciding how to go about developing a proof.

For statement forms created from two statement variables P and Q, a truth table must include a column for P and another column for Q. There will also be four rows in the body of the truth table because there are four different combinations of truth values that can be assigned to P and Q: when P is assigned T, Q can be assigned either T or F; and when P is assigned F, Q can also be assigned either T or F. The truth table that follows has been constructed to conform to the assumptions stated in Axiom 3.16. Each row describes a particular assignment of truth values to the variables P and Q and then gives the resulting truth assignments to the statement forms heading the other columns.

P	Q	$P \wedge Q$	$P \vee Q$	$P \Rightarrow Q$	$P \Leftrightarrow Q$
T	T	T	T	T	T
T	F	F	T	F	F
F	T	F	T	T	F
F	F	F	F	T	T

Example 3.17 We will construct a truth table for the statement form $(P \wedge Q) \vee {\sim}P$. First, we set up the truth table based on the standard format required when there are two statement variables P and Q:

P	Q	$P \wedge Q$	${\sim}P$	$(P \wedge Q) \vee {\sim}P$
T	T			
T	F			
F	T			
F	F			

Note that besides including a column at the very right for the form $(P \wedge Q) \vee {\sim}P$, we have also included intermediate columns for the forms $P \wedge Q$ and ${\sim}P$ out of which the form $(P \wedge Q) \vee {\sim}P$ is created. This is usually a good idea anytime we want to produce a truth table for a statement form that is complicated and can be broken down into smaller pieces. We then fill in the columns for $P \wedge Q$ and $\sim P$ based on our assumptions from Axiom 3.16:

P	Q	$P \wedge Q$	$\sim P$	$(P \wedge Q) \vee \sim P$
T	T	T	F	
T	F	F	F	
F	T	F	T	
F	F	F	T	

Once these intermediate columns have been completed, we use them to assign the truth values to $(P \wedge Q) \vee \sim P$ based on the fact that this disjunction (*or*) is false only when both components $P \wedge Q$ and $\sim P$ are false. Therefore,

- In the first row, because $P \wedge Q$ has been assigned T and $\sim P$ has been assigned F, $(P \wedge Q) \vee \sim P$ should be assigned T.

- In the second row, because $P \wedge Q$ has been assigned F and $\sim P$ has been assigned F, $(P \wedge Q) \vee \sim P$ should be assigned F.

- In the third and fourth rows, because $P \wedge Q$ has been assigned F and $\sim P$ has been assigned T, $(P \wedge Q) \vee \sim P$ should be assigned T.

The completed truth table for $(P \wedge Q) \vee \sim P$ is shown below:

P	Q	$P \wedge Q$	$\sim P$	$(P \wedge Q) \vee \sim P$
T	T	T	F	T
T	F	F	F	F
F	T	F	T	T
F	F	F	T	T

Practice Problem 1   Create a truth table for the statement form $P \Rightarrow (\sim Q \wedge P)$.

We can construct a truth table for any statement form, even when the form contains more than two statement variables.

Example 3.18   Below is a truth table for the statement form $(P \wedge Q) \Rightarrow R$.

P	Q	R	$P \wedge Q$	$(P \wedge Q) \Rightarrow R$
T	T	T	T	T
T	T	F	T	F
T	F	T	F	T
T	F	F	F	T
F	T	T	F	T
F	T	F	F	T
F	F	T	F	T
F	F	F	F	T

Note that there is a row corresponding with each different combination of truth values that can be assigned to P, Q, and R. Because there are eight different combinations of truth assignments, there are eight rows in the body of the table. (Do you see that all possible truth assignments to the three variables have been accounted for?) The resulting truth values to the statement form $P \wedge Q$ have then been included to make it easier to compute the truth values for $(P \wedge Q) \Rightarrow R$. Be sure you see how each of the entries in the last two columns of the table is determined. For instance, in the third row, because P has been assigned the truth value T and Q the truth value F, we know that the conjunction $P \wedge Q$ should be assigned F. Then, because a conditional with a false hypothesis is considered true, we see that $(P \wedge Q) \Rightarrow R$ will be assigned T.

Practice Problem 2   Create a truth table for the statement form $(P \vee \sim Q) \Leftrightarrow R$.

Sometimes it is possible to fill in a truth table more efficiently than simply proceeding row by row.

Example 3.18 (continued)   Consider again the statement form $(P \wedge Q) \Rightarrow R$. This form is really an implication (*if . . . then*) having a conjunction (*and*) for its hypothesis. Because an implication is true unless its hypothesis is true while its conclusion is false, and the only way for the hypothesis $P \wedge Q$ to be true is for both P and Q to be true, it becomes clear that there is only one way for $(P \wedge Q) \Rightarrow R$ to be false, and that is when the assignments to the variables P, Q, and R are as specified in the second row of the truth table we constructed previously. Hence, the remaining entries in the final column of the table must all be T.

Practice Problem 3   Create a truth table for $\sim(P \vee Q)$ by using the fact that a disjunction is false only when all of its components are false and the fact that a statement's negation is true only when the statement itself is false.

A statement may exhibit more than one form. The statement

If x is even and x is greater than 2, then x is not prime

has the form $P \Rightarrow Q$ because it results from this form when the variable P is replaced by the statement *x is even and x is greater than 2* and the variable Q is replaced by the statement *x is not prime*. But this statement also has the form $(P \wedge Q) \Rightarrow \sim R$ because it can be produced by replacing P with *x is even*, Q with *x is greater than 2*, and R with *x is prime* in this form. The given statement also possesses these other forms: $P \Rightarrow \sim Q$, $(P \wedge Q) \Rightarrow R$, and P (see Problem 3D).

Tautologies and Contradictions

Some statement forms always produce the same truth value no matter what statements are substituted for its statement variables. One that is always true is called a **tautology**, and one that is always false is called a **contradiction**.

Example 3.19   The following truth table reveals that the statement form $P \vee \sim P$ is a tautology and the statement form $P \wedge \sim P$ is a contradiction.

P	$\sim P$	$P \vee \sim P$	$P \wedge \sim P$
T	F	T	F
F	T	T	F

Both of these conclusions should seem reasonable because, for any statement, either it or its negation, but not both, will be true.

Practice Problem 4   Use a truth table to show that the statement form $P \Rightarrow (Q \Rightarrow P)$ is a tautology.

A statement having the form of a tautology or a contradiction may also be referred to as, respectively, a **tautology** or a **contradiction**. Hence, as the statement

$$x \in A \text{ or } x \notin A$$

has the form $P \vee {\sim}P$ of a tautology, the statement itself may be considered a tautology, and as the statement

$$x \in A \text{ and } x \notin A$$

has the form $P \wedge {\sim}P$ of a contradiction, this statement may itself be considered a contradiction.

Answers to Practice Problems

1.

P	Q	${\sim}Q$	${\sim}Q \wedge P$	$P \Rightarrow ({\sim}Q \wedge P)$
T	T	F	F	F
T	F	T	T	T
F	T	F	F	T
F	F	T	F	T

2.

P	Q	R	${\sim}Q$	$P \vee {\sim}Q$	$(P \vee {\sim}Q) \Leftrightarrow R$
T	T	T	F	T	T
T	T	F	F	T	F
T	F	T	T	T	T
T	F	F	T	T	F
F	T	T	F	F	F
F	T	F	F	F	T
F	F	T	T	T	T
F	F	F	T	T	F

(continues)

Answers to Practice Problems (continued)

3. In order for $\sim(P \vee Q)$ to be true, it must be the case that $P \vee Q$ is false. But the only way for $P \vee Q$ to be false is if P is false and Q is false. Thus, the following truth table assigns T to $\sim(P \vee Q)$ when both P and Q are false and F to $\sim(P \vee Q)$ otherwise.

P	Q	$\sim(P \vee Q)$
T	T	F
T	F	F
F	T	F
F	F	T

4. The truth table

P	Q	$Q \Rightarrow P$	$P \Rightarrow (Q \Rightarrow P)$
T	T	T	T
T	F	T	T
F	T	F	T
F	F	T	T

shows that $P \Rightarrow (Q \Rightarrow P)$ is a tautology because every entry in the column corresponding with this form is a T.

3.3   Logical Equivalence

Questions to guide your reading of this section:

1. In the context of two mathematical statements, what is the difference between saying they are *mathematically equivalent* and saying they are *logically equivalent*?

2. How are truth tables used to determine whether two statement forms are logically equivalent?

3. Under what circumstances are two statements considered to be logically equivalent?

When do two statements that are written in different ways actually express the same idea? Sometimes it is simply the meaning of certain words or notations.

We may use our knowledge of geometry, namely that the diameter of any circle is twice that circle's radius, to conclude that the two statements

The radius of a circle is 4

and

The diameter of a circle is 8

are "equivalent." They say the same thing using different words. Because the equivalence relies upon the mathematical meanings of *radius*, *diameter*, and *circle*, this equivalence is a *mathematical equivalence*.

But sometimes the equivalence of two differently worded statements has only to do with their underlying logical structure, not with the actual content, mathematical or otherwise, of the statements themselves. Such an equivalence is referred to as a *logical equivalence*.

Example 3.20   The statements

$$x \in A \ or \ x \in B$$

and

$$x \in B \ or \ x \in A,$$

though not literally identical, express exactly the same information. Moreover, their equivalence has nothing to do with what they say mathematically. Rather, it results from the underlying logical equivalence of the forms they possess, $P \vee Q$ and $Q \vee P$. Our experience with the logical connective *or* suggests that the order in which the components of a disjunction are assembled is irrelevant. Intuitively, any pair of statements possessing these forms would appear to be equivalent. For instance, the statements

I will go to the beach or I will go to the movies

and

I will go to the movies or I will go to the beach.

are also equivalent to each other.

The equivalence of each pair of statements in Example 3.20 is the result of our belief that the statements in each pair possess *forms* that are logically equivalent. Intuitively, two statement forms are logically equivalent if they *mean the same thing logically*; that is, they are both true at the same time and they are both false at the same time.

Example 3.20 (continued)   Note that in the following truth table,

P	Q	$P \vee Q$	$Q \vee P$
T	T	T	T
T	F	T	T
F	T	T	T
F	F	F	F

the columns under $P \vee Q$ and $Q \vee P$ match row by row. The forms $P \vee Q$ and $Q \vee P$ are true at the same time and they are false at the same time. Hence, we may conclude that $P \vee Q$ is logically equivalent to $Q \vee P$.

Definition 3.21   Two statement forms are **logically equivalent** provided their truth tables coincide; that is, for any assignment of truth values to the statement variables appearing in the two statement forms, the resulting truth assignments to the forms are identical. We then define two *statements* to be **logically equivalent** if they have forms that are logically equivalent.

Example 3.20 (continued)   Because the statement forms $P \vee Q$ and $Q \vee P$ are logically equivalent, any statements possessing these forms are also logically equivalent. For instance, the statements

$$x \in A \ or \ x \in B$$

and

$$x \in B \ or \ x \in A$$

are logically equivalent.

Practice Problem 1   Show that the statements

     If n is even, it follows that n + 1 is odd and that n^2 is even

and

     If n is even, then n + 1 is odd, and, furthermore, if n is even, then n^2 is even

are logically equivalent.

Example 3.22   Consider the statement

It is not the case that the natural number x is both prime and greater than 100.

Does this statement mean

The number x is not prime and it is not greater than 100?

Or does it mean

The number x is not prime or it is not greater than 100?

Answering these questions is really a matter of determining whether the form $\sim(P \wedge Q)$ of the original statement is logically equivalent to either of the forms, $\sim P \wedge \sim Q$ and $\sim P \vee \sim Q$, possessed by the potential interpretations. You will investigate this situation in Problems 3G.1 and 3H.

Example 3.23   The following truth table for the statement forms $P \Rightarrow Q$ and $Q \Rightarrow P$ shows that these forms are not logically equivalent. In other words, a conditional statement is not logically equivalent to its converse.

P	Q	$P \Rightarrow Q$	$Q \Rightarrow P$
T	T	T	T
T	F	F	T
F	T	T	F
F	F	T	T

Specifically, note that the truth values under $P \Rightarrow Q$ and $Q \Rightarrow P$ do not match in the second row. When a mismatch occurs in even a single row, we must conclude that the forms being compared are not logically equivalent. In this case, because the truth values assigned to $P \Rightarrow Q$ and $Q \Rightarrow P$ also do not match in the third row, we could have used this row instead of the second row to provide evidence for our conclusion that these two forms are not logically equivalent.

Practice Problem 2   Show that the statement forms $(P \wedge Q) \Rightarrow R$ and $(P \Rightarrow R) \wedge (Q \Rightarrow R)$ are not logically equivalent.

Example 3.24   The statement forms $P \Rightarrow Q$ and $\sim Q \Rightarrow \sim P$ are logically equivalent. You are asked to demonstrate this in Problem 3G.1(b). Hence, *any conditional statement is logically equivalent to its contrapositive.* For instance, the statement

$$\text{If } x \in A, \text{ then } x \in B$$

is logically equivalent to the statement

$$\text{If } x \notin B, \text{ then } x \notin A.$$

Answers to Practice Problems

1. First note that the statement

   If n is even, it follows that n + 1 is odd and that n^2 is even

   has the form $P \Rightarrow (Q \wedge R)$, and the statement

   If n is even, then n + 1 is odd, and, furthermore,
   if n is even, then n^2 is even

   has the form $(P \Rightarrow Q) \wedge (P \Rightarrow R)$, with P holding a place for *n is even*, Q holding a place for *n + 1 is odd*, and R holding a place for n^2 *is even*. Examining the truth table

P	Q	R	$Q \wedge R$	$P \Rightarrow (Q \wedge R)$	$P \Rightarrow Q$	$P \Rightarrow R$	$(P \Rightarrow Q) \wedge (P \Rightarrow R)$
T	T	T	T	T	T	T	T
T	T	F	F	F	T	F	F
T	F	T	F	F	F	T	F
T	F	F	F	F	F	F	F
F	T	T	T	T	T	T	T
F	T	F	F	T	T	T	T
F	F	T	F	T	T	T	T
F	F	F	F	T	T	T	T

we note that the columns corresponding with the forms $P \Rightarrow (Q \wedge R)$ and $(P \Rightarrow Q) \wedge (P \Rightarrow R)$ perfectly match row by row, making these forms logically equivalent. Thus, any statements possessing these forms are logically equivalent, in particular, the two statements originally given to us.

2. The truth table that follows shows the statement forms $(P \wedge Q) \Rightarrow R$ and $(P \Rightarrow R) \wedge (Q \Rightarrow R)$ are not logically equivalent because the columns corresponding with them do not match in every row. For instance, in the fourth row of the table, where P is assigned T and both Q and R are assigned F, the resulting assignment to $(P \wedge Q) \Rightarrow R$ is T but the resulting assignment to $(P \Rightarrow R) \wedge (Q \Rightarrow R)$ is F.

P	Q	R	$P \wedge Q$	$(P \wedge Q) \Rightarrow R$	$P \Rightarrow R$	$Q \Rightarrow R$	$(P \Rightarrow R) \wedge (Q \Rightarrow R)$
T	T	T	T	T	T	T	T
T	T	F	T	F	F	F	F
T	F	T	F	T	T	T	T
T	F	F	F	T	F	T	F
F	T	T	F	T	T	T	T
F	T	F	F	T	T	F	F
F	F	T	F	T	T	T	T
F	F	F	F	T	T	T	T

3.4 Arguments and Validity

Questions to guide your reading of this section:

1. How do mathematical arguments differ from some other kinds of arguments?

2. What is meant by the *premises* of an argument? the *consequence*?

3. How can we use a truth table to determine if an argument form is *valid*?

4. How can we determine whether an argument is *valid*?

We have already indicated that the more formal approach to mathematics we are now undertaking requires that explanations be based on a process known as *deductive reasoning*. But exactly how does one go about reasoning deductively? And how can we be sure that the deductions we make are legitimate?

Before answering these questions, let us first make clear how mathematical arguments may differ from some other kinds of arguments. Depending on the context in which it arises, a nonmathematical argument may be accepted based only on a certain amount of persuasive or credible evidence, some of which may be fairly subjective or circumstantial. Political arguments, especially those made in support of a favored candidate or policy, often include elements that could be viewed as subjective because they may be based on varying sets of values, interests, or standards. An argument of the form

You really should vote for an independent candidate. Here's why...

may be compelling, but so could an argument encouraging support of a non-independent candidate. Such arguments, no matter how well constructed, are always open to charges of bias or partisanship.

Similarly, the "weight" of evidence needed to support a claim of guilt in a criminal trial is not "beyond *all possible* doubt" but "beyond *a reasonable* doubt." This latter standard is subjective because its interpretation can vary from one person to another.

Even a scientific argument along the lines of

Because of the overwhelming physical evidence,
it is clear that humans descended from apes

ultimately relies on gathering "enough" circumstantial evidence to defend its claim. The conclusion is still subject to revision if new contradictory evidence is discovered at a later date.

But some arguments are more than just persuasive or credible. They are logically indisputable.

Example 3.25   Here is an argument that should leave you with no doubts as to its acceptability:

To graduate, I must take a math course or a biology course. I do not intend to take a math course. Therefore, I will have to take a biology course.

Note that the conclusion, "I will have to take a biology course," appears to be inevitable under the stated assumptions that "I must take a math course or a biology course" and "I do not intend to take a math course." In fact, it is really the underlying logical structure of the statements comprising this argument, not their subject matter, that makes it seem indisputable: Given a mandatory choice between two options, if we can eliminate one of them, logic tells us that we now have no choice but to accept the other.

Example 3.25 illustrates the type of argument that is based on deductive reasoning. In such an argument, the truth of a particular statement is "deduced" or "drawn forth" from other given statements. All mathematical arguments are of this type.

Definition 3.26 For our purposes, an **argument**, which can be symbolized as

$$P_1, P_2, ..., P_n \therefore C,$$

consists of a set of statements $P_1, P_2, ..., P_n$, called the **premises** of the argument, that are assumed to be true, along with a statement C, called the **consequence** of the argument, whose truth has *allegedly* been deduced from the given premises. The symbol $\therefore$ is read *therefore*.

Example 3.25 (continued) The premises of the argument

To graduate, I must take a math course or a biology course. I do not intend to take a math course. Therefore, I will have to take a biology course.

are

To graduate, I must take a math course or a biology course.

and

I do not intend to take a math course.

The consequence is

I will have to take a biology course.

Practice Problem 1 Identify the premises and consequence of the following argument: *Whenever it rains I carry my umbrella. Right now I am not carrying my umbrella. Therefore, it is not raining.*

The question of whether to accept an argument boils down to whether the argument's consequence can be viewed as logically inevitable under the assumption that all of the argument's premises are true.

Example 3.27 Here is a mathematical argument having the *same form* as the argument given in Example 3.25:

Premises: The integer n is either even or odd.

The integer n is not even.

Consequence: The integer n is odd.

Note that anytime the premises of this argument are both true, the consequence must also be true. On an intuitive level, the inability of the consequence to be false when both premises are true is what makes us want to accept this argument.

As was suggested in Example 3.25, the acceptance of an argument should not depend on the actual informational content of the premises and consequence, only on their underlying logical structure.

Example 3.28   The same logical reasoning is at work in the arguments given in both Examples 3.25 and 3.27. Note that both arguments have the form

$$P \vee Q, \sim P \therefore Q.$$

Intuitively, we believe this *argument form* is acceptable, regardless of what statements replace the statement variables P and Q. This is because, as the following truth table reveals, any assignment of truth values to P and Q that makes both premises $P \vee Q$ and $\sim P$ true (this only happens in the third row, where P is assigned F and Q is assigned T) also makes the consequence Q true.

P	Q	$P \vee Q$	$\sim P$
T	T	T	F
T	F	T	F
F	T	T	T
F	F	F	T

The term *valid* is used to describe those arguments and argument forms that are acceptable. Building off of our work in Examples 3.25, 3.27, and 3.28, we formulate the following definitions.

Definition 3.29   An argument form is **valid** provided that every assignment of truth values to the variables appearing in it that makes all of the premises true also makes the consequence true.

An argument is then taken to be **valid** provided that it possesses an argument form that is valid.

When $P_1, P_2, ..., P_n \therefore C$ is a valid argument form or valid argument, we say that C is a **logical consequence** of $P_1, P_2, ..., P_n$.

Example 3.30   Consider again the argument form

$$P \vee Q, \sim P \therefore Q.$$

To determine whether this argument form is valid, we look at those rows in the truth table presented in Example 3.28 in which both premises $P \vee Q$ and $\sim P$ are true. If the consequence Q is also true in each such row, the argument form is valid. If there is even one row in which both premises are true and the consequence is false, the argument form is invalid. In this case, the only row in which both premises are true is the third row, and in this row the consequence is also true. Thus, the argument form is valid, and we may conclude that Q is a logical consequence of $P \vee Q$ and $\sim P$.

Then, as the arguments in Examples 3.25 and 3.27 both possess the valid argument form $P \vee Q, \sim P \therefore Q$, they themselves are valid.

Note that in stating the argument in Practice Problem 2, we use *thus* in place of *therefore*. Other synonyms for *therefore* are *hence, it follows that, whence,* and *consequently*.

Practice Problem 2   Show that the following argument is valid: *Anytime an integer is divisible by 4 or 6, the integer is even. The integer k is divisible by 4. Thus, k is even.*

Recall that a statement may exhibit more than one form. Thus, an *argument* may exhibit more than one *argument form*. It turns out (see Problem 3K.4) that if one of the forms an argument exhibits is valid, so are all of its other forms. This means that to show an argument is invalid (i.e., not valid), it is enough to find a single invalid argument form possessed by the argument.

Example 3.31   The following argument seems invalid because of our intuitive belief that just because P implies Q, it does not necessarily follow that Q implies P:

$$\text{If } x \in A, \text{ then } x \in B. \quad \text{Therefore, if } x \in B, \text{ then } x \in A. \tag{1}$$

This argument has the form

$$P \Rightarrow Q \therefore Q \Rightarrow P \tag{2}$$

where P is $x \in A$ and Q is $x \in B$. To verify the argument (1) is not valid, we verify that the form (2) it possesses is not valid. To do this, we construct the following truth table:

P	Q	$P \Rightarrow Q$	$Q \Rightarrow P$
T	T	T	T
T	F	F	T
F	T	T	F
F	F	T	T

Note that in the third row, the premise $P \Rightarrow Q$ is true, but the consequence $Q \Rightarrow P$ is false. Thus, the argument form (2) is not valid, and it then follows that any argument possessing this form, for instance (1), is not valid.

Practice Problem 3   Show that the following argument is invalid: *Anytime an integer is divisible by both 4 and 6, it is divisible by 12. The integer k is divisible by 4. Consequently, k is divisible by 12.*

Answers to Practice Problems

1. The premises are

    Whenever it rains I carry my umbrella

    and

    Right now I am not carrying my umbrella.

    The consequence is

    It is not raining.

2. Note that the argument has the form $(P \vee Q) \Rightarrow R, P \therefore R$, where P is *The integer k is divisible by 4*, Q is *The integer k is divisible by 6*, and R is *The integer k is even*. The truth table

P	Q	R	$P \vee Q$	$(P \vee Q) \Rightarrow R$
T	T	T	T	T
T	T	F	T	F
T	F	T	T	T
T	F	F	T	F
F	T	T	T	T
F	T	F	T	F
F	F	T	F	T
F	F	F	F	T

reveals the argument form to be valid because, in every row in which both premises $(P \vee Q) \Rightarrow R$ and P are true (this occurs only in the first and third rows), the consequence R is also true. Thus, any argument possessing this form is valid.

3. Note that the argument has the form $(P \wedge Q) \Rightarrow R$, $P \therefore R$, where P is *The integer k is divisible by 4*, Q is *The integer k is divisible by 6*, and R is *The integer k is divisible by 12*. The truth table

P	Q	R	$P \wedge Q$	$(P \wedge Q) \Rightarrow R$
T	T	T	T	T
T	T	F	T	F
T	F	T	F	T
T	F	F	F	T
F	T	T	F	T
F	T	F	F	T
F	F	T	F	T
F	F	F	F	T

reveals the argument form to be invalid because, in the fourth row both premises $(P \wedge Q) \Rightarrow R$ and P are true, but the consequence R is false. Thus, any argument possessing this form is invalid.

3.5   Statements Involving Quantifiers

Questions to guide your reading of this section:

1. What are the two types of *quantifiers* used most often in mathematics? How can each of them be translated and how is each one symbolized?

2. Under what circumstances should a statement involving *universal* quantification be considered true? a statement involving *existential* quantification?

3. How do we form and simplify the negation of a statement involving a quantifier?

Recall that an open statement is a sentence that includes one or more variables within it and that becomes a statement when values are assigned to the variables. A generic open statement involving a single variable x can be denoted symbolically by $P(x)$, which is read "P of x." If the open statement includes more variables, they are all listed in the parentheses, so, for instance, $P(x, y, z)$ represents an open statement with three variables x, y, and z.

Example 3.32   Let $P(x)$ denote the open statement $x = x^2$. Then $P(1)$ denotes the true statement $1 = 1^2$ and $P(1/2)$ the false statement $1/2 = (1/2)^2$.

Example 3.33   The open statement $A \subseteq B$ can be denoted by $P(A, B)$. Then $P(\mathbf{N}, \mathbf{R})$ is really just the true statement $\mathbf{N} \subseteq \mathbf{R}$.

Example 3.34   When more than one open statement appears in a given situation, we must use different leading letters to represent them. For instance, if the open statements $x = x^2$ and $x \neq 0$ both show up in a problem we are working on, we might represent the former by $P(x)$ and the latter by $Q(x)$.

Practice Problem 1   Let $P(x, A, B)$ represent the open statement $x \in A \cap B$. Determine what statement is represented by $P(1/7, \mathbf{Q}^+, \mathbf{Z})$ and whether it is true.

Quantifiers

A *quantifier* is a type of logical phrase that specifies *how many*. There are two types of quantifiers that are of particular interest to us, the *universal* quantifier and the *existential* quantifier.

Definition 3.35   The **universal quantifier** is denoted symbolically by $\forall$ and can be translated as any of the following:

for all   *for any*   *for every*   *for each*.

As these translations suggest, the purpose of the universal quantifier is to enable us to express a statement that applies to *all* objects of some particular type.

The **existential quantifier** is denoted symbolically by ∃ and can be translated as any of the following:

there exists   *there is*   *there is at least one*   *for some.*

The purpose of the existential quantifier is to enable us to express a statement asserting the *existence* of an object of a certain type satisfying certain conditions.

We can form new statements by quantifying variables that appear within open statements. Following are several examples illustrating the creation of such statements. Note that in each case the quantified variables are really serving as *placeholders for generic objects of a certain type*, and, therefore, the statement can be expressed without mentioning the letters used as variable names. For this reason, quantified variables are sometimes called **bound variables**, being "bound" or "tied" to their quantifiers.

Example 3.36   *For every real number x, $x = x^2$ (i.e., $\forall x \in \mathbf{R}, x = x^2$).*

This statement says that *every real number is equal to its square*. It is false because not every real number is equal to its square; for instance, $3 \neq 3^2$.
Technically,

$$\forall x \in \mathbf{R}, x = x^2$$

is actually shorthand for the more cumbersome

$$\forall x, \ \textit{if } x \in \mathbf{R}, \textit{ then } x = x^2.$$

Nearly all mathematicians make use of this shorthand, not only because it is more concise but also because it is a convenient way of making clear the domain of a variable that is bound to a quantifier.
Note that a literal reading of $\forall x \in \mathbf{R}, x = x^2$ as

For every x is a member of **R***, $x = x^2$*

does not make sense. It is important to understand that a form of notational shorthand is being used and this necessitates altering the literal translation to something along the lines of

For every real number x, $x = x^2$

or

For every x that is a member of $\mathbf{R}$, x = x^2.

Also note that there is nothing special about the use of the letter x in the statement $\forall x \in \mathbf{R}$, $x = x^2$. We could replace x with another letter and the resulting statement would say the same thing. That is, $\forall y \in \mathbf{R}$, $y = y^2$ expresses exactly the same information as $\forall x \in \mathbf{R}$, $x = x^2$. What is really important is that the *same* letter be used in all three blanks in the template

$$\forall \_ \in \mathbf{R}, \ \_ = \_^{2}$$

not *which* letter is used. The same letter must be used throughout because our intention is that the same object replace each occurrence of the letter. The placeholder principle that is at work here is exactly the same one used in describing a set via set-builder notation (see Section 2.1) or in defining a function via function notation (see Chapter 5). Note, however, as indicated near the beginning of this example, that we may completely suppress the use of any variables and express the statement entirely in words as

Every real number is equal to its square.

Practice Problem 2 Write a symbolic version of the statement *Each integer is less than its additive inverse.*

Example 3.37 *For some real number x, x = x^2* (i.e., $\exists x \in \mathbf{R}$, $x = x^2$).

This statement, which is true, says that *there is at least one real number that is equal to its square* (there are actually two such real numbers, 0 and 1). Again, note the ability to express the statement without using a variable.

In this case, the notational shorthand

$$\exists x \in \mathbf{R}, x = x^2$$

is really an abbreviation for

$$\exists x, x \in \mathbf{R} \text{ and } x = x^2.$$

Problem 3M.5 considers in more detail the distinction in translating statements of the form $\forall x \in A$, $P(x)$ and statements of the form $\exists x \in A$, $P(x)$.

Be careful not to literally translate $\exists x \in \mathbf{R}$, $x = x^2$ into

There exists x is a member of $\mathbf{R}$, x = x^2

as this is awkward at best. Rather, we should say something like

For some real number x, x = x²

or

There exists a member x of the set **R** *such that x = x²*

or even

There exists x ∈ **R** *such that x = x².*

In fact, the most common translation of an existentially quantified statement

$$\exists x,\ P(x)$$

is as

There exists x such that P(x)

with the phrase *such that* inserted to make the statement read more smoothly. Again, the choice of the letter x is not relevant. The statement $\exists t \in$ **R**, $t = t^2$ is no different in meaning from the statement $\exists x \in$ **R**, $x = x^2$.

Practice Problem 3 Write the symbolic statement

$$\exists k \in \mathbf{N},\ 3k = k + 6$$

in words with no variables and no symbols other than 3 and 6.

The following axiom formalizes the circumstances under which each of $\forall x,\ P(x)$ and $\exists x,\ P(x)$ should be considered true. The axiom validates our conclusions in Examples 3.36 and 3.37 that $\forall x \in$ **R**, $x = x^2$ is false and $\exists x \in$ **R**, $x = x^2$ is true.

Axiom 3.38 (*The Truth of Statements Involving Quantifiers*) Given an open statement $P(x)$ involving the variable x:

- $\forall x,\ P(x)$ is true iff $P(a)$ is true for every object a in the domain of x.
- $\exists x,\ P(x)$ is true iff $P(a)$ is true for at least one object a in the domain of x.

Practice Problem 4 Determine whether the given statement is true or false.

(a) $\forall n \in \mathbf{N}, n < n^2$. (b) $\exists n \in \mathbf{N}, n + 1 > 2n$. (c) $\exists n \in \mathbf{R}^+, n + 1 > 2n$.

Example 3.39 When the same quantifier appears two or more times in a row, we usually write the quantifier only once and then list the variables to which the type of quantification is being applied. For instance,

$$\forall x \in \mathbf{R}, \forall y \in \mathbf{R}, x + y = y + x,$$

which communicates the true statement that when adding any two real numbers, the order in which the sum is calculated does not matter, would usually be written as

$$\forall x, y \in \mathbf{R}, x + y = y + x.$$

This convention is consistent with the wording of the statement as

For all real numbers x and y, x + y = y + x

in which there is only one occurrence of the phrase *for all* but it applies to both variables x and y.

When a statement involves both universal and existential quantification, we do not want to arbitrarily change the order in which the quantifiers appear as this may lead to a new statement whose meaning is different from the original.

Example 3.40 The statement

$$\forall A, \exists B, A \subseteq B,$$

that is,

For every set A, there is a set B such that $A \subseteq B$,

is true because any given set A does indeed have a superset, for instance, the set A itself. However, if we were to switch the order in which the quantifiers appear, we would get

$$\exists B, \forall A, A \subseteq B,$$

which can be translated as

There exists a set B such that for every set A, $A \subseteq B$,

or simply

There is a set that is a superset of all sets.

Because there is no single set that contains all other sets, the new statement is false.

Sometimes universal quantification is made implicit when no confusion can result.

Example 3.41   Think about the statement

If x is negative, then x^2 is positive.

You have probably made this sort of statement yourself. The intended meaning is that the square of *any* negative real number is positive, even though there is no explicit universal quantification in the actual statement. Thus, symbolically, this statement could be expressed as $\forall x < 0,\ x^2 > 0$.

Negating Statements Containing Quantifiers

In mathematics, we often need to negate statements involving quantifiers. For example, to understand what the statement $A \nsubseteq B$ means, we really need to negate the statement

For every x, if $x \in A$, then $x \in B$

that defines $A \subseteq B$ and that involves universal quantification.

Example 3.42   Let us try to decide what the negation of the statement

All the students in our class plan to teach math   (1)

should be. Note that, formally, the negation is the statement

It is not the case that all the students in our class plan to teach math.

This formal negation is rather awkwardly worded. We would like to find a statement that is logically equivalent to it but that is worded more clearly. One approach is to think about what would be required to be able to say that our original statement is false. Intuitively, it would be false if we can find at least one student in our class who does not plan to teach math. In other words, we might believe that the statement

There is a student in our class who does not plan to teach math

is logically equivalent to the negation of the original statement (1).

Thinking about Example 3.42 leads us to formulate the following axiom describing how to form the negations of statements containing quantifiers.

Axiom 3.43 (*Negating Statements Involving Quantifiers*)   For any open statement $P(x)$ involving a variable x,

$$\sim[\forall x, P(x)] \text{ is logically equivalent to } \exists x, \sim P(x)$$

and

$$\sim[\exists x, P(x)] \text{ is logically equivalent to } \forall x, \sim P(x).$$

Example 3.42 (continued)   Let $P(x)$ denote the open statement

x plans to teach math

where x can represent any student in our class. The statement $\forall x, P(x)$ is

All the students in our class plan to teach math

and, according to Axiom 3.43, its negation $\exists x, \sim P(x)$ is

There is a student in our class who does not plan to teach math.

The statement $\exists x, P(x)$ is

There is a student in our class who plans to teach math

and, according to Axiom 3.43, its negation $\forall x, \sim P(x)$ is

All of the students in our class do not plan to teach math,

which can be expressed less awkwardly as

None of the students in our class plan to teach math.

Example 3.44   The negation of $A \subseteq B$ is $A \nsubseteq B$. Because

$$A \subseteq B \text{ iff for every } x, \text{ if } x \in A, \text{ then } x \in B,$$

it follows that

$$A \nsubseteq B \text{ iff there exists } x \text{ such that } x \in A \text{ and } x \notin B$$

because the statement forms $\sim(P \Rightarrow Q)$ and $P \wedge \sim Q$ are logically equivalent (see your work on Problem 3G.1n).

Example 3.45   The negation of

There is a positive real number

is

Every real number is not positive,

which might itself be better expressed by writing

No real number is positive.

Practice Problem 5   Use Axiom 3.43 to form the negation of each statement. Write the negation as clearly as possible.

(a) The square of any integer is nonnegative.

(b) There is an odd integer whose opposite is even.

We now have enough logical apparatus in place to begin discussing how proofs are planned and written. This is the subject of the next chapter.

Answers to Practice Problems

1. The statement is $1/7 \in \mathbf{Q}^+ \cap \mathbf{Z}$, which is false.

2. One possibility is

$$\forall n \in \mathbf{Z}, n < -n$$

though a different letter could be used in place of n.

(*continues*)

Answers to Practice Problems (continued)

3. One possibility is

There is a natural number for which 3 times the number is equal to 6 more than the number.

4. (a) is false because 1 is a natural number and 1 is not less than 1^2.
   (b) is false because $n + 1 > 2n$ is untrue for every natural number n. But (c) is true because, for instance, $1/2$ is a positive real number and $1/2 + 1 > 2 \cdot (1/2)$.

5. (a) There is an integer whose square is negative.
   (b) The opposite of any odd integer is not even. That is, no odd integer has an even opposite.

Chapter 3 Problems

3A. Statements

1. Determine whether the given sentence is a statement. If it is a statement, tell whether it is simple.

   (a) The number 12 is odd.

   (b) Baseball was the most-attended professional sport in the United States in 2006.

   (c) Run!

   (d) If it rains, I'll hang out at the mall. (Assume it is clear who is speaking.)

   (e) Sue and Ted married each other. (Assume it is clear who Sue and Ted are.)

2. Write each of the following symbolic statements constructed from statements U, V, and W as English sentences. Be careful not to write an ambiguous translation. For example, if we translate $U \wedge (V \vee W)$ as "U and V or W," the grouping present in the symbolic statement has been lost. Note also that it is not appropriate to write "U and $(V$ or $W)$" because parentheses are not used this way in written English.

   (a) $U \wedge (V \vee W)$.   **(b)** $U \Rightarrow (V \vee W)$.

   (c) $\sim U \Leftrightarrow (V \wedge W)$.   **(d)** $(U \Rightarrow V) \Rightarrow W$.

3. Determine whether the given open statement is true for all values of the variable x, some but not all values of x, or none of the values of x. Take the domain of x to be $\mathbf{R}$.

    (a) $\sin(x) + \cos(x) = 1$.   **(b)** $\sin^2(x) + \cos^2(x) = 1$.   **(c)** $x^2 < 0$.

3B. Conditional Statements

1. Each of the following is a conditional statement. If it is not already expressed in *if...then* form, put it into this form. Then identify the hypothesis and conclusion of the statement.

    (a) If it is raining out, then I won't go to the beach.

    (b) I can pass the test as long as it is not too hard.

    (c) When it rains, it pours.

    (d) I'll come see you whenever I'm in town.

    (e) I'll drive provided that my car is back from the shop.

2. Based on your work in (1), what other words or phrases can play the same role as *if* in expressing a conditional statement?

3. Write the converse and the contrapositive of each statement in (1).

3C. Truth Tables

1. Construct truth tables for the following statement forms.

    (a) $P \Rightarrow {\sim}P$.   **(b)** $P \Rightarrow (Q \Rightarrow P)$.   **(c)** $P \Rightarrow [(P \Rightarrow Q) \Rightarrow P]$.

    (d) $(P \Rightarrow Q) \Rightarrow P$.   **(e)** $({\sim}P \Rightarrow Q) \Rightarrow [({\sim}P \Rightarrow {\sim}Q) \Rightarrow P]$.

    (f) $(P \vee Q) \wedge ({\sim}P \wedge {\sim}Q)$.   **(g)** $P \wedge (Q \vee R)$.

    (h) $P \Rightarrow (Q \Rightarrow R)$.   **(i)** $P \Leftrightarrow (Q \vee {\sim}P)$.

2. Is the statement form $P \Rightarrow P \Rightarrow P$ ambiguous as written? Use a truth table to provide evidence defending your conclusion.

3D. A Statement May Possess Multiple Forms

Show that the statement

If x is even and x is greater than 2, then x is not prime

possesses the given form by replacing each variable in the form with a statement so as to create the original statement.

    1. $P \Rightarrow {\sim}Q$.   **2.** $(P \wedge Q) \Rightarrow R$.   **3.** P.

3E. Tautologies and Contradictions

Which of the statement forms in Problem 3C.1 are tautologies? Which are contradictions?

3F. Necessary Versus Sufficient

1. Fill in each blank with *necessary* or *sufficient* so as to make a true statement.

   (a) To pass Calculus II, it is _____ that I understand the mathematics developed in Calculus I.

   (b) To win the game, it is _____ that the other team not show up.

   (c) To win the game, it is _____ that our team show up.

   (d) For a positive integer to be composite, it is _____ that it have 10 factors.

2. Each of the completed statements in (1) is a conditional statement. Translate each into *if...then* form.

3. The conditional statement $P \Rightarrow Q$ may be translated as

$$\textit{If P, then Q.}$$

or

$$\textit{P implies Q.}$$

Based on your answers to (1) and (2) above, fill in the blank with the correct word, *necessary* or *sufficient*, so that the resulting statement is an accurate translation of $P \Rightarrow Q$.

$$P \text{ is } \text{_____} \text{ for } Q.$$
$$Q \text{ is } \text{_____} \text{ for } P.$$

3G. Logically Equivalent Statement Forms

1. Use a truth table to determine whether the statement forms in each pair are logically equivalent to one another.

   (a) $P \wedge Q$ and $Q \wedge P$. (b) $P \Rightarrow Q$ and $\sim Q \Rightarrow \sim P$.

   (c) $P \Rightarrow Q$ and $\sim P \vee Q$. (d) $P \Leftrightarrow Q$ and $(P \Rightarrow Q) \wedge (Q \Rightarrow P)$.

   (e) $P \vee (Q \wedge R)$ and $(P \vee Q) \wedge R$.

   (f) $P \vee (Q \wedge R)$ and $(P \vee Q) \wedge (P \vee R)$.

(g) $P \wedge (Q \vee R)$ and $(P \wedge Q) \vee (P \wedge R)$.   **(h)** $\sim(\sim P)$ and P.

(i) $P \Rightarrow P$ and P.   **(j)** $\sim(P \wedge Q)$ and $\sim P \wedge \sim Q$.

(k) $\sim(P \wedge Q)$ and $\sim P \vee \sim Q$.   **(l)** $\sim(P \vee Q)$ and $\sim P \vee \sim Q$.

(m) $\sim(P \vee Q)$ and $\sim P \wedge \sim Q$.   **(n)** $\sim(P \Rightarrow Q)$ and $P \wedge \sim Q$.

(o) $\sim(P \Leftrightarrow Q)$ and $(P \wedge \sim Q) \vee (Q \wedge \sim P)$.

2. Explain why it makes sense to say that statement forms P and Q are logically equivalent if and only if the statement form $P \Leftrightarrow Q$ is a tautology.

3. Must any two tautologies be logically equivalent? What about any two contradictions? Explain your conclusions.

3H. DeMorgan's Laws

Use the conclusions you reached in Problem 3G.1(j,k,l,m) to fill in each blank with the correct logical connective, $\wedge$ or $\vee$ (each is used once).

1. $\sim(P \wedge Q)$ is logically equivalent to $\sim P$ ___ $\sim Q$.
2. $\sim(P \vee Q)$ is logically equivalent to $\sim P$ ___ $\sim Q$.

3I. A Common Error

At first glance, many people believe that the statement forms $\sim(P \Rightarrow Q)$ and $P \Rightarrow \sim Q$ are logically equivalent because they think that if P does not imply Q, it must be that P implies the negation of Q. In this problem, you will show that this reasoning is flawed.

1. Consider the statements

   If a number is prime, then it is odd

   and

   If a number is prime, then it is not odd.

   (a) Provide evidence showing each of these statements is false.

   (b) Because the first statement is false, would its negation be true or false? (Do not bother trying to form the negation of the statement; just use what you know about how the truth values of a statement and its negation are related to each other.)

   (c) Loosely speaking, two statements are logically equivalent if they have the same truth value, both true or both false. Based on (a) and (b), explain why the negation of the first statement could not be logically equivalent to the second statement. (This provides intuitive evidence that $\sim(P \Rightarrow Q)$ and $P \Rightarrow \sim Q$ are not logically equivalent.)

2. Use a truth table to provide conclusive evidence showing that the forms $\sim(P \Rightarrow Q)$ and $P \Rightarrow \sim Q$ are not logically equivalent.

3J. Applying Logical Algebra to Membership Conditions

1. Revisiting Problem 2N.1: For each set in this problem, find the simplest way to express the notion that x is a member of the set using only the statements $x \in A$, $x \notin A$, $x \in B$, and $x \notin B$, along with the logical connectives *and* and *or*.

2. Revisiting Problem 2N.3: For each set in this problem, find the simplest way to express the notion that x is a member of the set using only the statements $x \in A$, $x \notin A$, $x \in B$, $x \notin B$, $x \in C$, and $x \notin C$, along with the logical connectives *and* and *or*.

3. Rewrite the given statement as an equivalent statement that makes use of only the statements $x \in A$, $x \notin A$, $x \in B$, and $x \notin B$, along with one of the logical connectives *and* and *or*. Try to find the simplest such statement possible.

   (a) $x \notin A \cap B$. (b) $x \notin A \cup B$. (c) $x \notin A - B$.

4. The definition of the symmetric difference of two sets was given in Problem 2P. Which of the following statements is a correct translation of the statement $x \notin A \Delta B$?

   (a) $x \notin A - B$ or $x \notin B - A$. (b) $x \notin A - B$ and $x \notin B - A$.

3K. Valid Argument Forms

1. Use a truth table to show that each argument form is valid.

   (a) $P \Rightarrow Q, P \therefore Q$. (b) $P \Rightarrow Q, \sim Q \therefore \sim P$. (c) $P \wedge Q \therefore P$.

   (d) $P \Rightarrow Q, Q \Rightarrow R \therefore P \Rightarrow R$. (e) $P \Leftrightarrow Q, Q \Leftrightarrow R \therefore P \Leftrightarrow R$.

   (f) $P \Rightarrow Q, Q \Rightarrow P \therefore P \Leftrightarrow Q$. (g) $P \Leftrightarrow Q, P \therefore Q$.

   (h) $P \Leftrightarrow Q \therefore \sim P \Leftrightarrow \sim Q$.

2. Use a truth table to show that each argument form is invalid.

   (a) $P \vee Q \therefore P$. (b) $P \Rightarrow Q, Q \therefore P$. (c) $P \Rightarrow Q, \sim P \therefore \sim Q$.

3. Explain why it makes sense to say that the argument form $P_1, P_2, \ldots,$ $P_n \therefore C$ is valid if and only if the statement form $(P_1 \wedge P_2 \wedge \ldots \wedge P_n) \Rightarrow C$ is a tautology.

4. Explain why, if one form that an argument exhibits is valid, all forms the argument exhibits must be valid. (HINT: Consider the form exhibited by an argument in which all statement variables correspond with simple statements in the argument itself. Note that any other form the argument exhibits just replaces some combinations of variables and statement connectives with other statement variables. Try to explain why such a substitution process would still yield an argument form that is valid if the original argument form was itself valid.)

3L. Valid Arguments

1. Determine whether or not the given argument is valid. Explain how you reach your conclusions.

   (a) It is known that if x is a rational number, then $x^2 + 1$ is a rational number. It is also known that x is a rational number. Therefore, we may conclude that $x^2 + 1$ is a rational number.

   (b) *Premises:*   If $x = a + b$, then x is positive.

       $x \leq 0$.

       Consequence: $x \neq a + b$.

   (c) We are given that if $A \subseteq B$, then $C \subseteq D$. It turns out that $A \not\subseteq B$. Hence, $C \not\subseteq D$.

   (d) Suppose that if the natural number a is even, so is the natural number b. Furthermore, suppose that if b is even, then a is odd. Therefore, we conclude that if a is even, a is also odd.

2. Suppose that P and Q are logically equivalent statements. Explain why it must follow that $P \therefore Q$ is a valid argument.

3M. Statements Involving Quantifiers

1. We have studied three different ways of creating statements from an open statement. Specifically, given an open statement $P(x)$, we can

   • replace the variable x with an object a within its domain to obtain the statement $P(a)$;

   • apply universal quantification to the variable x to obtain the statement $\forall x, P(x)$;

   • apply existential quantification to the variable x to obtain the statement $\exists x, P(x)$.

   Let $Q(s)$ be the open statement

   $$s < s^2,$$

with the domain of the variable s being **N**. Create each of the statements

$$Q(1) \qquad Q(3) \qquad \forall s, Q(s) \qquad \exists s, Q(s)$$

from this open statement. Then tell which of the statements you create are true and which are false.

2. Let $P(x)$ be the open statement

x is a math major.

Write a coherent English sentence that translates the given symbolic statement and that does not make use of a letter or other symbol as a variable. Assume that the domain of x consists of the members of your class.

(a) $\forall x, P(x)$. **(b)** $\exists x, P(x)$. **(c)** $\exists x, \sim P(x)$.
(d) $\forall x, \exists y, P(x) \Rightarrow \sim P(y)$.

3. Correctly label each sentence as a statement or an open statement. Take the domain of each variable to be **R**.

(a) $x^2 \leq 0$. **(b)** $\forall x, x^2 \leq 0$. **(c)** $\exists x, x^2 \leq 0$. **(d)** $\forall x, \exists y, x = y^2$.
(e) $\forall x, x = y^2$.

4. Decide whether the statements in each pair are logically equivalent. Assume that $\mathbf{R}^+$ is the domain of each variable.

(a) $\forall x, x^2 > x$ and $\forall y, y^2 > y$.
(b) $\exists \theta, \cos(\theta) = 0$ and $\exists u, \cos(u) = 0$.
(c) $\exists \theta, \cos(\theta) = 0$ and $\exists u, \cos(\theta) = 0$.
(d) $\forall \varepsilon, \exists \delta, \forall x, |x - 5| < \delta \Rightarrow |x^2 - 25| < \varepsilon$ and
$\forall \delta, \exists \varepsilon, \forall t, |t - 5| < \varepsilon \Rightarrow |t^2 - 25| < \delta$.
(e) $\forall \varepsilon, \exists \delta, \forall x, |x - 5| < \delta \Rightarrow |x^2 - 25| < \varepsilon$ and
$\forall \delta, \exists \varepsilon, \forall x, |x - 5| < \delta \Rightarrow |x^2 - 25| < \varepsilon$.

5. Examples 3.36 and 3.37 pointed out that whereas $\forall x \in A, P(x)$ translates to

$$\forall x, \text{ if } x \in A, \text{ then } P(x),$$

$\exists x \in A, P(x)$ translates to

$$\exists x, x \in A \text{ and } P(x).$$

You may wonder why $\exists x \in A$, $P(x)$ cannot be translated as

$$\exists x, \text{ if } x \in A, \text{ then } P(x). \tag{1}$$

To see why, argue that (1) can be true even when there is no member t of the set A for which $P(t)$ is true, which would contradict the second part of Axiom 3.38.

6. You may recall from calculus that every differentiable function is continuous, but it is not the case that every continuous function is differentiable. Moreover, whereas some functions are differentiable, others are not. And whereas some functions are continuous, others are not. (For this problem, it is not important that you remember what these properties actually mean mathematically.) Let $D(f)$ represent the open statement

    The function f is differentiable.

    and let $C(f)$ represent the open statement

    The function f is continuous.

    Write a coherent English sentence that translates the given symbolic statement and does not make use of a letter or other symbol as a variable; do your best to avoid awkward language in your translations. Then use the information given in the first three sentences of this problem to determine which of these statements are true and which are false.

    (a) $\forall f, D(f) \wedge C(f)$.

    (b) $\forall f, D(f) \Rightarrow C(f)$.

    (c) $\forall f, C(f) \Rightarrow D(f)$.

    (d) $\forall f, C(f) \Rightarrow {\sim}D(f)$.

    (e) $\exists f, C(f) \wedge {\sim}D(f)$.

7. We have seen that interchanging the positions of universal and existential quantifiers in a statement can have an effect on the resulting statement's meaning and truth. Explain the difference in meaning between the two statements in each of the following pairs of statements.

    (a) $\exists x \in \mathbf{R}, \forall y \in \mathbf{R}^{+}, x < y$ and $\forall y \in \mathbf{R}^{+}, \exists x \in \mathbf{R}, x < y$

    (b) $\forall a \in \mathbf{R}, \exists n \in \mathbf{Z}, a - 1 < n \leq a$ and $\exists n \in \mathbf{Z}, \forall a \in \mathbf{R}, a - 1 < n \leq a$

8. Both of the statements in (7a) above are true, but their meanings are different. What is the largest value of the variable x appearing in the first of these statements that you may cite to argue that this statement

is true? Why is it not necessary to ever cite this number to convince someone of the truth of the second statement?

9. Only one of the statements in (7b) above is true. Which one? Try to defend your conclusion.

10. The quantification in the statement

$$\text{If } x > 1, \text{ then } x^2 > 1.$$

has been expressed implicitly.

(a) Rewrite the statement so that the quantification is made explicit and the domain of the variable x is also made clear.

(b) The given statement is true. Form the converse of the statement. Is the converse true? Provide evidence to support your conclusion.

11. Consider the sentences

$$\text{If } x = 3, \text{ then } x^2 = 9. \tag{1}$$

and

$$\text{If } x = 4, \text{ then } x^2 = 9. \tag{2}$$

Most people view (1) as being true and (2) as being false. But technically these sentences are really open statements, so in reality the truth of either one of them may depend on the number assigned to the variable x.

(a) Find a real number value for x that will make (2) true.

(b) Explain why there is no real number value for x that will make (1) false.

(c) It can be argued that the reason most people view (1) as being true and (2) as being false is that these sentences are really statements in which universal quantification over the set $\mathbf{R}$ has been made implicit. Assuming that this is the case, explain why (1) should be regarded as true and (2) should be regarded as false.

12. Consider the statements

$$\exists a \in \mathbf{R}, \forall b \in \mathbf{R}, a + b = b.$$

and

$$\forall b \in \mathbf{R}, \exists a \in \mathbf{R}, a + b = 0.$$

(a) In which statement can the value of a change as the value of b changes? In which statement must the value of a remain the same no matter what the value of b?

(b) In the statement where the value of a must remain the same, what is the actual value of a?

(c) In the statement where the value of a can change, what is the value of a that is associated with $b = 2/3$? with $b = -2/3$? with $b = 0$?

3N. Negating Statements Involving Quantifiers

1. Consider the statement

    All roads lead to Rome.

    Which of the following is the negation of this statement?

    (a) *No roads lead to Rome.*

    (b) *At least one road does not lead to Rome.*

2. Consider the statement

    At least one road leads to Rome.

    Which of the following is the negation of this statement?

    (a) *At least one road does not lead to Rome.*

    (b) *No roads lead to Rome.*

3. Each of the following statements involves one or more quantifiers. For each statement:

    - Write the statement in symbolic terms using the appropriate quantifiers $\forall$ and $\exists$, along with a symbolic representation of the appropriate open statement. As appropriate, introduce letters as variables. For example, the statement in (a) below can be expressed symbolically as $\exists B, B \neq A$, where the letter B has been introduced as a variable whose domain consists of sets.

    - Then write the formal negation of the symbolic statement you have written and simplify it using Axiom 3.43.

    - Then correctly translate the simplified negation you have formed back into a coherent English sentence. Does this sentence appear to you to be the negation of the original English sentence that was given?

    (a) There is a set that is not equal to the set A.

    (b) The set B is a subset of every set.

    (c) Every set is a subset of the set C.

    (d) Every set is nonempty.

    (e) There is a nonempty subset of every set.

    (f) Every rational number is a real number.

    (g) There is a prime number that is not odd.

    (h) The square of every real number is at least 0.

    (i) The sum of 0 and any real number is that real number.

    (j) For each real number, there is a real number that can be added to it to yield a sum of 0.

    (k) There is no largest real number.

4. Write the negation of each statement as clearly as possible.

    (a) No one can win all the time.

    (b) Everybody needs somebody's help sometime.

    (c) Where there's smoke there must be a fire.

5. Two of the statements (a)–(e) in Problem 3M.6 are negations of each other. Which two?

6. Many mathematical definitions involve the use of quantifiers. You may or may not recall the following definitions (it is not important for this problem if you do not remember them or if you are having difficulty understanding exactly what they say mathematically):

    (a) A function f is **odd** if for every allowable input x, $f(-x) = -f(x)$.

    (b) The **limit** of a function f as x approaches a is the number L if for every $\varepsilon > 0$, there exists $\delta > 0$ such that for every $x \in \mathbf{R}$, if $0 < |x - a| < \delta$, then $|f(x) - L| < \varepsilon$.

    (c) A sequence (s_n) **converges** to the number p if for every $\varepsilon > 0$, there exists $k \in \mathbf{N}$ such that for any $n \in \mathbf{N}$, if $n \geq k$, then $|s_n - p| < \varepsilon$.

Note that to understand what it means for a function to *not* be odd, for a function to *not* have a certain number as its limit, or for a sequence to *not* converge to a certain number, we must be able to formulate the negation of each of the statements following *if* in the above definitions. Form these negations and simplify the resulting statements as much as possible using Axiom 3.43 and, where appropriate, the fact that $\sim(P \Rightarrow Q)$ is logically equivalent to $P \wedge \sim Q$.

7. Write a statement that asserts the existence of an object based on the assumed truth of the given statement.

    (a) The sets A and B are not equal.

    (b) The set A is nonempty.

8. The definition of disjoint sets was given in Problem 2R. Use this definition to write a translation of the following statement that involves existential quantification.

The sets A and B are not disjoint.

Chapter 4

Planning and Writing Proofs

In the previous chapter, we learned what it means to reason *deductively* from given information. This deductive process is at the heart of creating mathematical arguments, and it is in this chapter that we begin to learn how to structure and develop mathematical proofs. We will see that the way a proof is planned and subsequently written up has a lot to do with the *form* exhibited by the statement we intend to prove. Our attention to this matter here will assist us as we attempt to organize both our thinking and our writing whenever we are faced with the task of creating a proof.

4.1 The Proof-Writing Context

Questions to guide your reading of this section:

1. On what kind of reasoning does a proof rely?

2. What name is given to a statement for which we have developed a valid proof?

3. What axioms concerning sets and real numbers serve as the basic mathematical assumptions upon which we will build our arguments?

4. In making a mathematical conjecture, what sorts of statements could we take as implicit premises?

5. In what ways can a mathematical conjecture's explicit premises be stated?

6. Besides premises, what other statements can appear within a proof?

We write proofs to support the truth of the conjectures we formulate as a result of our mathematical investigations. These conjectures are created within the context of certain assumptions, called **axioms**, we have agreed to. In essence, a **proof** is a written record of a series of valid deductive arguments that together explain the logical inevitability of a conjecture. Once a valid proof for a conjecture has been found, the conjecture becomes a **theorem**. In this chapter, we show you how to develop proofs and how to structure your writing of proofs into forms consistent with accepted mathematical practice.

A proof for a mathematical statement is typically planned by analyzing a *form* the statement exhibits. We will provide you with strategies for proving statements that possess certain standard forms. These strategies will make some parts of the proof-writing process almost automatic. Otherwise, writing a mathematical proof is like many other types of writing. It requires planning, insight, reflection, revision, and editing. The end product itself will consist of complete sentences organized into paragraphs. Our writing will incorporate mathematical terminology and notation, and it will be important that we use such terms and notation correctly so that our sentences are well structured and convey our ideas clearly and precisely. We will also need to be sure we are able to justify each statement in our proofs, in many cases providing explicit justification for each deduction as we write.

The process of learning how to develop and write your own proofs will actually go a long way toward helping you to comprehend and become more comfortable with mathematical language. As your ability to write mathematics improves, so will your ability to read critically and analyze mathematical text, and vice versa. Both skills, reading and writing, are essential to learning and doing mathematics. Any proof you read or write has the potential for helping you to gain deeper insight into mathematical ideas, and that is really the ultimate payoff for anyone who enjoys doing mathematics.

We will be writing proofs in the context of set theory and the real number system. The basis for our deductions about sets is Axiom 2.39, and the basis for our deductions about real numbers is Axiom 2.46: These axioms comprise our assumptions about sets and real numbers. Our attention to the real number system is significant as much of your prior work in mathematics fundamentally relies on basic properties of the algebra of the real numbers. But our consideration of sets is also important because the language and notations of set theory are used throughout mathematics, especially in the communication of abstract mathematical concepts. Set theory also provides a relatively neutral arena for developing proof-writing skills. Even with our work in Chapter 2, you are probably much less familiar with the basic properties of sets than you are with properties of numbers. We hope that your more limited experience with sets will help you to better appreciate the need for formal arguments.

It is usually customary to omit explicit mention of the application of set-theoretic axioms in most proofs, in essence because these axioms are used so often. In contrast, at least at first, we will usually take care to cite the application of the axioms for the real numbers. This is primarily because although they represent familiar mathematical properties, numbers have many other properties just as familiar to us that we are not assuming. For the most part, these other facts can be deduced from the axioms.

From Valid Arguments to Proofs

Recall that an argument consists of a set of statements that are assumed to be true, the so-called *premises* of the argument, along with a statement whose truth is to be logically deduced from the premises, the *consequence* of the argument. A proof for a mathematical conjecture is really a write-up of the deductive arguments used to move legitimately from the conjecture's premises to its ultimate consequence. Axioms, definitions, previously proved theorems, and tautologies may all be regarded as **implicit premises** in the sense that they are not usually stated as part of a conjecture, but they are assumed to be true in the context within which the conjecture has been proposed. Any valid application of relevant implicit premises may appear in the proof of a mathematical conjecture.

Example 4.1   Among the other assumptions stated in Axiom 2.46, we are assuming that **R** is closed under addition. Thus, if we are working on a proof and have discovered that x is a real number, we could conclude that $x + 1$ is also a real number because of the closure of **R** under addition.

Example 4.2   The statement

$$p \in J \cap K \text{ if and only if both } p \in J \text{ and } p \in K$$

is allowable within a proof because it simply quotes the definition of set intersection.

Example 4.3   Suppose we have already proved the Pythagorean Theorem (stated in Chapter 1). Then, as a previously established theorem, it could be applied within a subsequent proof.

Example 4.4   The statement

The value of the real number x is either zero or nonzero

could appear in a proof because it has the form of the tautology $P \vee {\sim}P$.

A conjecture usually also incorporates additional **explicit premises**, premises specifically included within the statement of the conjecture. At the very beginning of a proof, any explicit premises of the conjecture we are proving are usually assumed to be true.

Example 4.5   Consider the conjecture

Let m and n be even integers. Then $m + n$ is even.

The sentence

Let m and n be even integers

imposes the explicit premise that

m and n are even integers.

This premise would usually be assumed true at the beginning of a proof of the conjecture.

Because the same conjecture can be expressed in a variety of ways, it is important to be able to identify any explicit premises on which a conjecture relies.

Example 4.5 (continued) The conjecture

Let m and n be even integers. Then m + n is even.

can also be written as

If m and n are even integers, then m + n is even.

We may still view the statement

m and n are even integers

as an explicit premise whose truth is being assumed.

In fact, whenever a conjecture has been expressed as a conditional statement, the hypothesis may be viewed as an explicit premise.

Example 4.6 In the conjecture

$$\textit{If } x \in A - B, \textit{ then } x \in A \cup B$$

the hypothesis

$$x \in A - B$$

may be taken as an explicit premise.

Example 4.7 Now consider the conjecture

Suppose a, b, c, and d are real numbers. If a < b and c < d then a + c < b + d.

There are several explicit premises in this conjecture; namely,

a, b, c, and d are real numbers, a < b, and c < d.

In attempting to prove the conjecture, all of these premises would usually be assumed true.

Practice Problem 1   Identify the explicit premises in the following conjecture:

Let f be a function. If f is differentiable, then f is continuous.

Back in Section 3.4, we learned how to draw forth "new" true statements from given premises using the notion of *logical consequence*. This deductive process is at the heart of creating a proof, and any such deductions, provided they result from valid arguments, may appear in a proof.

Example 4.6 (continued)   Consider once again the conjecture

If $x \in A - B$, then $x \in A \cup B$.

The explicit premise

$$x \in A - B$$

can be taken as true, as can the implicit premise

$x \in A - B$ if and only if both $x \in A$ and $x \notin B$,

which is just the definition of set difference. Then, as the argument form

$$P, P \Leftrightarrow Q \therefore Q$$

is valid (see Problem 3K.1g), we may deduce from these two premises that

$x \in A$ and $x \notin B$.

This new statement is true and may appear in a proof of the conjecture, as it was obtained as a logical consequence of a valid argument whose premises are true.

Practice Problem 2   Suppose that both $A \subseteq B$ and $x \in A$ are explicit premises in a conjecture we are trying to prove.

(a) What logical consequence can be drawn forth from these explicit premises?

(b) What implicit premise did you rely on to make your deduction in (a)?

Thus, the proof of a theorem may include the theorem's explicit premises, any application of relevant implicit premises, and any statement that can be deduced from the premises via a valid argument. We are now ready to consider how we might go about proving the different kinds of statements that regularly appear in mathematical settings.

Answers to Practice Problems

1. There are two explicit premises: *f is a function* and *f is differentiable.*

2. (a) $x \in B.$

(b) Our deduction relies on the implicit premise

$A \subseteq B$ *if and only if, for every x, if $x \in A$, then $x \in B$,*

which is just the definition of subset.

4.2   Proving an *If... Then* Statement

Questions to guide your reading of this section:

1. In the standard strategy used to prove an *if... then* statement, what do we assume and what do we try to show?

2. What are some of the stylistic conventions we should follow when writing up a proof?

The majority of statements proved in mathematics take the form of an *if... then* statement $P \Rightarrow Q$. We have learned that the only way $P \Rightarrow Q$ can be false is if the hypothesis P is true and the conclusion Q is false. Thus, if we assume the truth of the hypothesis P and are then able to show that in this circumstance the conclusion Q is also true, we would be able to deduce the truth of the conditional statement $P \Rightarrow Q$ itself. This is the standard approach to proving an *if...then* statement.

Proof Strategy 4.8 (*Proving an "If...Then" Statement*)   To prove the *if...then* statement $P \Rightarrow Q$, assume the hypothesis P and deduce the conclusion Q.

Note that this proof strategy is aligned with our observation in Section 4.1 that the hypothesis of an *if...then* statement we are trying to prove may be taken as an explicit premise and, therefore, would usually be assumed true at the beginning of a proof of the statement.

Example 4.6 (continued)   To prove

$$\text{If } x \in A - B, \text{ then } x \in A \cup B,$$

we could assume the hypothesis $x \in A - B$ and then try to deduce the conclusion $x \in A \cup B$.

Example 4.9   To prove

$$\text{If } a, b, \text{ and } c \text{ are integers, then so is } a + bc,$$

we could assume a, b, and c are integers and try to deduce that $a + bc$ is an integer.

Practice Problem 1   Suppose we want to use our standard strategy for proving an *if...then* statement to prove

$$\text{If } p \in A \text{ and } A \cup B \subseteq C, \text{ then } p \in C.$$

(a)   What could we assume?

(b)   What would we try to deduce?

Once we have clearly distinguished what we should assume from what we need to deduce, we must develop an argument that gets us from the former to the latter.

Example 4.6 (continued)   How can we deduce what we want to show, $x \in A \cup B$, from our assumption that $x \in A - B$? It quickly becomes apparent that certain implicit premises, namely the definitions of the union and difference of sets, will play a critical role in achieving our goal. Recalling and studying these definitions

$$x \in A \cup B \text{ iff either } x \in A \text{ or } x \in B$$
$$x \in A - B \text{ iff both } x \in A \text{ and } x \notin B$$

we may observe that our premise $x \in A - B$ requires that x be a member of A, and this is enough to conclude that $x \in A \cup B$, as membership in a union requires membership in only one of the sets over which the union is formed. We have just identified the key argument to be made in our proof.

Now the deductions in a proof must be sequenced to form a clear path from what we already know to what we are trying to show. In addition, a proof should incorporate appropriate documentation to justify these deductions.

Example 4.6 (continued) In writing up our proof of the conjecture

$$\textit{If } x \in A - B, \textit{ then } x \in A \cup B,$$

we must be sure to properly sequence our deductions so that there is a clear flow from our initial assumption ($x \in A - B$) to what we are ultimately trying to show ($x \in A \cup B$). Along the way, we want to include appropriate justification to support our deductions. Here is what we might write:

Assume $x \in A - B$. We will show $x \in A \cup B$. By the definition of set difference, as $x \in A - B$, it follows that $x \in A$ and $x \notin B$. But, because $x \in A$, we may now conclude, by the definition of union, that $x \in A \cup B$.

Note that our proof begins with the assumption of the explicit premise $x \in A - B$ (this is what we *know*). The use of relevant implicit premises, specifically the definitions of the difference and union of sets, is documented so that anyone reading our proof can follow our line of reasoning. The proof ends once we have deduced the conclusion $x \in A \cup B$ (what we wanted to *show*) of the original *if...then* statement that was our conjecture.

It is also important to note what is *not* usually included in the write-up of a proof. First, the logical maneuvering that underlies the arguments in a proof is usually not explicitly pointed out.

Example 4.6 (continued) For instance, we did not mention that because $P \wedge Q \therefore P$ is a valid argument form, we are able to deduce "$x \in A$" from "$x \in A$ and $x \notin B$." The reader is expected to be knowledgeable of elementary logic and to fill in for herself the logical underpinnings of the arguments presented in a proof.

Also, the reader is expected to be familiar with the definitions of any terms that arise within a proof. It is how these definitions are being used in the proof that should be pointed out, not statements of the definitions themselves. The same goes for axioms and previously proved theorems; their application, not their formal statement, is what should be presented in a proof.

Example 4.6 (continued)   We never explicitly wrote out the definitions of the difference and union of sets within the proof we wrote. That is, we did not include a sentence such as, "Now the definition of set difference tells us that $x \in A - B$ if and only if both $x \in A$ and $x \notin B$." Instead, our proof focused on how this definition, which the reader of the proof is expected to be familiar with, can be used to deduce "$x \in A$ and $x \notin B$" from the explicit premise "$x \in A - B$."

Stylistic Conventions in Writing Proofs

The write-up of a proof should adhere to certain stylistic conventions that the mathematical community has established.

Example 4.6 (continued)   This is how we would normally display the theorem and proof we have developed in this example:

Theorem: If $x \in A - B$, then $x \in A \cup B$.

Proof: Assume $x \in A - B$. We will show $x \in A \cup B$. By the definition of set difference, as $x \in A - B$, it follows that $x \in A$ and $x \notin B$. But, because $x \in A$, we may now conclude, by the definition of union, that $x \in A \cup B$. $\square$

To begin, note that we clearly labeled the statement we proved using the word *Theorem*. Fundamentally, a proof explains, via deductive reasoning, why a certain conjecture should be considered true; that is, why the conjecture is really a theorem. Thus, once we believe we can write a proof for a conjecture, we begin to refer to the conjecture as a theorem and label it accordingly. Other labels that are sometimes used in place of *Theorem* include *Proposition* and *Claim*. You should always clearly identify a statement you intend to prove by giving it one of these labels.

Next observe that we used the label *Proof* to indicate where the proof of our theorem begins. You should always label the beginning of a proof with the word *Proof*. We also used the symbol $\square$ to indicate the end of our proof and shall do so throughout this book.

Most of the time, when a proof is being written in accordance with a standard proof strategy, the very first sentence of the proof should give the reader a sense of how that strategy is being used. For instance, we used the sentence

$$\textit{Assume } x \in A - B$$

at the beginning of our proof to indicate that we are assuming the hypothesis of the *if...then* statement we are proving in order to carry out the standard strategy for proving an *if...then* statement. This sentence alerts the reader of our intention to use this strategy. Other words that would convey the same information as *assume* in this context would be *suppose, let,* and *consider.* Also, notice that the second sentence in our proof

$$We\ will\ show\ x \in A \cup B$$

reminds us of the ultimate consequence we hope to deduce. This sentence is really optional, but it can help us to keep our proof focused and on track.

Observe also that a sequence of deductions is being made in our proof. As we move from one deduction to another, we used words and phrases such as *we may conclude that, hence,* and *it follows that* to indicate that a deduction is being made. Other words that could have been used are *then, therefore,* and *thus.* These words and phrases are interchangeable in this context.

The writer of a proof is expected to provide some level of justification for the deductions being made. Note carefully how the words *as* and *because* are used to reference justification for particular deductions in our proof. For instance, the third sentence of the proof states

$$By\ the\ definition\ of\ set\ difference,\ as\ x \in A - B,\ it\ follows\ that\ x \in A\ and\ x \notin B.$$

This sentence actually makes the deduction

$$x \in A\ and\ x \notin B.$$

The clause

$$as\ x \in A - B$$

explains why this deduction is valid. Another word that could replace *as* in this context is *because.* However, the word *if* is not an appropriate replacement here because the sentence

$$By\ the\ definition\ of\ set\ difference,\ if\ x \in A - B,\ it\ follows\ that\ x \in A\ and\ x \notin B$$

does not make a deduction. Instead, this sentence simply states a *conditional* observation of what *could* be deduced *if* a certain hypothesis were true. The word *if* makes it sound like we do not already know that $x \in A - B$, when, in fact, we do. In a proof, we really need to *make* deductions, not just state what deductions *could* be made.

The amount and nature of the justification appearing in a proof varies depending on the audience for whom the proof is written. For now, we recommend you think of yourself and your fellow students as your audience and that you attempt to provide thorough documentation of every move you make in the proofs you write. As you gain experience, you will sometimes be able to provide less justification because of what you can expect from your audience. You can use the write-ups of proofs that appear in this book as a guide for both how and when to justify your conclusions.

> **Practice Problem 2**   Use your answers from Practice Problem 1 to write a proof of the statement *If $p \in A$ and $A \cup B \subseteq C$, then $p \in C$.*

Example 4.9 (continued)   We now prove the statement

$$\textit{If a, b, and c are integers, then so is } a + bc$$

using the plan we already developed in accordance with the standard strategy for proving a conditional statement. Our proof relies on the assumption we have made (see Axiom 2.46) that the set $\mathbf{Z}$ is closed under both addition and multiplication, meaning that whenever we add or multiply two integers, the result is an integer.

Theorem: If a, b, and c are integers, then so is $a + bc$.
Proof: Let a, b, and c be integers. We will show $a + bc$ is an integer. Because $\mathbf{Z}$ is closed under multiplication and both b and c are integers, it follows that bc is an integer. Then, because $\mathbf{Z}$ is closed under addition and both a and bc are integers, it follows that $a + bc$ is an integer.   □

> **Practice Problem 3**   The **square** of a real number a, denoted a^2, is defined so that $a^2 = a \cdot a$. Prove that if x and y are rational numbers, so is $x^2 + y^2$.

Sometimes a conditional statement we want to prove is not expressed in *if... then* form. In such a situation, it is usually a good idea to rewrite the statement so that the hypothesis and the conclusion have been clearly identified.

Example 4.10   If we want to prove the statement

$$\textit{The square of an odd integer is odd,}$$

we might first rewrite it as

$$\textit{If an integer is odd, then its square is also odd.}$$

Furthermore, when a statement we want to prove does not provide names for the objects to which it refers, we usually must create names for these objects within our proof.

Example 4.10 (continued)   We could write the beginning of the proof of

If an integer is odd, then its square is also odd

as follows:

    Theorem: If an integer is odd, then its square is also odd.
    Proof: Suppose n is an odd integer. We will show n^2 is odd. [*argument under construction*]   □

Essentially, we have implicitly rewritten the theorem as

If the integer n is odd, then n^2 is also odd

so that its statement includes a name for the odd integer that is mentioned. We then just applied our standard strategy for proving an *if...then* statement to start our proof. Of course, we could have used a different letter in place of n.

We inserted the parenthetical remark

[argument under construction]

to indicate where our proof needs further development. We will use this remark throughout the text when part of a proof we are writing is not yet complete. This particular proof will be completed in Section 4.6.

Practice Problem 4   Consider the statement

The sum of a rational number and an irrational number is irrational.

(a) Rewrite this statement in *if...then* form and so that letters naming the numbers to be added have been introduced.

(b) In applying the standard strategy for proving an *if...then* statement to prove the statement you wrote in (a), what would we assume and what would we try to deduce?

Answers to Practice Problems

1. (a) $p \in A$ and $A \cup B \subseteq C$.   (b) $p \in C$.

2. *Theorem:* If $p \in A$ and $A \cup B \subseteq C$, then $p \in C$.

   Proof: Assume $p \in A$ and $A \cup B \subseteq C$. As $p \in A$, it follows, by the definition of union, that $p \in A \cup B$. Then, because $A \cup B \subseteq C$, the definition of subset allows us to conclude that $p \in C$. $\square$ [*As is typical in writing a proof, you might have chosen somewhat different wording to indicate the deductions being made or to reference justification for the deductions.*]

3. *Theorem:* If x and y are rational numbers, then so is $x^2 + y^2$.

   Proof: Consider rational numbers x and y. Because **Q** is closed under multiplication, we may conclude that both x^2 and y^2 are rational. Then, because **Q** is closed under addition, it follows that $x^2 + y^2$ is rational. $\square$

4. (a) If a is a rational number and b is an irrational number, then $a + b$ is irrational. [*Other letters could have been chosen in place of a and b.*]

   (b) We would assume a is a rational number and b is an irrational number. We would try to deduce that $a + b$ is irrational.

4.3 Proving a *For All* Statement

Questions to guide your reading of this section:

1. Regarding the basic strategy used to prove *for all* statements: What is the strategy and how is it implemented? What is the "hat/envelope" analogy used to model the strategy? Why is it important for the object chosen at the beginning of a proof that uses this strategy to remain anonymous except for being of the required type?

2. How do we usually prove a subset relationship $A \subseteq B$?

3. How do we go about disproving a false statement that involves universal quantification?

Recall that the universal quantifier $\forall$ may be translated as *for all, for every, for each,* or *for any.* Thus, a statement of the form $\forall x, P(x)$ is true only when the property $P(x)$ is true for *all* objects within the domain of the variable x. This suggests the following proof strategy.

Proof Strategy 4.11 (*Proving a "For All" Statement*)   To prove $\forall x, P(x)$, consider an anonymous x within the domain of the variable and show the statement $P(x)$ is true for this arbitrarily selected x.

Example 4.12   The *for all* statement

$$\text{For any real number } a,\ 0a = 0$$

may be expressed as the *if...then* statement

$$\text{If } a \text{ is a real number, then } 0a = 0$$

To prove this conditional statement, our standard strategy for proving an *if...then* statement tells us we could assume a is a real number and then try to show $0a = 0$. Note the agreement of this approach with the new Proof Strategy 4.11: We are essentially beginning our proof by considering an anonymous a within the domain $\mathbf{R}$ of the variable and then will try to show $P(a)$, where in this case $P(a)$ is the statement $0a = 0$.

It is very important to remember that our standard strategy for proving a *for all* statement requires us to maintain the anonymity of the "selected" object throughout our proof.

Example 4.12 (continued)   In attempting to prove

$$\text{For any real number } a,\ 0a = 0,$$

we begin by considering an anonymous real number a. Thus, we know a is a real number, but we do not know which one. Because we want ultimately to prove a property is true for *any* real number, we could not assign a particular real number value to a or make additional assumptions about a. For instance, we could not make $a = 5$ and then continue the rest of our argument using 5 in place of a; if we did so, we would only be proving $0 \cdot 5 = 0$, not that 0 multiplied by *any* real number yields 0. Nor could we assume that a is positive because then we would only be proving that 0 multiplied by a *positive* real number yields 0.

The process of maintaining the anonymity of the "selected" object when applying Proof Strategy 4.11 to prove $\forall x, P(x)$, and the reasoning behind this process, may be modeled by imagining that we have a hat into which we have put all the objects within the domain of the variable x. We thoroughly shuffle the objects in the hat, select one of them at random, and give the object a name, say x (usually the name is the same letter that was used as the variable). Once selected, we do not look to see which object we have chosen because we do not want to make any further assumptions about it that could prejudice the generality of our argument. In fact, it is helpful to imagine placing the selected object in an envelope that is then sealed and marked with whatever name it was given. In this way, the object remains anonymous except for the knowledge that it is within the domain of the original variable. Next, we attempt to prove that for the selected object x, the statement $P(x)$ is true. If we are able to do this, we then conclude that because the object was chosen at random and could have been any of the objects within the domain of the variable x, our conclusion that $P(x)$ is true is actually valid for *all* such objects.

Example 4.13   If we want to prove the *for all* statement

$$\text{For all real numbers } a, \ b, \ c, \text{ and } d, \ a(b + cd) = (ac)d + ba,$$

we could start by considering anonymous real numbers a, b, c, and d (in terms of the "hat/envelope" model, imagine four hats, each containing all of the real numbers, and four envelopes, one marked a, one marked b, one marked c, and the last marked d). We would then try to show that the equation $a(b + cd) = (ac)d + ba$ holds true for these numbers.

Practice Problem 1   If we are trying to prove

$$\text{For every prime number } p, \text{ the number } p + 7 \text{ is not prime}$$

using our standard strategy for proving a *for all* statement, we should begin by considering an anonymous prime number p. Why can't we then just assume $p = 11$ and observe that $p + 7 = 11 + 7 = 18$, which is not prime?

Proof Strategy 4.11 also helps us structure our write-up of the proof of a *for all* statement.

Example 4.13 (continued)   We have already begun to develop a plan for proving

$$\text{For all real numbers } a, \ b, \ c, \text{ and } d, \ a(b + cd) = (ac)d + ba$$

in accordance with Proof Strategy 4.11. In implementing our plan, here is how we might begin writing the proof:

Theorem: For all real numbers a, b, c, and d, $a(b + cd) = (ac)d + ba$.
Proof: Consider real numbers a, b, c, and d. We will show that $a(b + cd) = (ac)d + ba$. *[argument under construction]*   □

The first sentence

Consider real numbers a, b, c, and d

of our proof indicates that anonymous real numbers a, b, c, and d are being selected at random in order to carry out the standard strategy for proving a *for all* statement. As was the case in writing up the proof of an *if...then* statement, we are using language that conveys to our reader our intent to use a standard proof-writing strategy. Other sentences that we could have used instead include

Let a, b, c, and d be real numbers
Suppose a, b, c, and d are real numbers

and

Assume a, b, c, and d are real numbers.

We could also write the first sentence as, for instance,

*Consider a, b, c, d ∈ **R***

to make use of the efficiency of set-theoretic notation.

Practice Problem 2   Write the first two sentences of a proof of the statement

For every prime number p, the number p + 7 is not prime

that uses the standard strategy for proving a *for all* statement.

After we have used Proof Strategy 4.11 to begin structuring our write-up of the proof of a *for all* statement, we need to figure out how to demonstrate what needs to be shown.

Example 4.13 (continued)   We are trying to show the equation

$$a(b + cd) = (ac)d + ba$$

is true. This is really just a matter of applying certain axioms for the real numbers (see Axiom 2.46) to rewrite $a(b + cd)$ as $(ac)d + ba$.

Theorem: For all real numbers a, b, c, and d, $a(b + cd) = (ac)d + ba$.
·*Proof:* Consider real numbers a, b, c, and d. We will show that $a(b + cd) = (ac)d + ba$. Observe that

$$a(b + cd) = ab + a(cd) = a(cd) + ab = (ac)d + ab = (ac)d + ba,$$

where the first equality holds because of the distributive property, the second because of the commutativity of addition, the third because of the associativity of multiplication, and the fourth because of the commutativity of multiplication. $\square$

Practice Problem 3   Prove that for all real numbers a and b, $a(1 + ba) = a + a^2 b$.

Proving Set Inclusion

The definition of subset tells us that to prove $A \subseteq B$, we need to prove that

for every x, if $x \in A$, then $x \in B$,

which can be accomplished by applying our standard strategy for proving a *for all* statement. Specifically, we would need to consider an anonymous x that is a member of A and show x is also a member of B.

Proof Strategy 4.14 (*Proving Set Inclusion*)   To prove $A \subseteq B$, consider an anonymous member x of the set A and show x is a member of the set B. (A different letter may be used in place of x.)

For example, to prove $A \cap B \subseteq A$, we would consider an anonymous x in $A \cap B$ and then show x is in A. In this case, the definition of intersection moves us immediately from our assumption to what we want to show. Our proof appears after the following statement cataloging several elementary subset relationships.

Theorem 4.15   For any sets A and B,

    (i) $A \cap B \subseteq A$ and $A \cap B \subseteq B$;

    (ii) $A \subseteq A \cup B$ and $B \subseteq A \cup B$;

    (iii) $A - B \subseteq A$.

Proof: Assume A and B are sets. We will prove (i) and leave (ii) and (iii) for you to prove in Practice Problem 4.

First, we prove $A \cap B \subseteq A$. Let $x \in A \cap B$. Then, by the definition of intersection, $x \in A$ and $x \in B$, so, in particular, $x \in A$.

Now, to prove $A \cap B \subseteq B$, let $x \in A \cap B$. Again, by the definition of intersection, it follows that $x \in A$ and $x \in B$, so, in particular, $x \in B$.   □

Practice Problem 4   Prove (ii) and (iii) of Theorem 4.15.

A Standard Proof Strategy May Not Always Be the Best Option

The standard strategy for carrying out a certain type of proof may not always be the most appropriate or efficient way to pursue a proof. For example, we can prove the empty set is a subset of every set, but the standard strategy for proving a set inclusion would require us to begin the proof with an anonymous member of the empty set, which is impossible because the empty set, by definition, has no members. Instead, our proof of this result will go back to the definition of subset and make use of the fact that a conditional statement with a false hypothesis is true.

Theorem 4.16   For any set A, $\varnothing \subseteq A$.

Proof: Consider a set A. Note that for any x, the *if...then* statement

$$\text{If } x \in \varnothing, \text{ then } x \in A$$

has a false hypothesis and, therefore, must be considered true. Thus, by definition of subset, $\varnothing \subseteq A$.   □

Disproving a False *For All* Statement

In mathematics, we expect to formulate conjectures that we will attempt to prove, but we realize that some of them will turn out to be false. So there will be times when we need to establish that a *for all* statement is false. Recall that if $P(a)$ is false for a particular value a within the domain of the variable x, the *for all* statement $\forall x, P(x)$ is false; such a value a is referred to as a **counterexample** to $\forall x, P(x)$. Thus, as soon as we produce a single counterexample to a statement involving universal quantification, we know that the statement cannot be true. The existence of the counterexample is said to **disprove** the statement.

Proof Strategy 4.17 (*Disproving a "For All" Statement*)   To show the statement $\forall x, P(x)$ is false (i.e., to **disprove** it), find an object a within the domain of the variable x for which the statement $P(a)$ is false; such an object a is called a **counterexample** to $\forall x, P(x)$.

Example 4.18   The number 0 serves as a counterexample to the statement

The square of any real number is positive

because 0 is a real number and $0^2 = 0$ is not positive. By exhibiting this counterexample, we are disproving the statement.

Example 4.19   Any integer that is not a natural number provides a counterexample to the statement $\mathbf{Z} \subseteq \mathbf{N}$. For example, -3 is a counterexample, as -3 is an integer but is not a natural number.

Practice Problem 5   Disprove the statement *Every prime number is odd.*

Answers to Practice Problems

1.   The prime p will no longer be anonymous if we specify its value. Our standard strategy for proving a *for all* statement requires that the prime p remain anonymous throughout the proof so that our argument will apply generally to *any* prime p.

2.   *Theorem:* For every prime number p, $p + 7$ is not prime.

   Proof: Let p be a prime number. We will show $p + 7$ is not prime. [*argument under construction*] $\square$

3.   *Theorem:* For all real numbers a and b, $a(1 + ba) = a + a^2b$.

   Proof: Let a and b be real numbers. Then

$$a(1 + ba) = a \cdot 1 + a(ba) = a + a(ba) = a + a(ab) = a + (aa)b = a + a^2b,$$

   where the first equality holds because of the distributive property, the second because 1 is the multiplicative identity, the third because multiplication is commutative, the fourth because multiplication is associative, and the last because of the definition of a^2. $\square$

(*continues*)

Answers to Practice Problems (continued)

4. *Theorem:* For any sets A and B, $A \subseteq A \cup B$, $B \subseteq A \cup B$, and $A - B \subseteq A$.

Proof: Let A and B be sets.

First we prove $A \subseteq A \cup B$. Suppose $x \in A$. It logically follows that $x \in A$ or $x \in B$. Thus, by the definition of union, we may conclude that $x \in A \cup B$.

Now we prove $B \subseteq A \cup B$. Suppose $x \in B$. It logically follows that $x \in A$ or $x \in B$. Thus, by the definition of union, we may conclude that $x \in A \cup B$.

Finally, we prove $A - B \subseteq A$. Suppose $x \in A - B$. Then, by the definition of set difference, $x \in A$ and $x \notin B$. So, in particular, $x \in A$. $\square$

5. Note that 2 is a prime number that is not odd, so 2 serves as a counterexample to the given *for all* statement.

4.4 The *Know/Show* Approach to Developing Proofs

Questions to guide your reading of this section:

1. What is the *Know/Show* approach to developing proofs? In attempting to prove a mathematical conjecture, how do we identify what we *know*? How do we identify what we need to *show*?

2. In carrying out the *Know/Show* approach, why do our updates of what we know represent "moving forward" in our proof, but our updates of what we want to show represent "moving backward?"

3. On what property will an argument relating additive and multiplicative concepts usually rely?

In this text, we are applying an approach to developing proofs commonly referred to as **Know/Show**. This approach emphasizes the distinction between what we already *know* (i.e., the premises upon which a conjecture rests), from what we want to *show* (i.e., the ultimate consequence to be drawn forth or deduced from what we already know). In many cases, either before or after presenting the write-up of a proof, we will provide some discussion of our plan for how the proof was developed. Often, especially in this chapter, our discussion will include a display of what we know and what we want to show as follows: *We know* _____ and *we want to show* _____.

The most elementary application of the *Know/Show* approach reflects our standard strategy for proving an *if...then* statement $P \Rightarrow Q$: The hypothesis P can be considered something that we *know* and the conclusion Q what we want to *show*.

Example 4.6 (continued)   When we proved

$$\text{If } x \in A - B, \text{ then } x \in A \cup B$$

in Section 4.2, we assumed $x \in A - B$ and deduced $x \in A \cup B$. We would have identified what we know and what we want to show in this situation as follows: *We know* $x \in A - B$ and *we want to show* $x \in A \cup B$.

However, the *Know/Show* approach is applicable whenever we are trying to develop a proof, not just in the context of proving *if...then* statements.

Example 4.20   We can apply *Know/Show* to prove that for any sets A, B, and C, $A \cap (B - C) \subseteq B - (C \cap A)$. Because we want to prove a *for all* statement, we begin by applying Proof Strategy 4.11 and consider anonymous sets A, B, and C. Then, as the property we want to establish is a set inclusion, we apply Proof Strategy 4.14 and let x be an anonymous member of the set $A \cap (B - C)$ and try to show x is in the set $B - (C \cap A)$. In other words, *we know* $x \in A \cap (B - C)$ and *we want to show* $x \in B - (C \cap A)$.

In any proof, what we *know* and what we need to *show* must be treated very differently because of the simple fact that what we *know* is already true, whereas what we want to *show* can only be regarded as true once it has been logically deduced via a valid argument.

What we *know* is what we can build from. With anything we *know* we should ask questions like: *Where can we go from here? What can we deduce from this known information?*

Example 4.20 (continued)   Here, *we know* $x \in A \cap (B - C)$. What can we deduce from this information? Applying the definition of intersection, we can determine that $x \in A$ and $x \in B - C$. Can we make any further deductions? Well, from our new knowledge that $x \in B - C$, we can apply the definition of set difference to determine that $x \in B$ and $x \notin C$. Thus, we now know that $x \in A$ and $x \in B$ and $x \notin C$.

What we need to *show* is what we have to build toward. With anything we need to *show* we should ask questions like: *What will lead us here? How could we reach this conclusion?*

Example 4.20 (continued)   Here, *we want to show* $x \in B - (C \cap A)$. How can we obtain this conclusion? One way would be if we knew $x \in B$ and $x \notin C \cap A$, because we could then use the definition of set difference to reach the desired conclusion. This leads us to ask how we might conclude that $x \notin C \cap A$. Well, if we knew that either $x \notin C$ or $x \notin A$, the definition of intersection would allow us to conclude that $x \notin C \cap A$. So, we could reach our ultimate conclusion if we are able to deduce that $x \in B$ and either $x \notin C$ or $x \notin A$.

Note that, as just illustrated, what we know and what we want to show change as definitions and proof strategies are applied. Each time we revise what we know, we are actually "moving forward" in our proof by making a legitimate deduction that is based on information we are assuming to be true. However, each time we revise what we want to show, we are really "moving backward" from the very end of our proof toward the middle.

Example 4.20 (continued)   We can diagram our updates of what we know as follows, with the implication arrows indicating that each update is deduced from the previous one.

$$\textit{We know: } x \in A \cap (B - C).$$
$$\Downarrow$$
$$\textit{We know (update 1): } x \in A \text{ and } x \in B - C.$$
$$\Downarrow$$
$$\textit{We know (update 2): } x \in A \text{ and } x \in B \text{ and } x \notin C.$$

But in diagramming our updates of what we want to show, the implication arrows indicate that each update leads to the previous one, with the original version of what we want to show occurring at the very end of the chain.

$$\textit{We want to show (update 2): } x \in B \text{ as well as either } x \notin C \text{ or } x \notin A.$$
$$\Downarrow$$
$$\textit{We want to show (update 1): } x \in B \text{ and } x \notin C \cap A.$$
$$\Downarrow$$
$$\textit{We want to show: } x \in B - (C \cap A).$$

After we have revised what we know and what we want to show as much as we are able, we need to find a way to get from the most updated version of what we know to the most updated version of what we want to show.

Example 4.20 (continued)   Here, we need to figure out how to get from

$$\textit{We know (update 2): } x \in A \text{ and } x \in B \text{ and } x \notin C.$$

to

$$\textit{We want to show (update 2): } x \in B \text{ as well as either } x \notin C \text{ or } x \notin A.$$

But logic does this for us because

- at this stage, we know $x \in B$, so this component of what we want to show is already taken care of; and
- knowing that $x \notin C$ allows us to conclude that either $x \notin C$ or $x \notin A$.

Once we have a complete "flow of consequences" leading from the original version of what we know to the original version of what we want to show, we are in a position to write up our argument. But we must make sure the steps in our written argument are sequenced to reflect the "flow."

Example 4.20 (continued)   We can now build a path of deductions leading from the original version of what we know in this situation to the original version of what we want to show:

$$\textit{We know: } x \in A \cap (B - C).$$
$$\Downarrow$$
$$\textit{We know (update 1): } x \in A \text{ and } x \in B - C.$$
$$\Downarrow$$
$$\textit{We know (update 2): } x \in A \text{ and } x \in B \text{ and } x \notin C.$$
$$\Downarrow$$
$$\textit{We want to show (update 2): } x \in B \text{ as well as either } x \notin C \text{ or } x \notin A.$$
$$\Downarrow$$
$$\textit{We want to show (update 1): } x \in B \text{ and } x \notin C \cap A.$$
$$\Downarrow$$
$$\textit{We want to show: } x \in B - (C \cap A).$$

Our proof is written to reflect the deductive sequence represented in the above diagram.

Theorem: For any sets A, B, and C, $A \cap (B - C) \subseteq B - (C \cap A)$.
Proof: Let A, B, and C be sets. Consider $x \in A \cap (B - C)$. We will show $x \in B - (C \cap A)$. Because $x \in A \cap (B - C)$, the definition of intersection allows us to conclude that $x \in A$ and $x \in B - C$. Then, as $x \in B - C$, the definition of set difference allows us further to conclude that $x \in B$ and $x \notin C$. Now, as $x \notin C$, the definition of intersection tells us that $x \notin C \cap A$. So, as $x \in B$ and $x \notin C \cap A$, it follows, using the definition of set difference, that $x \in B - (C \cap A)$. $\square$

Practice Problem 1   Use the *Know/Show* approach to develop a proof that for any sets A, B, and C, $C \cap (A - B) \subseteq A - (B - C)$.

In Example 4.20, the updates of what we know were produced separately from the updates of what we want to show. Usually, though, these updates influence one another, so they should occur in tandem. In addition, it is almost always better to focus first on what we want to show because doing so usually gives us some ideas for how we might apply the information that we know and also helps us to structure the writing of our proof so that it is aligned with our standard proof strategies.

Suppose, for instance, that we want to prove that for all sets A, B, and C, if $A \subseteq B$ and $B \subseteq C$, then $A \subseteq C$. Then *we know $A \subseteq B$ and $B \subseteq C$, and we want to show $A \subseteq C$.*

Focusing first on what we want to show, the standard strategy for proving set inclusion tells us we should consider an anonymous x in A and try to show that x is in C. Note that our focus on what we want to show, rather than what we already know, at the outset of our planning helps us to implement a standard proof strategy. Moreover, we can now update what we know and what we want to show as follows: *We know $A \subseteq B$, $B \subseteq C$, and $x \in A$, and we want to show $x \in C$.*

Now, as we know $x \in A$ and $A \subseteq B$, the definition of subset allows us to conclude that $x \in B$. Updating further, *we know $A \subseteq B$, $B \subseteq C$, $x \in A$, and $x \in B$, and we want to show $x \in C$.*

Then, knowing that $B \subseteq C$ and $x \in B$, the definition of subset applied once more allows us to reach our desired conclusion that $x \in C$. Our proof follows the development we have just traced.

Theorem 4.21   For all sets A, B, and C, if $A \subseteq B$ and $B \subseteq C$, then $A \subseteq C$.

Proof: Consider sets A, B, and C, and suppose $A \subseteq B$ and $B \subseteq C$. We will show $A \subseteq C$. Let $x \in A$. As $A \subseteq B$ and $x \in A$, the definition of subset allows us to conclude that $x \in B$. Then, as $B \subseteq C$ and we now know that $x \in B$, the definition of subset allows us to conclude that $x \in C$. $\square$

Practice Problem 2   Prove that if $A \subseteq B$, then $A - C \subseteq B - C$.

Significance of the Distributive Property

An examination of Axiom 2.46 reveals that among our axioms for the real numbers, only the distributive property $a(b + c) = ab + ac$ involves both addition and multiplication. Therefore, when a statement we intend to prove relates additive and multiplicative concepts, it is reasonable to expect that the proof will make use of the distributive property. To illustrate, we will prove that for any real number a, $0a = 0$. Relative to *Know/Show*, *we know* a is a real number, and *we want to show* $0a = 0$. Knowing that a is a real number means a satisfies all of the properties listed in Axiom 2.46. We can use any of these axioms within our proof, but which of them might be relevant will only become apparent as we think about what we want to show; namely, $0a = 0$.

Now there are really only two mathematical notions concerning the real numbers that are to be found within the equation $0a = 0$, the notion of 0 and the notion of multiplication. It is reasonable to ask what our axioms for the real numbers tell us about these notions. Well, what distinguishes the number 0 is its status as the additive identity: Whenever 0 is added to a real number, the end result is that same real number. So, the equation $0a = 0$ we are attempting to derive relates the "additive" notion of 0 to the operation of multiplication. Thus, we anticipate using the distributive property in our proof.

Our analysis thus far, then, points us toward an application of the distributive property, but most likely one that includes the expression $0a$ appearing in our equation $0a = 0$. Fundamentally, there are only two such applications,

$$0(a + x) = 0a + 0x \tag{1}$$

and

$$a(0 + x) = a0 + ax, \tag{2}$$

where x can be any real number [as multiplication is commutative, there should be no cause for concern about the appearance of $a0$ rather than $0a$ in (2)]. How do we decide if either of these equations might help us? Note that Equation (2) is much more promising than Equation (1) because (2) allows for a simplification based on the notion that 0 is the additive identity $(0 + x = x)$, whereas (1) allows for no such simplification. So, we will see where (2) leads us.

Simplifying the left side of (2) using the fact that $0 + x = x$ gives us

$$ax = a0 + ax. \tag{3}$$

Intuitively, if we "cancel out" the ax appearing on both sides of (3), we will be left with $0 = a0$, which is essentially what we wanted to show.

You will notice in our write-up of the proof of Theorem 4.22 that we use the number 0 in place of x. Most mathematicians prefer to bring as few objects into their proofs as possible, so rather than introduce x as another real number along with 0 and a, because it appears from the above analysis that x could be any real number, we have chosen 0.

Theorem 4.22   For any real number a, $0a = 0$.

Proof: Consider a real number a. We will show $0a = 0$. By the distributive property,

$$0a + 0a = (0 + 0)a. \tag{4}$$

Then, because 0 is the additive identity, we know that $0 + 0 = 0$, and we may rewrite Equation (4) as

$$0a + 0a = 0a.$$

Again using the fact that 0 is the additive identity, we may rewrite this equation as

$$0a + 0a = 0a + 0. \tag{5}$$

Now, as $0a \in \mathbf{R}$ because of the closure of $\mathbf{R}$ under multiplication, we know $0a$ has an additive inverse, $-(0a)$, with the property that $-(0a) + 0a = 0$. Therefore, adding $-(0a)$ to both sides of (5), we obtain

$$-(0a) + (0a + 0a) = -(0a) + (0a + 0),$$

from which it follows, using the associativity of addition, that

$$[-(0a) + 0a] + 0a = [-(0a) + 0a] + 0,$$

which, because $-(0a) + 0a = 0$, can then be simplified to

$$0 + 0a = 0 + 0,$$

and then to

$$0a = 0,$$

by once again applying the fact that 0 is the additive identity. □

Practice Problem 3 Prove that for any real number t, $t + t = 2t$. You may freely use the fact that $1 + 1 = 2$ in your proof.

Answers to Practice Problems

1. We would begin by using our standard strategies for proving a *for all* statement and proving a set inclusion to write the beginning of our proof.

   Theorem: For all sets A, B, and C, $C \cap (A - B) \subseteq A - (B - C)$.
   Proof: Suppose A, B, and C are sets. Assume $x \in C \cap (A - B)$. We will show $x \in A - (B - C)$. [*argument under construction*] □

   We now apply the *Know/Show* approach to identify and update what we know.

   We know: $x \in C \cap (A - B)$.

   Applying the definition of intersection allows us to make a deduction.

We know (updated): $x \in C$ and $x \in A - B$.

Applying the definition of set difference allows us to make a further deduction.

We know (further update): $x \in C$ and $x \in A$ and $x \notin B$.

Having extracted as much information from what we know as is possible, we examine what we want to show.

We want to show: $x \in A - (B - C)$.

According to the definition of set difference, to make this deduction we would need to show the following:

We want to show (updated): $x \in A$ and $x \notin B - C$.

Now, to deduce $x \notin B - C$, the definition of set difference tells us it would be enough to show either $x \notin B$ or $x \in C$. Thus:

We want to show (further update): $x \in A$ and either $x \notin B$ or $x \in C$.

Note that our final update of what we know tells us that $x \in A$ and $x \in C$, and this information is enough to obtain the final update of what we want to show, so we are ready to finish writing our proof, being careful to sequence our deductions in the appropriate order (i.e., working forward through our updates of what we know and backward through our updates of what we want to show).

Theorem: For all sets A, B, and C, $C \cap (A - B) \subseteq A - (B - C)$.
Proof: Suppose A, B, and C are sets. Assume $x \in C \cap (A - B)$. We will show $x \in A - (B - C)$. As $x \in C \cap (A - B)$, the definition of intersection allows us to conclude that $x \in C$ and $x \in A - B$. Then, as $x \in A - B$, the definition of set difference tells us that $x \in A$ and $x \notin B$. Because $x \in C$, it follows using the definition of set difference that $x \notin B - C$. Finally, applying the definition of set difference once more, as $x \in A$ and $x \notin B - C$, we may conclude that $x \in A - (B - C)$. □

2.   At first *we know* $A \subseteq B$, and *we want to show* $A - C \subseteq B - C$.

Because we want to establish a set inclusion, we use the standard strategy for doing so, taking an anonymous $x \in A - C$ and showing that $x \in B - C$. So now *we know* $A \subseteq B$ and $x \in A - C$, and *we want to show* $x \in B - C$.

Applying the definition of set difference allows us to deduce from $x \in A - C$ that $x \in A$ and $x \notin C$. This same definition tells us that in order to show $x \in B - C$, we could show $x \in B$ and $x \notin C$. Thus, *we know* $A \subseteq B$, $x \in A$, and $x \notin C$, and *we want to show* $x \in B$ and $x \notin C$.

We can use what we know at this stage to get what we want to show at this stage.

(*continues*)

Answers to Practice Problems (continued)

Theorem: If $A \subseteq B$, then $A - C \subseteq B - C$.

Proof: Suppose $A \subseteq B$, and let $x \in A - C$. The definition of set difference allows us to conclude that $x \in A$ and $x \notin C$. Because $x \in A$ and $A \subseteq B$, the definition of subset tells us that $x \in B$. So, as we now have $x \in B$ and $x \notin C$, it follows, by the definition of set difference, that $x \in B - C$. □

3. Rather than present just the finished version of the proof, we will walk you through its development. If you were not able to complete the proof on your own, you may be able to do so after reading only part of the following. First, you should clearly identify the statement to be proved:

Theorem: For any real number t, $t + t = 2t$.

Then you should apply our standard strategy for proving a *for all* statement to begin writing the proof:

Proof: Consider a real number t. We will show that $t + t = 2t$. [*argument under construction*] □

Thus, according to *Know/Show*, we have the following: *We know* t is a real number, and *we want to show* $t + t = 2t$.

Now we have to do some thinking about how we can show $t + t = 2t$. Because the equation we are attempting to derive involves both addition and multiplication, we expect to make use of the distributive property. (If you did not try to use the distributive property, stop reading and see if you can figure out the rest of the proof on your own now.) High school math teachers usually explain why $t + t = 2t$ using the distributive property and the fact that $t = 1t$, as this gives $t + t = 1t + 1t = (1 + 1)t = 2t$. This is what we must include in our proof, along with appropriate justification.

Theorem: For any real number t, $t + t = 2t$.

Proof: Consider a real number t. We will show that $t + t = 2t$. Because 1 is the multiplicative identity, we know that $1t = t$. Thus, using the distributive property and the fact that $1 + 1 = 2$, we have $t + t = 1t + 1t = (1 + 1)t = 2t$. □

4.5 Existence and Uniqueness

Questions to guide your reading of this section:

1. What are some of the mathematical situations in which we can assert the existence of an object?

2. What are the two standard approaches to proving that a mathematical object with a certain property actually exists?

3. How do we prove that there is only one object possessing a certain property, in other words, that the object possessing the property is unique?

The ability to bring an object into existence is critical in creating certain proofs. In Section 4.4, when we proved $0a = 0$ for any real number a, our proof used the assumption that every real number has an additive inverse to assert the existence of an additive inverse $-(0a)$ for the real number $0a$. The existence of $-(0a)$ and the fact that $-(0a)$ added to $0a$ yields 0 was crucial to our argument.

As of now, we have at least five ways that an object's existence can be asserted within a proof we are writing:

- If we know a is a real number, we may assert the existence of its additive inverse $-a$ (i.e., if $a \in \mathbf{R}$, then there exists $-a \in \mathbf{R}$ with the property that $a + (-a) = 0$).

- If we know a is a nonzero real number, we may assert the existence of its multiplicative inverse $1/a$ (i.e., if $a \in \mathbf{R} - \{0\}$, then there exists $1/a \in \mathbf{R}$ with the property that $a \cdot (1/a) = 1$).

- If we know a set A is *nonempty*, we may assert the existence of an object, say x, that is a member of A (i.e., if $A \neq \varnothing$, then there must exist $x \in A$).

- If we know a set A is *not* a subset of a set B, we may assert the existence of an object, say x, that is a member of A and that is not a member of B (i.e., if $A \nsubseteq B$, then there exists $x \in A$ such that $x \notin B$).

- If we know a set A is *not* equal to a set B, we may assert the existence of an object, say x, that is a member of one of the sets A or B but not the other (i.e., if $A \neq B$, then there exists x such that either $x \in A$ and $x \notin B$ or else $x \in B$ and $x \notin A$).

Note that while we may assert the existence of an object when one set is *not* a subset of another set or when two sets are *not* equal, knowing two sets *are* equal or that one set *is* a subset of another will not generally allow us to assert the existence of an object because some or all of the sets under discussion could be empty.

Practice Problem 1 In each case, tell whether the given information allows us to assert the existence of an object. If it does, bring this object into existence and describe any properties the object must satisfy.

(a) $x \in \mathbf{R}$. (b) $y \in \mathbf{R}^+$. (c) $A - B \neq \varnothing$.

(d) $C \subseteq A \cap B$. (e) $C \nsubseteq A \cap B$.

Example 4.23 In the following proof, we use the assumption that a is a nonzero real number to assert the existence of its multiplicative inverse $1/a$. The desired conclusion follows immediately once we multiply both sides of the given equation $a^2 = a$ by $1/a$ and simplify.

Theorem: If a is a nonzero real number for which $a^2 = a$, then $a = 1$.

Proof: Suppose a is a nonzero real number for which $a^2 = a$. Because a is nonzero, a has a multiplicative inverse $1/a$. Multiplying both sides of $a^2 = a$ by $1/a$ yields

$$(1/a) \cdot a^2 = (1/a) \cdot a.$$

Then, as by definition $a^2 = a \cdot a$, we have

$$(1/a) \cdot (a \cdot a) = (1/a) \cdot a.$$

Applying the associativity of multiplication gives us

$$[(1/a) \cdot a] \cdot a = (1/a) \cdot a.$$

But, as $1/a$ is a multiplicative inverse of a, this equation may be simplified to

$$1 \cdot a = 1.$$

Finally, because 1 is the multiplicative identity, we may conclude that

$$a = 1. \quad \square$$

Proving Existence

We can establish the existence of an object having a particular property by either looking for an already existing object that happens to have the property or creating a "new" object from existing objects in such a way that the new object possesses the property.

Proof Strategy 4.24 (*Proving Existence*) To show that an object satisfying a certain property actually exists, either

1. find an object whose existence is already known and show that the object possesses the desired property; or

2. define or construct an object in terms of other objects whose existence is already known and then show that the newly defined/constructed object possesses the desired property.

When it is possible to bring an object into existence using one of the premises under which our argument will be created, this object often turns out to be precisely the one we are looking for to satisfy some particular property.

Example 4.25 We will prove that if $C - A \not\subseteq C - B$, then $B \not\subseteq A$. Thus, *we know* $C - A \not\subseteq C - B$, and *we want to show* $B \not\subseteq A$. To show $B \not\subseteq A$, we must find a member of B that is not in A. Because our premise allows us to assert the existence of an object that is in $C - A$ but not in $C - B$, it is reasonable to ask ourselves whether this object could be a member of B but not a member of A. The definition of set difference tells us that this is indeed the case.

Theorem: If $C - A \not\subseteq C - B$, then $B \not\subseteq A$.
Proof: Suppose $C - A \not\subseteq C - B$. Then there exists $v \in C - A$ [*we do not always need to use the letter x as the name of an object whose existence is being asserted*] such that $v \notin C - B$. According to the definition of set difference, because $v \in C - A$, we may conclude that $v \in C$ and $v \notin A$, and because $v \notin C - B$, we may also conclude that $v \notin C$ or $v \in B$. Our conclusions that $v \in C$ and either $v \notin C$ or $v \in B$ allow us to deduce that $v \in B$. Hence, as $v \in B$ and $v \notin A$, we may conclude, by the definition of subset, that $B \not\subseteq A$. $\square$

Practice Problem 2 Prove that if $A \neq \varnothing$, then, for any set B, $A \cup B \neq \varnothing$.

Example 4.25 shows how to apply (1) of Proof Strategy 4.24, in which an object we already know exists winds up being the object we want to show satisfies a certain property. Sometimes, though, the object whose existence we are attempting to demonstrate must be created from other objects whose existence is known. Solving algebraic equations provides us with one situation in which this approach, (2) of Proof Strategy 4.24, may be applied.

Recall that a statement that two mathematical expressions represent the same thing is called an **equation**. Usually, the expressions are symbolic and the **equals sign**, =, is written between them to form the equation. A **solution** to an equation involving a variable is a value within the domain of the variable that, when substituted for the variable in the equation, makes the equation true.

You are used to working with equations involving real numbers. The equation $x + 4 = 10$ has 6 as its only solution, whereas the equation $x^2 = 100$ has two solutions, 10 and −10, and the equation $x + 1 = x$ has no solutions.

In this text, you have also worked with equations involving sets. If we regard A as a variable representing a set, one solution to the equation $A \cap \mathbf{N} = \{1, 2\}$ is $A = \{x \in \mathbf{R} \mid x < 3\}$ and another is $A = \{1, 2\}$, and there are many others.

To verify a certain object is a solution to an equation, we must replace the variable in the equation with the object and make sure the resulting equation (which now has no variables in it) is true.

Example 4.26   When asked to "solve" the equation

$$4x + 7 = -5,$$

you would probably subtract 7 from both sides to get

$$4x = -12$$

and then divide both sides by 4 to get

$$x = -3.$$

But logically, this just shows that *if* the equation $4x + 7 = -5$ has a solution, then the solution must be -3. In other words, we have found that -3 is the only candidate for a solution to the equation. To verify that -3 is a solution to $4x + 7 = -5$, we must substitute -3 for the variable x in this equation and show that the resulting equation

$$4 \cdot (-3) + 7 = 5$$

is true. Of course, because $4 \cdot (-3) + 7 = -12 + 7 = 5$, it does follow that -3 is a solution to $4x + 7 = -5$.

The reason this is important is because a *candidate for a solution* to an equation may not always be an *actual solution*. For example, to "solve" the equation

$$\frac{x}{x+1} - \frac{2}{x^2 - 1} = 1 \tag{1}$$

we could multiply through by the least common denominator $x^2 - 1$, which gives us

$$x(x - 1) - 2 = x^2 - 1,$$

apply the distributive property to get

$$x^2 - x - 2 = x^2 - 1,$$

and subtract x^2 from both sides and add 2 to both sides to get

$$-x = 1,$$

from which we may conclude that

$$x = -1.$$

But substituting -1 for x in the original Equation (1) makes both fractions on the left side undefined (remember that division by 0 is undefined), so -1 is not a solution to (1). In fact, (1) has no real number solutions.

In the proof of Theorem 4.27 below, note how we explicitly construct our solution $(1/a) \cdot [c + (-b)]$ to the equation $ax + b = c$ out of objects we already have in hand; namely, the numbers a, b, and c.

Theorem 4.27 Let a, b, and c be real numbers with $a \neq 0$. Then there exists a solution to the equation $ax + b = c$.

Proof: Assume a, b, and c are real numbers with $a \neq 0$. As $a \neq 0$, a has a multiplicative inverse $1/a$. Let $x = (1/a) \cdot [c + (-b)]$. Then

$$
\begin{aligned}
ax + b &= a((1/a) \cdot [c + (-b)]) + b \\
&= [a \cdot (1/a)] \cdot [c + (-b)] + b && \text{(multiplication is associative)} \\
&= 1 \cdot [c + (-b)] + b && \text{($1/a$ is the multiplicative inverse of a)} \\
&= [c + (-b)] + b && \text{(1 is the multiplicative identity)} \\
&= c + [(-b) + b] && \text{(addition is associative)} \\
&= c + 0 && \text{($-b$ is the additive inverse of b)} \\
&= c, && \text{(0 is the additive identity)}
\end{aligned}
$$

which demonstrates that $(1/a) \cdot [c + (-b)]$ is a solution to $ax + b = c$. $\square$

The third sentence of this proof,

$$\textit{Let } x = (1/a) \cdot [c + (-b)],$$

is where our construction of the solution is taking place; we are creating a new number using the given numbers a, b, and c along with our knowledge of addition, multiplication, opposites, and reciprocals, and then assigning the new number the name x (it is common to name a proposed solution to an equation using the same letter appearing as the variable in the equation). Such a sentence will usually appear in an existence proof in which an object is constructed out of other objects.

When an object is constructed from other objects in a proof, the reasoning behind the construction is often not immediately apparent. Here, our choice of $(1/a) \cdot [c + (-b)]$ as a solution to $ax + b = c$ comes about only after some preliminary "scratchwork" in which we "solve" $ax + b = c$ for x (by adding $-b$ to both sides of the equation and then multiplying through by $1/a$), thus obtaining a candidate for the solution. But our proof does not include the process by which we discover our candidate, only the verification that the candidate really is a solution.

Note also that we provided parenthetical justification for each step in the series of calculations forming the heart of our proof. Sometimes this sort of detailed justification of the steps in an algebraic simplification may be left out of a proof as long as, for the most part, steps are not being combined. By laying out all of our steps, it should be possible for a reader familiar with the mathematical properties being applied to fill in for himself the specific justification for each step.

> **Practice Problem 3** Prove that every nonempty set has at least two subsets.

Proving Uniqueness

There are also times when we want to show there is only one object having a certain property; in other words, that the object having the property is *unique*.

Proof Strategy 4.28 (*Proving Uniqueness*) Suppose that an object of a certain type satisfying a certain property exists. To prove this object is unique, assume there are two objects with the property and show they are the same.

We demonstrate the use of Proof Strategy 4.28 to establish the uniqueness of the empty set by assuming there are two empty sets A and B and then demonstrating that they are actually equal to each other. That is, *we know A and B are both empty sets, and we want to show $A = B$*.

Theorem 4.29 Any two empty sets are equal.

Proof: Suppose A and B are both empty sets. By the definition of empty set, neither A nor B has any members, so that, for all x, $x \in A$ if and only if $x \in B$. Thus, by the definition of set equality, $A = B$. $\square$

Note that in our proof, both simple statements $x \in A$ and $x \in B$ in the biconditional

$$x \in A \text{ if and only if } x \in B$$

are false for any x, thus making the biconditional itself true for any x. This is why the definition of set equality allows us to conclude that $A = B$.

Practice Problem 4   Let a, b, and c be real numbers with $a \neq 0$. Prove that the solution to the equation $ax + b = c$ is unique.

In Axiom 2.46, we make it sound like 0 is the only additive identity, and 1 is the only multiplicative identity. This is indeed the case.

Theorem 4.30 (*Uniqueness of the Additive and Multiplicative Identities*)   The number 0 is the only additive identity for $\mathbf{R}$, and the number 1 is the only multiplicative identity for $\mathbf{R}$.

   Proof: We show that 0 is the unique additive identity for $\mathbf{R}$ and leave the proof that 1 is the unique multiplicative identity for $\mathbf{R}$ as Practice Problem 5 later.
   Let z be a real number that is an additive identity for $\mathbf{R}$, meaning $z + a = a$ for every $a \in \mathbf{R}$. In particular, this tells us that $z + 0 = 0$. But we also have $z + 0 = z$ because 0 is an additive identity. Because each of z and 0 is equal to the real number $z + 0$, it follows that they must be equal to each other; that is, $z = 0$. This allows us to conclude that 0 is the only additive identity for $\mathbf{R}$. $\square$

The proof of Theorem 4.30 illustrates a variation on the application of Proof Strategy 4.28 in demonstrating uniqueness. Because we already know 0 is an additive identity for $\mathbf{R}$, our proof only needed to entertain the possibility of "another" additive identity z. We then considered what would happen if we added these two additive identities together to form $z + 0$. The fact that each of z and 0 is an additive identity led to $z + 0$ having to be equal to both 0 and z, from which we concluded that $z = 0$.

Practice Problem 5   Prove that 1 is the only multiplicative identity for $\mathbf{R}$.

Not only are the additive and multiplicative identities unique, but also each real number has a unique additive inverse, and each nonzero real number has a unique multiplicative inverse.

Theorem 4.31 (*Uniqueness of Additive and Multiplicative Inverses*)   For each real number a, the number $-a$ is the only additive inverse for a. Also, for each nonzero real number a, the number $1/a$ is the only multiplicative inverse for a.

Proof: We show the uniqueness of additive inverses and leave the uniqueness of multiplicative inverses as an exercise (see Problem 4I.2).

Consider $a \in \mathbf{R}$ and suppose that b is an additive inverse for a, so that $b + a = 0$. Observe that

$$b = b + 0 = b + [a + (-a)] = [b + a] + (-a) = 0 + (-a) = -a. \quad \square$$

The key to figuring out the proof of Theorem 4.31 is to realize that because both $-a$ and b are additive inverses of a, the expression $b + a + (-a)$ can be simplified to either b or $-a$, forcing these "two" additive inverses of a to be the same.

Note that the series of equalities at the very end of our proof was left unjustified. Rather than supporting each individual equality, we have chosen to write out each step of the simplification process, leaving it up to the reader to fill in the justification of each step for herself.

Answers to Practice Problems

1. (a) We may conclude that x has an additive inverse $-x$ with the property that $x + (-x) = 0$. If $x \neq 0$, we would also be able to conclude that x has a multiplicative inverse $1/x$ with the property that $x \cdot (1/x) = 1$.

   (b) We may conclude y has an additive inverse $-y$ such that $y + (-y) = 0$ and a multiplicative inverse $1/y$ such that $y \cdot (1/y) = 1$.

   (c) We may conclude that there exists $x \in A$ such that $x \notin B$.

   (d) Knowing only that $C \subseteq A \cap B$, we may not assert the existence of any objects because the sets C and $A \cap B$ might be empty.

   (e) We may conclude that there exists $x \in C$ such that $x \notin A \cap B$.

2. *Claim:* If $A \neq \varnothing$, then, for any set B, $A \cup B \neq \varnothing$.

   Proof: Assume $A \neq \varnothing$ and consider an anonymous set B. Because $A \neq \varnothing$, there exists $x \in A$. So, by the definition of union, it follows that $x \in A \cup B$, making $A \cup B \neq \varnothing$. $\quad \square$

3. *Claim:* Every nonempty set has at least two subsets.

   Proof: Consider a nonempty set A. We must show that A has at least two subsets. We have already established that the empty set $\varnothing$ is a subset of every set, so, in particular, $\varnothing \subseteq A$. Because A is nonempty, there exists $x \in A$. Now we can construct $\{x\}$, which is a set different from $\emptyset$, as $\{x\}$ has a member, namely x, that is not a member of $\emptyset$. And $\{x\} \subseteq A$, as the only member of $\{x\}$ is x and we know that $x \in A$. Thus, A does have at least two subsets, $\emptyset$ and $\{x\}$. $\square$

4.   *Claim:* If a, b, and c are real numbers with $a \neq 0$, then the solution to the equation $ax + b = c$ is unique.

Proof: Let a, b, and c be real numbers with $a \neq 0$, and consider the equation $ax + b = c$. We have already shown that this equation has a solution. Assume s and t are both solutions to this equation. Then, $as + b = c$ and $at + b = c$. Hence, $as + b = at + b$. Adding $-b$ to both sides of this last equation produces $as = at$. Then, multiplying through by the multiplicative inverse $1/a$ of a, which exists because $a \neq 0$, we may conclude that $s = t$. $\square$

5.   *Claim:* 1 is the only multiplicative identity for **R**.

Proof: Suppose there is another multiplicative identity for **R**, say m. Because 1 is a multiplicative identity, $1 \cdot m = m$. But as m is a multiplicative identity, we also have $1 \cdot m = 1$. Therefore, as both m and 1 are equal to $1 \cdot m$, it follows that $m = 1$. $\square$

4.6   The Role of Definitions in Creating Proofs

Questions to guide your reading of this section:

1.   Why does the *definition* for a mathematical term often play a key role in the creation of a proof for a statement that involves the term?

2.   What form must an *even* integer take by definition? an *odd* integer?

3.   What equation defines the *additive inverse* of a real number a? What equation defines the *multiplicative inverse* of a nonzero real number a?

4.   How is the operation of *subtraction* defined? the operation of *division*?

5.   What is the precise meaning of $a < b$ according to the definition of "$<$"?

The definition of a mathematical term often plays a central role in the proof of a statement involving the term. When a statement that we *know* to be true involves a defined term, we may use the term's definition to make a deduction. When a statement that we want to *show* is true involves a defined term, we may reach the desired conclusion by showing that the corresponding statement in which the term's definition has been applied is true.

Example 4.32   In Example 4.6, we proved that if $x \in A - B$, then $x \in A \cup B$. By applying the definition of set difference to our assumption that $x \in A - B$, we were able to deduce that $x \in A$ and $x \notin B$. By applying the definition of union to what we wanted to show, namely $x \in A \cup B$, we realized we could reach this conclusion by showing that $x \in A$ or $x \in B$.

These observations yield the following proof strategy.

Proof Strategy 4.33 (*Definitions and Know/Show*) Suppose the *if and only if* statement $P \Leftrightarrow Q$ formally defines P. Then

1. If we *know* P, we may conclude Q.
2. If we want to *show* P, we need only show Q.

Even and Odd Integers

An integer n is **even** if there is an integer k such that $n = 2k$ and is **odd** if there is an integer k such that $n = 2k + 1$. For example, the integer 26 is even because $26 = 2 \cdot 13$ and 13 is an integer, whereas the integer -7 is odd because $-7 = 2 \cdot (-4) + 1$ and -4 is an integer. Note that the integer 5 is not even because there is no integer k for which $5 = 2k$ (this equation will only be true when $k = 5/2$, which is not an integer).

Example 4.10 (continued) Back in Section 4.2, we began to develop a proof for the theorem

If n is an odd integer, then n^2 is odd.

Thus, *we know n is an odd integer* and *we want to show n^2 is odd.*

Using Proof Strategy 4.33 in the context of the definition of odd integer, we may revise this information as follows: *We know $n = 2k + 1$ for some integer k,* and *we want to show $n^2 = 2j + 1$ for some integer j.* In this revision of what we know and what we want to show, we have used different letters, k and j, to reflect the fact that there is a good chance that n and n^2 are different integers. We may now observe that, based on what we know at this point,

$$n^2 = (2k+1)^2 = 4k^2 + 4k + 1 = 2(2k^2 + 2k) + 1,$$

so that it appears $2k^2 + 2k$ may take the role of the integer j we are looking for. Our proof takes all of this analysis into account.

Theorem: If n is an odd integer, then n^2 is odd.
Proof: Suppose n is an odd integer. We will show n^2 is odd. Because n is odd, $n = 2k + 1$ for some integer k. Now

$$n^2 = (2k+1)^2 = 4k^2 + 4k + 1 = 2(2k^2 + 2k) + 1,$$

and because the closure of **Z** under both multiplication and addition guarantees that $2k^2 + 2k$ is an integer, we may now conclude that, by definition, n^2 is odd. □

Note that our proof makes no mention of j. Remember that the letter j was only being used as a placeholder for an integer we were trying to find. In contrast, the letter k was used as the actual name of an integer that was obtained from our assumption that n was indeed odd. When we calculated n^2 in terms of k, we discovered that the appropriate choice for our placeholder j was the integer $2k^2 + 2k$, itself created from the integer k.

Practice Problem 1   Use the definition of even integer to prove that the sum of two even integers is even.

Additive and Multiplicative Inverses

Although we usually write $-a$ for the *additive inverse* of a real number a, it is the equation $a + (-a) = 0$ that actually defines $-a$. That is, by definition, b is the **additive inverse** of a iff $a + b = 0$. When proving results about additive inverses, we typically must make use of this conceptual definition. For example, we use it in proving that the additive inverse $-(-a)$ of $-a$ is a in Theorem 4.34 below. The theorem can be viewed as a type of "cancellation" property because it allows us to eliminate two minus signs when they are applied one right after the other.

Theorem 4.34   For any real number a, $-(-a) = a$.

   Proof: Let $a \in \mathbf{R}$. We must show a is the additive inverse of $-a$. By definition, this means we must show that $(-a) + a = 0$. But, as $-a$ is the additive inverse of a, by definition we know that $a + (-a) = 0$. Then, because addition is commutative, we may conclude that $(-a) + a = 0$, which, as indicated previously, means the additive inverse of $-a$ is a, that is, $-(-a) = a$. □

Practice Problem 2   As you are no doubt aware from prior mathematical work, the additive inverse of a sum is the sum of the additive inverses.

   (a) According to the definition of additive inverse, what equation must hold true in order for $(-a) + (-b)$ to be the additive inverse of $a + b$?

   (b) Use your answer to (a) to prove that $(-a) + (-b)$ is the additive inverse of $a + b$.

The next theorem we present allows us to establish the usual "sign rules" for multiplication of positive and negative numbers. Because we are relating the additive notion of additive inverses to the operation of multiplication, it is not surprising that our proof will use the distributive property. Note how we show that $(-a)b$ is the additive inverse of ab by demonstrating that $ab + (-a)b = 0$.

Theorem 4.35 (*Sign Rules for Multiplication*) For all real numbers a and b, $(-a)b = -(ab)$ and $(-a)(-b) = ab$.

Proof: We shall prove the first equality and leave the other one as an exercise (see Practice Problem 3). Consider anonymous $a,\, b \in \mathbf{R}$. Observe that

$$ab + (-a)b = [a + (-a)]b = 0b = 0,$$

that is, $ab + (-a)b = 0$. Thus, the real number $(-a)b$ must be the additive inverse of ab, as ab has only one additive inverse, and $(-a)b$ adds to ab to produce the additive identity 0. Thus, $(-a)b = -(ab)$. □

> **Practice Problem 3** We have already proved that a minus sign appearing in a factor of a product may be moved out in front to apply to the entire product. We have also proved that the opposite of the opposite of a real number is that number. Use these facts to prove that for all real numbers a and b, $(-a)(-b) = ab$.

A **corollary** to a theorem is a theorem that is a near immediate consequence of the original theorem. We now state and prove a corollary to Theorem 4.35.

Corollary 4.36 For any real number a, $(-1)a = -a$.

Proof: Let $a \in \mathbf{R}$. Then $(-1)a = -(1a) = -a$, where the first equality results by applying the already proved fact that a minus sign may be moved out of a factor and placed in front of the product. □

Recall that every nonzero real number a has a unique **multiplicative inverse** b with the property that $a \cdot b = 1$. Although we usually write $1/a$ for the multiplicative inverse of a, it is the conceptual idea that the product of a number and its multiplicative inverse is 1 that defines the notion. Several properties of multiplicative inverses are developed in the problems at the end of this chapter.

Subtraction and Division

The existence of additive and multiplicative inverses permits us to define subtraction and division. The operation "−" of **subtraction** is defined so that for any real numbers a and b, $a - b = a + (-b)$, whereas the operation "÷" of **division** is defined so that for any real number a and any nonzero real number b, $a \div b = a \cdot (1/b)$. In other words, to subtract we add the opposite, and to divide we multiply by the reciprocal. For instance, $12 - 5 = 12 + (-5)$ and $12 \div 5 = 12 \cdot (1/5)$.

We now prove several fundamental properties of subtraction; note carefully the use of the definition of subtraction in our proofs. Additional properties of subtraction, as well as properties of division, can be found in Problem 4J.4.

Theorem 4.37   For any real number a, $a - 0 = a$.

Proof: Suppose a is a real number. Observe that

$$a - 0 = a + (-0) = a + 0 = a,$$

where the first equality is justified because of the definition of subtraction, the second because, as $0 + 0 = 0$, 0 is the additive inverse of 0, and the third because 0 is the additive identity. □

Theorem 4.38   For any real number a, $a - (-a) = a + a$.

Proof: Consider a real number a. Then

$$a - (-a) = a + (-(-a)) = a + a,$$

where the first equality holds because of the definition of subtraction and the second because we have already proved that the additive inverse of $-a$ is a. □

> **Practice Problem 4**   Use the definition of subtraction and the already proven fact that the additive inverse of a sum is the sum of the additive inverses to prove that for all real numbers a, b, and c, $a - (b + c) = a - b - c$.

Inequalities

When one number is less than another, subtracting the smaller from the larger will yield a positive number that represents the amount that must be added to the smaller number to obtain the larger. For instance, 3 is less than 5, and the difference $5 - 3 = 2$ is the amount that must be added to 3 in order to obtain 5. This observation leads to the following definition.

Definition 4.39 Given real numbers a and b, we say a is **less than** b, denoted $a < b$, if $b - a$ is positive. When $a < b$, we may also write $b > a$ and say that b is **greater than** a. Furthermore, we write $a \le b$ if either $a < b$ or $a = b$, and we may write $b \ge a$ in place of $a \le b$.

For example, $3 < 5$ because $5 - 3 = 2$ and 2 is positive. In place of $3 < 5$ we may write $5 > 3$. Also, because $3 < 5$, we have $3 \le 5$ and $5 \ge 3$. It is also true that $3 \le 3$ and $3 \ge 3$ as $3 = 3$.

We now prove that every positive real number is greater than 0. In developing our proof, we have to think about what the definition of ">" tells us about the statement $a > 0$:

$$a > 0 \ \textit{iff} \ a - 0 \ \textit{is positive.}$$

Thus, to show $a > 0$, according to the definition of ">", we must show $a - 0$ is positive. To do this we will use Theorem 4.37, which states that $a - 0 = a$.

Theorem 4.40 For every positive real number a, $a > 0$.

Proof: Let a be a positive real number. We must show $a > 0$. By definition of ">" it suffices to show that $a - 0$ is positive. But because we have already proved that $a - 0 = a$ and because we have assumed that a is positive, it follows that $a - 0$ is positive. □

Practice Problem 5 Recall that, by definition, a real number a is said to be **negative** if its additive inverse $-a$ is positive. Prove that every negative real number is less than 0.

Theorem 4.41 (*Algebra with Inequalities*) Let a, b, c, and d be real numbers.

1. If $a < b$ and $c < d$, then $a + c < b + d$.
2. If $a < b$ and $c > 0$, then $ac < bc$.
3. If $a < b$ and $c < 0$, then $ac > bc$.

We establish (1) of Theorem 4.41 and leave (2) and (3) as exercises (see Practice Problem 6 and Problem 4J.6a). Note that in developing a proof for (1), *we know* $a < b$ and $c < d$, and *we want to show* $a + c < b + d$. Decoding the inequalities according to the definition of "<" tells us that *we know* $b - a$ and $d - c$ are both positive, and *we want to show* $(b + d) - (a + c)$ is positive. Thus, the problem to be solved can then be stated as follows: If we know $b - a$ and $d - c$ are positive, how do we argue that $(b + d) - (a + c)$ is positive? Once we recognize that $(b + d) - (a + c)$ can be obtained by adding $b - a$ and $d - c$ (note how the result of Practice Problem 4 is being used), we have our solution.

Proof of Theorem 4.41 (1): Suppose $a < b$ and $c < d$. Then, by definition of "$<$", $b - a$ and $d - c$ are both positive. Hence, as $\mathbf{R}^+$ is closed under addition, it follows that $(b - a) + (d - c)$ is positive, and because $(b - a) + (d - c) = (b + d) - (a + c)$, we may then conclude that $(b + d) - (a + c)$ is positive. Thus, once again applying the definition of "$<$", we determine that $a + c < b + d$. $\square$

Practice Problem 6   Prove Theorem 4.41(2).

Answers to Practice Problems

1. *Claim:* If m and n are even integers, then $m + n$ is even.

    Proof: Suppose m and n are even integers. Then there exist integers j and k such that $m = 2j$ and $n = 2k$. It follows that $m + n = 2j + 2k = 2(j + k)$, where $j + k$ is an integer, being the sum of two integers. Thus, by definition, $m + n$ is even. $\square$

2. **(a)** $(a + b) + [(-a) + (-b)] = 0$

    (b)   *Claim:* The additive inverse of $a + b$ is $(-a) + (-b)$.

    Proof: Consider real numbers a and b. By definition, to show $(-a) + (-b)$ is the additive inverse of $a + b$, we must show $(a + b) + [(-a) + (-b)] = 0$. Observe that

    $$(a + b) + [(-a) + (-b)] = [a + (-a)] + [b + (-b)] = 0 + 0 = 0,$$

    where the first equality is justified by the commutativity and associativity of addition and the second equality because $-a$ and $-b$ are, respectively, the additive inverses of a and b. Thus, the additive inverse of $a + b$ is $(-a) + (-b)$; that is, $-(a + b) = (-a) + (-b)$. $\square$

3. *Claim:* For all real numbers a and b, $(-a)(-b) = ab$.

    Proof: Let $a, b \in \mathbf{R}$. Then

    $$(-a)(-b) = -[a(-b)] = -[(-b)a] = -[-(ba)] = ba = ab$$

    (can you justify each equality here?). $\square$

4. *Claim:* For all real numbers a, b, and c, $a - (b + c) = a - b - c$.

    (continues)

Answers to Practice Problems (continued)

Proof: Consider real numbers a, b, and c. Then

$$a - (b + c) = a + (-(b + c)) = a + (-b) + (-c) = a - b - c,$$

where the first and third equalities are justified by the definition of subtraction and the second equality because we have already proved that the additive inverse of a sum is the sum of the additive inverses. □

5. *Claim:* For every negative real number a, $a < 0$.

Proof: Let a be a negative real number. Then, by definition, $-a$ is positive. Now

$$0 - a = 0 + (-a) = -a,$$

where the first equality follows using the definition of subtraction and the second using the fact that 0 is the additive identity. Thus, as $0 - a = -a$ and we already know $-a$ is positive, it follows that $0 - a$ is positive. Hence, by definition, $a < 0$. □

6. *Claim:* If $a < b$ and $c > 0$, then $ac < bc$.

Proof: Suppose $a < b$ and $c > 0$. Then, by definition, $b - a$ is positive and $c - 0$, that is, c, is positive. Because the product of positive numbers is also positive, it now follows that $(b - a)c$ is positive. But $(b - a)c = bc - ac$, so we may conclude that $bc - ac$ is positive. Thus, by definition of "$<$", we now have $ac < bc$. □

4.7 Proving and Expressing a Mathematical Equivalence

Questions to guide your reading of this section:

1. How do we usually prove an *if and only if* statement $P \Leftrightarrow Q$?
2. What is the *double containment* strategy for proving two sets are equal?
3. How do we usually prove the equivalence of more than two statements?
4. What is the difference between saying P is *sufficient* for Q and saying P is *necessary* for Q?

When an *if and only if* statement $P \Leftrightarrow Q$ is true, we say that P and Q are **equivalent** and call $P \Leftrightarrow Q$ an **equivalence**. Every mathematical definition expresses an equivalence, but not all equivalences are definitions.

Example 4.42   A Venn diagram suggests that the only way the intersection of two sets A and B could be equal to the set A is for the set A to be a subset of the set B. Thus, we may formulate the conjecture that for any sets A and B,

$$A \cap B = A \text{ if and only if } A \subseteq B.$$

This alleged equivalence is not expressing a definition because the statement to either side of *if and only if* does not define the statement on the other side.

How do we go about proving a conjecture that expresses a mathematical equivalence? Back in Problem 3K.1(f), you showed that $P \Rightarrow Q$, $Q \Rightarrow P \therefore P \Leftrightarrow Q$ is a valid argument form, which means that the truth of $P \Leftrightarrow Q$ follows immediately from the truth of $P \Rightarrow Q$ and $Q \Rightarrow P$. This observation leads to the following standard approach to proving an *if and only if* statement.

Proof Strategy 4.43 (*Proving an "If and Only If" Statement*)   To prove the *if and only if* statement $P \Leftrightarrow Q$, prove both *if...then* statements $P \Rightarrow Q$ and $Q \Rightarrow P$.

Example 4.42 (continued)   Thus, to prove

$$A \cap B = A \text{ if and only if } A \subseteq B,$$

we need only prove the two *if...then* statements:

$$\text{If } A \cap B = A, \text{ then } A \subseteq B$$

and

$$\text{If } A \subseteq B, \text{ then } A \cap B = A.$$

In using Proof Strategy 4.43 to establish $P \Leftrightarrow Q$, we often label the argument for $P \Rightarrow Q$ using ($\Rightarrow$) and the argument for $Q \Rightarrow P$ using ($\Leftarrow$).

Example 4.44   In the following proof of

$$n \text{ is odd if and only if } n + 1 \text{ is even,}$$

we use ($\Rightarrow$) to indicate where we are proving

$$\text{If } n \text{ is odd, then } n + 1 \text{ is even}$$

and ($\Leftarrow$) to indicate where we are proving

$$\text{If } n + 1 \text{ is even, then } n \text{ is odd.}$$

Theorem: An integer n is odd if and only if $n + 1$ is even.

Proof: ($\Rightarrow$) Suppose n is an odd integer. Then $n = 2k + 1$ for some integer k. So

$$n + 1 = (2k + 1) + 1 = 2k + 2 = 2(k + 1),$$

where $k + 1$ is an integer as the sum of two integers is also an integer. Therefore, by definition, $n + 1$ is even.

($\Leftarrow$) Suppose $n + 1$ is an even integer. Then $n + 1 = 2k$ for some integer k. Thus,

$$n = 2k - 1 = (2k - 2) + 1 = 2(k - 1) + 1,$$

where $k - 1$ is an integer as the difference of two integers is also an integer. Therefore, by definition, n is odd. □

Practice Problem 1   Prove that an integer n is even if and only if $n - 3$ is odd.

Proving Set Equality

By definition, two sets are **equal** provided they have exactly the same members. But it would appear that, for any sets A and B,

$$A = B \text{ if and only if both } A \subseteq B \text{ and } B \subseteq A.$$

You are asked to prove this fact in Problem 4L.1. It is the basis for the most commonly used method for showing two sets are equal, sometimes referred to as **double containment** because each set is shown to be contained in the other.

Proof Strategy 4.45 (*Proving Two Sets Are Equal*)   To prove sets A and B are equal, prove both $A \subseteq B$ and $B \subseteq A$.

Note that one of the conditional statements we need to establish when applying Proof Strategy 4.43 in Example 4.42,

$$\text{If } A \subseteq B, \text{ then } A \cap B = A,$$

involves showing two sets are equal. Hence, we will apply Proof Strategy 4.45, double containment, in doing this. In writing up the proof, we use the label "($\subseteq$)" to indicate where we are proving the "forward" containment $A \cap B \subseteq A$ and the label "($\supseteq$)" to indicate where we are proving the "backward" containment $A \cap B \supseteq A$ (this is another standard stylistic convention in proof writing).

Theorem 4.46   For any sets A and B, $A \cap B = A$ if and only if $A \subseteq B$.

    Proof: Consider sets A and B.

        ($\Rightarrow$) Suppose $A \cap B = A$. We must show $A \subseteq B$. So consider an anonymous $x \in A$; we must demonstrate that $x \in B$. Because $x \in A$ and $A \cap B = A$, it follows that $x \in A \cap B$. Then, by the definition of intersection, we may conclude that $x \in B$.

        ($\Leftarrow$) Suppose $A \subseteq B$. We must show $A \cap B = A$.

        ($\subseteq$) Consider an anonymous $x \in A \cap B$. Then, by the definition of intersection, $x \in A$.

        ($\supseteq$) Consider an anonymous $x \in A$. Because $A \subseteq B$ by hypothesis, it follows that $x \in B$. Because $x \in A$ and $x \in B$, it follows, by the definition of intersection, that $x \in A \cap B$. $\square$

Practice Problem 2   Prove that for any sets A and B, $A - (A \cap B) = A - B$.

Proving the Equivalence of More Than Two Statements

Sometimes we have more than two statements that we believe to be equivalent to one another. Consider, for instance, the following theorem.

Theorem 4.47   For any real number a, the following are equivalent:

1. $a > 0$;
2. a is positive;
3. $-a$ is negative;
4. $-a < 0$.

According to Proof Strategy 4.43, to prove this theorem it looks as if we need to prove all twelve (!) of the following implications (1) $\Rightarrow$ (2), (2) $\Rightarrow$ (1), (1) $\Rightarrow$ (3), (3) $\Rightarrow$ (1), (1) $\Rightarrow$ (4), (4) $\Rightarrow$ (1), (2) $\Rightarrow$ (3), (3) $\Rightarrow$ (2), (2) $\Rightarrow$ (4), (4) $\Rightarrow$ (2), (3) $\Rightarrow$ (4), and (4) $\Rightarrow$ (3). However, because

$$P \Rightarrow Q, Q \Rightarrow R \therefore P \Rightarrow R \tag{$\dagger$}$$

is a valid argument form (see Problem 3K.1d), the truth of some of these implications follows from the truth of some of the others. For instance, once we have proved both (1) ⇒ (2) and (2) ⇒ (3), it immediately follows, using (†), that (1) ⇒ (3).

Practice Problem 3   Explain why it is enough to prove only the four implications (1) ⇒ (2), (2) ⇒ (3), (3) ⇒ (4), and (4) ⇒ (1) in order to prove Theorem 4.47.

In general, if we have a list of statements all of which are to be proved equivalent to one another, (†) tells us that it is enough to prove only that each statement listed implies the statement immediately following it in the list and that the last statement listed implies the first one (note that this forms a complete "loop" of implications). This is the approach we take in proving Theorem 4.47. The proof relies heavily on the definition of *less than* ($r < s$ iff $s - r$ is positive) and the definition of *negative* (r is negative iff $-r$ is positive).

Proof of Theorem 4.47: Let a be a real number.

(1) ⇒ (2): Assume $a > 0$. Then, by definition, $a - 0$ is positive. But because we have already proved that $a - 0 = a$, we may now conclude that a is positive.

(2) ⇒ (3): Assume a is positive. Now we have already proved that $-(-a) = a$, so it follows that $-(-a)$ is positive. Hence, by definition, $-a$ is negative.

(3) ⇒ (4): Assume $-a$ is negative. By definition, then, $-(-a) = a$ is positive. So as $0 - (-a) = 0 + [-(-a)] = 0 + a = a$ (*can you justify each of these equalities?*), it follows that $0 - (-a)$ is positive. So, by definition of "<", we may conclude that $-a < 0$.

(4) ⇒ (1): Assume $-a < 0$. Then, by definition of "<", $0 - (-a)$ is positive. But because $a - 0 = a = 0 + a = 0 + [-(-a)] = 0 - (-a)$, it follows that $a - 0$ is positive. Thus, by the definition of "<", $a > 0$. ☐

Necessary and Sufficient Conditions

Assuming A and B are sets, try to answer the following questions:

- Is the condition $A \subseteq B$ *necessary* in order for $A = B$? In other words, *must* A be a subset of B if the sets A and B are to be equal?

- Is the condition $A \subseteq B$ *sufficient* for concluding that $A = B$? In other words, if we know that A is a subset of B, would that be *enough* to conclude that the sets A and B are equal?

A little thought reveals the answers to be "yes" and "no," respectively. The subset relation $A \subseteq B$ *necessarily holds* if we know that $A = B$, but $A \subseteq B$ is *insufficient* by itself to conclude $A = B$ because $A \subseteq B$ allows for B to include elements that may not be members of A.

Generally, when the *if...then* statement $P \Rightarrow Q$ is true, we say that P is a **sufficient condition** for Q, and when the *if...then* statement $Q \Rightarrow P$ is true, we say that P is a **necessary condition** for Q. Note that saying P is a sufficient condition for Q is the same thing as saying Q is a necessary condition for P.

Example 4.48 Because the statement

$$\text{If } A = B, \text{ then } A \subseteq B$$

is true, we can say that $A \subseteq B$ is a necessary condition for $A = B$ and that $A = B$ is a sufficient condition for $A \subseteq B$. But because the statement

$$\text{If } A \subseteq B, \text{ then } A = B$$

is not generally true, we conclude that $A \subseteq B$ is not sufficient for $A = B$, and $A = B$ is not necessary for $A \subseteq B$.

Example 4.49 An integer n is **divisible by** an integer m if there is an integer k such that $n = mk$. Thus, for example, 42 is divisible by 6 because $42 = 6 \cdot 7$, and -85 is divisible by 17 because $-85 = 17 \cdot (-5)$.

Note that divisibility by 4 is *sufficient* for divisibility by 2 because in order for an integer to be divisible by 2, it is *enough* to know that the integer is divisible by 4. This is just a different way of saying that the conditional statement

$$\text{If the integer } n \text{ is divisible by } 4, \text{ then } n \text{ is divisible by } 2$$

is true.

However, divisibility by 4 is *not necessary* for divisibility by 2 because in order for an integer to be divisible by 2, it is *not required* that the integer be divisible by 4 (consider, for instance, the integer 10). This is really just another way to say that the conditional statement

$$\text{If the integer } n \text{ is divisible by } 2, \text{ then } n \text{ is divisible by } 4$$

is false.

If both $P \Rightarrow Q$ and $Q \Rightarrow P$ are true, we may say that P is **both necessary and sufficient for** Q, which is really just another way of saying that P **and** Q

are equivalent, as the truth of both $P \Rightarrow Q$ and $Q \Rightarrow P$ leads automatically to the truth of $P \Leftrightarrow Q$.

Example 4.50 In Problem 4L.1 you show that, for any sets A and B,

$$A = B \text{ if and only if both } A \subseteq B \text{ and } B \subseteq A.$$

Thus, the condition

$$A \subseteq B \text{ and } B \subseteq A$$

is both necessary and sufficient for the equality of the sets A and B. That is, the notion of two sets being equal is equivalent to the notion that each of the two sets is a subset of the other.

Practice Problem 4 Determine whether the given condition is necessary but not sufficient, sufficient but not necessary, both necessary and sufficient, or neither necessary nor sufficient for the integer n to be even.

(a) $n + 1$ is odd. (b) $n + 2$ is odd.

(c) $2n$ is even. (d) n is divisible by 6.

Expressing the Equivalence of Two Mathematical Statements

The equivalence of two mathematical statements can be expressed in a variety of ways. Back in Theorem 4.46, we learned that for any sets A and B, the statements $A \cap B = A$ and $A \subseteq B$ are equivalent; that is, they mean the same thing mathematically. This equivalence can be expressed in any of the following ways:

Theorem: For any sets A and B, the following are equivalent:

1. $A \cap B = A$;
2. $A \subseteq B$.

Theorem: For any sets A and B, $A \cap B = A$ if and only if $A \subseteq B$.
Theorem: For any sets A and B, the condition $A \subseteq B$ is both necessary and sufficient for $A \cap B = A$.
Theorem: For any sets A and B, if $A \subseteq B$, then $A \cap B = A$, and conversely. Note that in the final statement of the theorem, we have made use of the fact that $Q \Rightarrow P$ is the converse of $P \Rightarrow Q$.

Answers to Practice Problems

1. *Claim:* An integer n is even if and only if $n - 3$ is odd.
    Proof: Let $n \in \mathbf{Z}$.
    ($\Rightarrow$) Suppose n is even. Then $n = 2k$ for some integer k. So

    $$n - 3 = 2k - 3 = (2k - 4) + 1 = 2(k - 2) + 1,$$

    where $k - 2$, being a difference of integers, is itself an integer.
    Thus, by definition, $n - 3$ is odd.
    ($\Leftarrow$) Suppose $n - 3$ is odd. Then $n - 3 = 2k + 1$ for some integer k.
    It follows that $n = 2k + 4 = 2(k + 2)$, where $k + 2$, being a sum of
    integers, is itself an integer. Thus, by definition, n is even. $\square$

2. *Theorem:* For any sets A and B, $A - (A \cap B) = A - B$.
    Proof: Let A and B be sets.
    ($\subseteq$) Consider $x \in A - (A \cap B)$; we must show $x \in A - B$. We may
    use the definition of set difference to deduce from our hypothesis
    that $x \in A$ and $x \notin A \cap B$. Then, as $x \notin A \cap B$, it follows, by the
    definition of intersection, that $x \notin B$. So, as $x \in A$ and $x \notin B$, we may
    conclude, using the definition of set difference, that $x \in A - B$.
    ($\supseteq$) Consider $x \in A - B$; we must show $x \in A - (A \cap B)$. We may
    use the definition of set difference to deduce from our hypothesis that
    $x \in A$ and $x \notin B$. Then, as $x \notin B$, by the definition of intersection we
    may conclude that $x \notin A \cap B$. Thus, as $x \in A$ and $x \notin A \cap B$, it fol-
    lows, using the definition of set difference, that $x \in A - (A \cap B)$. $\square$

3. It is enough to prove only the implications $(1) \Rightarrow (2)$, $(2) \Rightarrow (3)$,
    $(3) \Rightarrow (4)$, and $(4) \Rightarrow (1)$ because all of the other implications fol-
    low from these. Specifically,

    * $(2) \Rightarrow (1)$ follows from $(2) \Rightarrow (3)$, $(3) \Rightarrow (4)$, and $(4) \Rightarrow (1)$.
    * $(1) \Rightarrow (3)$ follows from $(1) \Rightarrow (2)$ and $(2) \Rightarrow (3)$.
    * $(3) \Rightarrow (1)$ follows from $(3) \Rightarrow (4)$ and $(4) \Rightarrow (1)$.
    * $(1) \Rightarrow (4)$ follows from $(1) \Rightarrow (2)$, $(2) \Rightarrow (3)$, and $(3) \Rightarrow (4)$.
    * $(3) \Rightarrow (2)$ follows from $(3) \Rightarrow (4)$, $(4) \Rightarrow (1)$, and $(1) \Rightarrow (2)$.
    * $(2) \Rightarrow (4)$ follows from $(2) \Rightarrow (3)$ and $(3) \Rightarrow (4)$.
    * $(4) \Rightarrow (2)$ follows from $(4) \Rightarrow (1)$ and $(1) \Rightarrow (2)$.
    * $(4) \Rightarrow (3)$ follows from $(4) \Rightarrow (1)$, $(1) \Rightarrow (2)$, and $(2) \Rightarrow (3)$.

4. **(a)** Both necessary and sufficient.
    (b) Neither necessary nor sufficient.
    (c) Necessary but not sufficient.   **(d)** Sufficient but not necessary.

4.8   Indirect Methods of Proof

Questions to guide your reading of this section:

1. How do we prove a statement using the method of *proof by contradiction*? How do we apply this method when the statement to be proved is an *if...then* statement?

2. How do we prove an *if...then* statement by *contraposition*?

The proofs we have been writing up to this point have been planned in accordance with the forms of the statements we wanted to prove. We have developed standard strategies for writing proofs of *for all* statements, *if...then* statements, *if and only if* statements, and so forth. Proofs that are based on the direct application of such strategies are often referred to as **direct proofs**. Sometimes, though, it is useful, or even necessary, to use a so-called **indirect** method of proof. We will explore two types of indirect methods of proof, *proof by contradiction* and *proof by contraposition*.

Proof by Contradiction

Recall our fundamental logical assumption that any given statement P is either true or false, but not both. Another way to think about this assumption is that either P or its negation $\sim P$ is true, but not both of them. The method of proof known as **proof by contradiction** establishes the truth of a statement P by eliminating the possibility that the statement's negation $\sim P$ is true. This is done by assuming $\sim P$ is true and arguing until a contradiction is reached. The derivation of a contradiction eliminates the possibility that $\sim P$ is true; hence, the original statement P must be true.

Proof Strategy 4.51 (*Proof by Contradiction*)   To prove the statement P using proof by contradiction, assume its negation $\sim P$ is true and argue until deducing a contradiction.

In any proof by contradiction, the negation of what we ultimately are trying to prove becomes part of what we know. So, in general, if we want to prove the statement P using contradiction, we have the following: *We know $\sim P$* and *we want to show* any contradiction we can find.

Back in Section 4.3, we proved that the empty set is a subset of every set. We now give a proof by contradiction for this theorem.

Theorem 4.16 (revisited)   For every set A, $\varnothing \subseteq A$.

Proof: Suppose to the contrary that there is a set A such that $\varnothing \nsubseteq A$. By the definition of subset, it follows that there exists $x \in \varnothing$ such that $x \notin A$. But $x \in \varnothing$ contradicts the definition of the empty set as a set having no members.   □

Note how we alerted the reader of our plan to use proof by contradiction by writing *Suppose to the contrary. . .* at the very beginning of the proof. Other phrases commonly used to indicate that a proof will be done via contradiction include *Assume by way of contradiction that. . .* and *We proceed by contradiction*. Note also how our proof adheres to Proof Strategy 4.51: It begins by assuming that it is not the case that the empty set is a subset of every set (this assumption is the negation of what we are trying to prove), and it ends once a contradiction has been derived (in this case that the empty set has a member).

In Practice Problem 1, you will use contradiction to prove that no integer can be both even and odd. As a preliminary result, we first prove that 1 is odd but not even. In proving 1 is not even, we use proof by contradiction by assuming 1 is even and arguing until we obtain a contradiction.

Theorem 4.52   The integer 1 is odd, but is not even.

Proof: Note that because $1 = 2 \cdot 0 + 1$ and 0 is an integer, 1 is, by definition, odd. Now assume by way of contradiction that 1 is even. Then, by definition, $1 = 2 \cdot k$ for some integer k. However, by the definition of multiplicative inverse, because $1 = 2 \cdot k$, it follows that k is a multiplicative inverse of 2. But, because a real number cannot have more than one multiplicative inverse, and we know that $1/2$ is a multiplicative inverse of 2, it follows that $k = 1/2$, contradicting the earlier conclusion that k is an integer.   □

Practice Problem 1   Use the fact that 1 is not even, along with the method of proof by contradiction, to prove that no integer can be both even and odd.

Using Contradiction to Prove an *If. . . Then* Statement

Proof by contradiction is a strategy that is often useful in attempting to prove an *if. . . then* statement $P \Rightarrow Q$. We begin by assuming the hypothesis P is true, just as we would for a direct proof. However, we establish the conclusion Q is true indirectly by assuming $\sim Q$ is true and arguing until we reach a contradiction.

Proof Strategy 4.53 (*Proving an "If. . . Then" Statement by Contradiction*)   To prove the *if. . . then* statement $P \Rightarrow Q$ by contradiction, assume the hypothesis P is true and also assume the negation $\sim Q$ of the conclusion is true. Then argue until deducing a contradiction.

Applying the *Know/Show* approach to proving an *if. . . then* statement $P \Rightarrow Q$ by contradiction gives us the following: *We know P and $\sim Q$, and we want to show* any contradiction we can find.

There is often more than one way to develop a proof of a given statement. In the next example, we develop an argument that uses proof by contradiction to establish a result we first derived in our proof of Theorem 4.46.

Example 4.54   Here we will prove

$$\text{If } A \cap B = A, \text{ then } A \subseteq B.$$

We begin by assuming the hypothesis $A \cap B = A$. We want to show $A \subseteq B$, but rather than doing so directly, we will use contradiction. So besides assuming $A \cap B = A$, we will also assume that $A \nsubseteq B$. We will then argue until we reach a contradiction.

> *Theorem:* If $A \cap B = A$, then $A \subseteq B$.
>
> *Proof:* Suppose $A \cap B = A$ and suppose by way of contradiction that $A \nsubseteq B$. Then, by the definition of subset, there exists $x \in A$ such that $x \notin B$. The definition of intersection then tells us that $x \notin A \cap B$. Now, because $A \cap B = A$ by hypothesis, we may conclude that $x \notin A$, contradicting the earlier statement that $x \in A$.   □

Note the order in which our proof makes use of the two premises, $A \cap B = A$ and $A \nsubseteq B$. We worked with the premise $A \nsubseteq B$ first, as this allowed us to assert the existence of a particular object, which we called x, that is in A but not in B. The existence of this object essentially contradicted the set equality that the other premise, $A \cap B = A$, asserts. Generally speaking, if a premise allows you to assert the existence of an object, you will probably need to use this premise to get ahold of the object very early in the proof.

You probably recall that the only way for a product of two real numbers to be 0 is if one of the factors is 0. We now give a proof of this result using contradiction. Note how our proof uses one of DeMorgan's Laws, in this case the fact that the negation of an *or* statement $P \vee Q$ is the *and* statement $\sim P \wedge \sim Q$, in assuming the negation of the conclusion.

Theorem 4.55 (*When a Product Is Zero, One of the Factors Is Zero*)   For all real numbers a and b, if $ab = 0$, then $a = 0$ or $b = 0$.

> *Proof:* Consider real numbers a and b. Suppose $ab = 0$ and suppose to the contrary that $a \neq 0$ and $b \neq 0$. Because a and b are nonzero, they have multiplicative inverses, $1/a$ and $1/b$. It now follows from the equation $ab = 0$ that $(1/b)(1/a) \cdot ab = (1/b)(1/a) \cdot 0$. The left side of this latter equation can be simplified to 1 using the fact that the product of a number and its multiplicative inverse is 1 (applied twice), and the right side can be simplified to 0 using the fact that the product of a real number and 0 is 0. We may therefore conclude that $1 = 0$, a contradiction.   □

Practice Problem 2   Use proof by contradiction, along with the previously proved fact that 1 is not even, to prove that if n is even, then $n + 1$ is not even.

Because proof by contradiction involves assuming the negation of what we intend to show, we are in effect providing ourselves with "extra" hypothesis when we use this proof technique. In addition, using proof by contradiction means that we no longer have a specific statement we are trying to deduce and that our proof will be complete when we have derived *any* contradiction at all. These facets of the method of proof by contradiction often make it possible to prove a statement for which a direct proof is elusive. Proof by contradiction is also a reasonable option when the explicit premises of an argument do not seem to provide much information. Our recommendation is that, unless you are directed to a specific proof strategy, first try a direct method of proof based on the form the statement you want to prove exhibits; if you cannot come up with a direct proof, then try a proof by contradiction.

Proof by Contraposition

There is another indirect method of proof that is sometimes used to establish an *if...then* statement. It is called **proof by contraposition** and is based on the fact that a conditional statement $P \Rightarrow Q$ is logically equivalent to its contrapositive $\sim Q \Rightarrow \sim P$ (see Problem 3G.1b).

Proof Strategy 4.56 (*Proof by Contraposition*)   To prove the *if...then* statement $P \Rightarrow Q$ by contraposition, assume the negation $\sim Q$ of its conclusion and then deduce the negation $\sim P$ of its hypothesis.

Our *Know/Show* perspective in the setting of a proof of $P \Rightarrow Q$ that uses contraposition yields the following: *We know* $\sim Q$ and *we want to show* $\sim P$.

   Proof by contraposition is a valuable strategy because sometimes the proof of a conditional statement $P \Rightarrow Q$ turns out to be easier if it is done by contraposition.

Example 4.57   If we want to prove

$$\textit{If } A - B = A \cup B, \textit{ then } B = \varnothing$$

directly, we would assume $A - B = A \cup B$ and then try to show $B = \varnothing$. But it is probably not immediately clear how we could use the hypothesis that $A - B$ has exactly the same members as $A \cup B$ to argue that B has no members. We are not saying that the proof cannot be done directly, only that it may be somewhat difficult to think about in this way. However, if we use proof by contraposition, we would assume $B \neq \varnothing$ and then try to show $A - B \neq A \cup B$. Noting that any member of B would be a member of $A \cup B$ but would not be a member of $A - B$, we see that the assumption that B has at least one member leads relatively easily to the conclusion that $A - B \neq A \cup B$. Here is our proof.

   Theorem: If $A - B = A \cup B$, then $B = \emptyset$.
   Proof: We proceed by contraposition. Suppose $B \neq \emptyset$. Then there exists $x \in B$. By the definition of union $x \in A \cup B$, and by the definition of difference $x \notin A - B$. Hence, by the definition of set equality, $A - B \neq A \cup B$. □

Note how our proof's first sentence identifies for the reader the proof strategy we are using.

Practice Problem 3   Let a and b be real numbers. Use contraposition to prove that if it is not the case that $a < b$, then $a \geq b$.

Answers to Practice Problems

1. *Claim:* No integer can be both even and odd.

    Proof: Suppose to the contrary that there is an integer n that is both even and odd. Then there exist integers j and k such that $n = 2j$ and $n = 2k + 1$. It follows that $2j = 2k + 1$ so that $1 = 2(j - k)$, where $j - k$, being a difference of integers, is itself an integer. Thus, by definition, we may conclude that 1 is even, contradicting the fact that we already proved 1 is not even. □

2. *Claim:* If n is even, then $n + 1$ is not even.

    Proof: Suppose n is even. Also suppose, by way of contradiction, that $n + 1$ is even. Then there exist integers j and k for which $n = 2j$ and $n + 1 = 2k$. Thus,

    $$1 = (n + 1) - n = 2k - 2j = 2(k - j),$$

    where $k - j$ is an integer as $\mathbf{Z}$ is closed under subtraction. Thus, by definition, we may conclude that 1 is even, contradicting the fact that we have already proved 1 is not even. □

3. *Claim:* If it is not the case that $a < b$, then $a \geq b$.

    Proof (by contraposition): Let a and b be real numbers and assume that it is not the case that $a \geq b$. We will show that $a < b$. Because it is not the case that $a \geq b$, neither $a > b$ nor $a = b$, from which it follows that $a - b$ is neither positive nor equal to 0. So, by the law of trichotomy, we may conclude that $a - b$ is negative. Thus, by definition, $-(a - b) = b - a$ is positive. So, using the definition of $<$, we have $a < b$. □

4.9   Proofs Involving *Or*

Questions to guide your reading of this section:

1.   What is the usual approach to proving an *or* statement?

2.   When can it be helpful to set up *cases* in a proof? Once cases are introduced into a proof, what must we be sure the cases do collectively, and what must we do in each case?

Recall the argument we presented back in Example 3.25 in which graduation from a certain college requires a student to take a course in either math or biology. If a student wants to graduate and does not plan to take a math course, the student will have to take a biology course. This example illustrates the logical equivalence of $P \vee Q$ and $\sim P \Rightarrow Q$, which makes sense intuitively: In order for $P \vee Q$ to be true if P is not true, it must be that Q is true (see also Problem 3G.1c). This logical equivalence tells us that to prove the *or* statement $P \vee Q$, we may instead prove the *if... then* statement $\sim P \Rightarrow Q$.

Proof Strategy 4.58 (*Proving an "Or" Statement*)   To prove the *or* statement $P \vee Q$, assume $\sim P$ and deduce Q.

We use this proof strategy within the proof of Theorem 4.59 below. Specifically, we establish

$$t \in A \cap B \ or \ t \in A - B$$

by instead proving

If $t \notin A \cap B$, then $t \in A - B$.

Note also how the proof uses double containment, Proof Strategy 4.45, our standard strategy for proving set equality.

Theorem 4.59   For any sets A and B, $A = (A \cap B) \cup (A - B)$.

Proof: Consider any sets A and B.

($\subseteq$) Let $t \in A$. To show $t \in (A \cap B) \cup (A - B)$, we must show, by the definition of union, that $t \in A \cap B$ or $t \in A - B$. This can be done by showing that if $t \notin A \cap B$, then $t \in A - B$.

So suppose $t \notin A \cap B$. Then, because we already assumed that $t \in A$, it follows from the definition of intersection that $t \notin B$. Because $t \in A$ and $t \notin B$, we may deduce, by the definition of set difference, that $t \in A - B$.

($\supseteq$) Let $t \in (A \cap B) \cup (A - B)$. Then, according to the definition of union, $t \in A \cap B$ or $t \in A - B$. It then follows, using the definitions of intersection and set difference, that either $t \in A$ and $t \in B$ or else $t \in A$ and $t \notin B$. In either case, $t \in A$. $\square$

In Section 4.8, we used proof by contradiction to establish that the only way for a product of two real numbers to be 0 is if one of the factors is 0. We now give a proof of this result that uses our standard strategy for proving *or* statements, Proof Strategy 4.58.

Theorem 4.55 (revisited) For any real numbers a and b, if $ab = 0$, then $a = 0$ or $b = 0$.

Proof: Consider real numbers a and b. Suppose $ab = 0$ and $a \neq 0$. It suffices to show that $b = 0$. Because $a \neq 0$, we know a has a multiplicative inverse $1/a$. Then, because $ab = 0$, it follows that $(1/a)(ab) = (1/a)(0)$. This equation can be simplified to the desired conclusion $b = 0$ (*how precisely?*). ☐

We can diagram the development of our new proof of Theorem 4.55 using the *Know/Show* approach. Initially, *we know* $ab = 0$, and *we want to show* $a = 0$ or $b = 0$. But we can establish the *or* statement we want to show by proving the *if... then* statement

$$\text{If } a \neq 0, \text{ then } b = 0.$$

This leads us to revise what we know and what we want to show as follows: *We know* $ab = 0$ and $a \neq 0$, and *we want to show* $b = 0$. Our written proof follows this plan.

Practice Problem 1 Make use of Proof Strategy 4.58 to prove that if $x \in A \cup B$, then $x \in A$ or $x \in B - A$.

Proofs Involving Cases

There are times when reaching a certain conclusion is more easily achieved by considering two or more separate cases based on either our initial premises or an intermediate deduction following from the premises.

Example 4.60 For instance, within the proof of the following theorem, we deduce that $x \in A$ or $x \in B$ and continue the proof by arguing independently from each of these two possibilities. The proof is then completed by demonstrating that our desired conclusion can be reached in each case.

Theorem: If $A \subseteq B$, then $A \cup B \subseteq B$.
Proof: Suppose $A \subseteq B$ and consider $x \in A \cup B$. Then, by the definition of union, $x \in A$ or $x \in B$. Our argument will be complete when we have shown that $x \in B$.

Case 1: $x \in A$
Then, as $A \subseteq B$, it follows that $x \in B$.

Case 2: $x \in B$
Then, rather trivially, $x \in B$.

Thus, regardless of which case actually holds for the anonymously selected $x \in A \cup B$, we have concluded that $x \in B$, and our proof is finished.   □

In a proof involving cases, it is important to clearly identify each individual case so that the reader understands exactly what hypothesis each case is assuming. We did this in the proof in Example 4.60 by using the labels *Case 1* and *Case 2* and following each label with the particular hypothesis under which that case was operating. This is probably the most common approach to organizing a proof that uses cases.

Note also that the cases created must cover all possibilities. In the proof from Example 4.60, because we had already deduced that $x \in A$ *or* $x \in B$, it is clear that our two cases exhaust all possibilities.

Finally, in order for a proof involving cases to be complete, not only must the cases we set up exhaust all possibilities, in every one of these cases we must be able to deduce the ultimate conclusion we are trying to reach.

Proof Strategy 4.61 (*Using Cases in a Proof*)   If we want to prove P and we have been able to deduce an *or* statement, we can set up separate cases based on the components of the *or* statement and show that P can be proved in each case.

You are asked to verify formally the validity of this strategy in Problem 4Q.1.

Example 4.62   Here is another example of a proof that incorporates cases.

Theorem: If $A \subseteq C \cap D$ and $B - A \subseteq C - D$, then $A \cup B \subseteq C$.
Proof: Suppose $A \subseteq C \cap D$ and $B - A \subseteq C - D$. We must show $A \cup B \subseteq C$. To do this, let $x \in A \cup B$; we will show $x \in C$. We will consider cases based on whether or not $x \in A$.

Case 1: $x \in A$.
Because $A \subseteq C \cap D$, it follows that $x \in C \cap D$. Then, by the definition of intersection, we may conclude that $x \in C$.

Case 2: $x \notin A$.
Because $x \in A \cup B$ and $x \notin A$, it follows, by the definition of union, that $x \in B$. Now, as $x \in B$ and $x \notin A$, the definition of set difference tells us that $x \in B - A$. But we know $B - A \subseteq C - D$, so it follows that $x \in C - D$. Thus, once again applying the definition of set difference, we are able to conclude that $x \in C$.

Therefore, in either case, we are able to deduce $x \in C$. Because one of the two cases must hold by the Fundamental Assumption About Set Membership, our proof is complete. □

Note that we set up the cases in this proof a bit differently than in the proof presented in Example 4.60. At first, it seems that we should proceed just as we did in that example, creating two cases, one where $x \in A$ and the other where $x \in B$. However, setting up a case where $x \in B$ would still require us, within that case, to consider the two possibilities, $x \in A$ or $x \notin A$. This is because $B - A \subseteq C - D$ is the only part of our hypothesis that involves B, and by itself, it will only allow us to conclude that members of B that are not members of A are members of C. To handle members of B that are members of A, we must use the other part of our hypothesis, $A \subseteq C \cap D$. Once we realize this, it becomes more convenient to set up the cases based on whether or not x is in A.

We next prove that the square of any real number is nonnegative. Our proof uses Proof Strategy 4.61 as it involves three different arguments for the cases where the given real number is positive, negative, or 0.

Theorem 4.63 For every real number a, $a^2 \geq 0$.

Proof: Let $a \in \mathbf{R}$. By the law of trichotomy, $a > 0$, $a < 0$, or $a = 0$.

Case 1: $a > 0$.
Multiplying both sides of the inequality $a > 0$ by the positive number a produces a new inequality, $a \cdot a > 0 \cdot a$, which must also be true. Simplifying, we get $a^2 > 0$ so that $a^2 \geq 0$.

Case 2: $a < 0$.
Multiplying both sides of the inequality $a < 0$ by the negative number a produces a new inequality, $a \cdot a > 0 \cdot a$, which must also be true (keep in mind that multiplying through by a negative number reverses the inequality sign). Simplifying, we get $a^2 > 0$ so that $a^2 \geq 0$ in this case as well.

Case 3: $a = 0$.
In this case, $a^2 = a \cdot a = 0 \cdot 0 = 0$. Thus, $a^2 \geq 0$ when $a = 0$. □

Practice Problem 2 We have not yet established that every integer must be either even or odd, but you are no doubt familiar with this fact. In this problem, assume that this is true and use cases, based on whether n is even or odd, to prove that for every integer n, $n^2 + n$ is even.

Answers to Practice Problems

1. *Claim:* If $x \in A \cup B$, then $x \in A$ or $x \in B - A$.
    Proof: Assume $x \in A \cup B$ and $x \notin A$. It suffices to show that $x \in B - A$. By definition of union, knowing that $x \in A \cup B$ and $x \notin A$, we may conclude that $x \in B$. So, as $x \in B$ and $x \notin A$, the definition of set difference tells us that $x \in B - A$. $\square$

2. *Claim:* For every integer n, $n^2 + n$ is even.
    Proof: Let $n \in \mathbf{Z}$. Then n is either even or odd.

    Case 1: n is even.
    Then $n = 2k$ for some integer k, and it follows that

    $$n^2 + n = (2k)^2 + 2k = 4k^2 + 2k = 2(2k^2 + k),$$

    where $2k^2 + k$ is an integer as $\mathbf{Z}$ is closed under both multiplication and addition.
    Hence, by definition, $n^2 + n$ is even.

    Case 2: n is odd.
    Then $n = 2k + 1$ for some integer k, and it follows that

    $$n^2 + n = (2k + 1)^2 + (2k + 1) = 4k^2 + 6k + 2 = 2(2k^2 + 3k + 1),$$

    where $2k^2 + 3k + 1$ is an integer as $\mathbf{Z}$ is closed under both multiplication and addition. Hence, by definition, $n^2 + n$ is even. $\square$

4.10   A Mathematical Research Situation

Questions to guide your reading of this section:

1. How did we use Venn diagrams to help us develop the four conjectures in the research situation presented in this section?

2. How did we apply our standard proof strategies and the *Know/Show* approach to develop and write proofs of these conjectures?

We have now developed enough set theory to consider set-theoretic research situations. Our primary goal here is to show you how we can go about exploring the relationship between two sets that are themselves formed from other sets. To be specific, given any sets A, B, and C, we can form the two sets

$$(A - B) - C \quad \text{and} \quad A - (B - C).$$

We might wonder whether the location of the parentheses in these two expressions is of any significance as, after all, only one set operation, that of the difference of sets, is being used. In other words, are these sets always equal, no matter what sets A, B, and C we start with? This question initiates our investigation.

Carrying Out the Investigation: Building Conjectures from Venn Diagrams

We know from our work with sets in Chapter 2 that Venn diagrams can be a useful tool in comparing two sets, so it is reasonable to consider making such diagrams for the sets we wish to compare. The Venn diagram for $(A - B) - C$ is

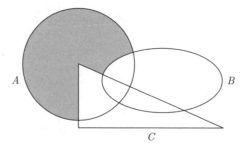

and the Venn diagram for $A - (B - C)$ is

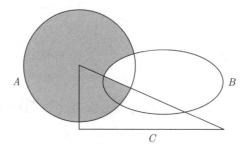

and these diagrams suggest that the sets are not always equal. Thus, we may formulate our first conjecture:

    Conjecture 1: There exist sets A, B, and C such that $(A - B) - C \neq A - (B - C)$.

However, the Venn diagrams we have created suggest that one of the sets is always a subset of the other. So, we can formulate a second conjecture.

    Conjecture 2: For any sets A, B, and C, $(A - B) - C \subseteq A - (B - C)$.

Looking again at the Venn diagrams, we note that the diagram for $A - (B - C)$ marks two regions that the Venn diagram for $(A - B) - C$ does not. Otherwise, the diagrams are identical. We may also observe that the two regions marked in the one diagram, but not in the other, appear to comprise the set $A \cap C$. Thus, upon reflection, it seems that if $A \cap C$ had no members, our two sets $(A - B) - C$ and $A - (B - C)$ would be equal (at least according to the diagrams). This leads us to two more conjectures:

Conjecture 3: If $A \cap C = \varnothing$, then $(A - B) - C = A - (B - C)$.
Conjecture 4: If $A - (B - C) \subseteq (A - B) - C$, then $A \cap C = \varnothing$.

Conjecture 3, if proved true, would establish $A \cap C = \varnothing$ as a sufficient condition for the equality of the sets $(A - B) - C$ and $A - (B - C)$, whereas Conjectures 2 and 4, if proved true, would establish $A \cap C = \varnothing$ as a necessary condition for the equality of these sets.

You might wonder why we are not stating Conjecture 4 as

$$\text{If } (A - B) - C = A - (B - C), \text{ then } A \cap C = \varnothing.$$

We certainly could. But, because we believe (see Conjecture 2) we can prove the subset relationship $(A - B) - C \subseteq A - (B - C)$ without any restriction (i.e., we believe, based on our Venn diagrams, that this set inclusion holds for all sets A, B, and C), our thinking is that it is only the reverse inclusion $A - (B - C) \subseteq (A - B) - C$ that will be needed to establish that $A \cap C = \varnothing$.

Proving Conjecture 1

Note that Conjecture 1 asserts the existence of sets A, B, and C with a certain property. To prove it, we will apply Proof Strategy 4.24, our standard strategy for proving existence. First, we will define specific sets A, B, and C for which we think it will turn out to be true that $(A - B) - C \neq A - (B - C)$. Then, we will have to show that our chosen sets do indeed possess this property. Based on the research we conducted after formulating Conjecture 1, it appears that we could choose any sets A, B, and C as long as A and C have at least one member in common, thus making $A \cap C$ nonempty. This is what we do at the beginning of our proof.

Conjecture 1: There exist sets A, B, and C such that $(A - B) - C \neq A - (B - C)$.

Proof: Let $A = \{1, 2, 3\}$, $B = \{2, 4, 6, 8\}$, and $C = \{2, 3, 4, 5\}$. Then $A - B = \{1, 3\}$ so that $(A - B) - C = \{1\}$. Also, $B - C = \{6, 8\}$ so that $A - (B - C) = \{1, 2, 3\}$. Thus, because, for instance, $2 \in A - (B - C)$, yet $2 \notin (A - B) - C$, we may conclude that $(A - B) - C \neq A - (B - C)$. $\square$

Having proved Conjecture 1, this statement is, of course, now a theorem.

Proving Conjecture 2

We use Proof Strategy 4.14, our standard strategy for establishing a set inclusion, in proving Conjecture 2.

Conjecture 2: For any sets A, B, and C, $(A - B) - C \subseteq A - (B - C)$.

Proof: Consider sets A, B, and C. Let $x \in (A - B) - C$. We must show $x \in A - (B - C)$. According to the definition of set difference, as $x \in (A - B) - C$, we may conclude that $x \in A$, $x \notin B$, and $x \notin C$. Now, because $x \notin B$, the definition of set difference tells us that $x \notin B - C$. Then, as we now have $x \in A$ and $x \notin B - C$, we may conclude that $x \in A - (B - C)$. $\square$

Because we have proved Conjecture 2, it is now also a theorem. Let us examine the proof relative to *Know/Show* to gain a better understanding of how it was developed. Initiating the standard approach to establishing a set inclusion gives us: *We know* $x \in (A - B) - C$, and *we want to show* $x \in A - (B - C)$.

Thinking first about what we want to show, the definition of set difference tells us *we want to show* $x \in A$ and $x \notin B - C$. And, because $x \notin B - C$ translates as $x \notin B$ or $x \in C$ (definition of set difference and one of DeMorgan's Laws), we now realize *we want to show* $x \in A$, as well as one of $x \notin B$ or $x \in C$.

Now, going back to what we know and applying the definition of set difference twice allows us to revise this premise so that *we know* $x \in A$, $x \notin B$, and $x \notin C$.

At this point, because we realize that we know both $x \in A$ and $x \notin B$, it is clear that we will be able to deduce our revised version of what we want to show. This is how we were able to pull together the proof we wrote for Conjecture 2 (go back now and re-read the proof to make sure you understand what we have just discussed).

Proofs: Discovering Versus Writing

We just commented on how we used the *Know/Show* approach to discover our proof of Conjecture 2. Whenever we apply *Know/Show* to think about how to develop a proof, we are really working both forward from what we know and backward from what we want to show. Our hope is that this will enable us to connect what we know to what we want to show. By finding such connections and determining how to apply them to the situation at hand, we wind up discovering how our argument should go.

Of course, once we have "discovered" our argument, we still need to write it up in the form of a proof. And a proof is almost always written so that there is a clear sense of "forward motion" in which each subsequent statement is deduced from statements appearing earlier in the proof. In other words, the order in which we present our ideas in a written proof must reflect the logical sequence in which deductions can be made, even though some of our preliminary work in discovering the fundamental connections that make our argument work will usually involve thinking backwards from what we ultimately want to show (we first addressed this issue in Section 4.4). We need to "think backwards" to try to determine how we might reach a certain conclusion, but we should never "write backwards" as this usually makes it difficult to maintain a clear distinction between what is known and what is being shown.

To illustrate, recall that in applying *Know/Show* to discover how our proof of Conjecture 2 was developed, we "thought backwards" from

$$We\ want\ to\ show:\ x \in A - (B - C)$$

to discover that this really meant

$$We\ want\ to\ show:\ x \in A,\ \text{as well as one of }x \notin B\ or\ x \in C.$$

The written proof itself, though, presents a clear sequence of deductions that begins with what is known and then moves step-by-step toward what needs to be shown. Specifically this sequence of deductions is as follows:

1. We began with what was known: $x \in (A - B) - C$.

2. From what was known, we used the definition of set difference to deduce that $x \in A$, $x \notin B$, and $x \notin C$.

3. We used our conclusion that $x \notin B$ and the definition of set difference to deduce that $x \notin B - C$ (but it probably would not have been clear that we would want to make this particular deduction if we had not thought about "working backwards" to revise what we ultimately wanted to show).

4. Having concluded that $x \in A$ and $x \notin B - C$, we applied the definition of set difference once more to deduce that $x \in A - (B - C)$, which was our ultimate goal.

Our written proof essentially explains how we can move from each step in the above chain to the next one.

Proving Conjecture 3

What follows is a direct proof of Conjecture 3. Be sure you see how we are using the relevant standard proof strategies and that you can follow all the details of the argument.

Conjecture 3: If $A \cap C = \varnothing$, then $(A - B) - C = A - (B - C)$.

Proof: Suppose $A \cap C = \varnothing$. Because we have already proved, in Conjecture 2, that $(A - B) - C \subseteq A - (B - C)$ for any sets A, B, and C, we need only show that $A - (B - C) \subseteq (A - B) - C$. So consider an anonymous $x \in A - (B - C)$. Then, by the definition of set difference, $x \in A$ and $x \notin B - C$. Now, using our hypothesis that $A \cap C = \varnothing$ and our deduction that $x \in A$, we may conclude that $x \notin C$. Then, as we have determined that $x \notin B - C$, and this requires that either $x \notin B$ or $x \in C$, we can use our conclusion that $x \notin C$ to deduce that $x \notin B$. Thus, we have $x \in A$, $x \notin B$, and $x \notin C$, and from this information, the definition of set difference allows us to reach our desired conclusion that $x \in A - (B - C)$. $\square$

Because we now have a proof for Conjecture 3, this conjecture is really a theorem. The proof itself can be developed by applying *Know/Show*. Our standard strategy for proving a conditional statement, Proof Strategy 4.8, along with our prior proof that $(A - B) - C \subseteq A - (B - C)$ for any sets A, B, and C, means that in this situation *we know* $A \cap C = \varnothing$ and *we want to show* $A - (B - C) \subseteq (A - B) - C$.

Focusing first on what we want to show, we decide to apply the standard approach to establishing a set inclusion, Proof Strategy 4.14, and consider an anonymous $x \in A - (B - C)$; our goal is to then show that $x \in (A - B) - C$. Thus, at this stage, *we know* $A \cap C = \varnothing$ and $x \in A - (B - C)$, and *we want to show* $x \in (A - B) - C$. Applying the definition of set difference allows us to update further so that *we know* $A \cap C = \varnothing$, $x \in A$, and also either $x \notin B$ or $x \in C$, and *we want to show* $x \in A$, $x \notin B$, and $x \notin C$. It is at this point that we are really able to identify how we can use what we know to establish what we want to show. We know right away that $x \in A$, so this part of what we want to show is done. Because we know that $x \in A$ and $A \cap C = \varnothing$, it is impossible for $x \in C$, so we must have $x \notin C$. Finally, as $x \notin C$, and we know that either $x \notin B$ or $x \in C$, logic permits us to conclude that $x \notin B$.

Before proceeding to consider Conjecture 4, we will present another proof of Conjecture 3, one that uses contradiction, Proof Strategy 4.51. We do so to give you more experience with this proof strategy, but also to remind you that there is usually more than one way to develop a proof for a conjecture. Make sure you can follow the implementation of the proof by contradiction strategy.

Conjecture 3: If $A \cap C = \varnothing$, then $(A - B) - C = A - (B - C)$.

Proof (by contradiction): Suppose $A \cap C = \varnothing$ and suppose $(A - B) - C \neq A - (B - C)$. Because we have already proved, in Conjecture 2, that $(A - B) - C \subseteq A - (B - C)$ for any sets A, B, and C, we may conclude that $A - (B - C) \nsubseteq (A - B) - C$. It follows immediately that there exists $x \in A - (B - C)$ such that $x \notin (A - B) - C$. Because $x \in A - (B - C)$, it follows that $x \in A$ and $x \notin B - C$, and the latter requirement tells us that either $x \notin B$ or $x \in C$. Also, because $x \notin (A - B) - C$, it follows that either $x \notin A - B$ or $x \in C$. Because we have already deduced that $x \in A$ and we have assumed that $A \cap C = \varnothing$, it follows that $x \notin C$. From this conclusion, because we know that either $x \notin B$ or $x \in C$, it follows that $x \notin B$, and because we already know that either $x \notin A - B$ or $x \in C$, it follows that $x \notin A - B$. Now, as $x \notin A - B$, the definition of set difference allows us to conclude that either $x \notin A$ or $x \in B$. But this provides us with a contradiction, as we have already deduced that $x \in A$ and $x \notin B$. $\square$

Proving Conjecture 4

We now present a proof for Conjecture 4 that uses contraposition (see Proof Strategy 4.56). Note that the object x whose existence is asserted based on the assumption that $A \cap C \neq \varnothing$ turns out to be an object demonstrating why $A - (B - C) \nsubseteq (A - B) - C$.

> *Conjecture 4:* If $A - (B - C) \subseteq (A - B) - C$, then $A \cap C = \varnothing$.

> *Proof:* Suppose $A \cap C \neq \varnothing$. We must show $A - (B - C) \nsubseteq (A - B) - C$. Because $A \cap C \neq \varnothing$, there exists $x \in A \cap C$, and it follows from the definition of intersection that $x \in A$ and $x \in C$. Now observe that as $x \in C$, the definition of set difference tells us that $x \notin (A - B) - C$ and $x \notin B - C$. But, as we now have determined that $x \in A$ and $x \notin B - C$, we may conclude that $x \in A - (B - C)$. This conclusion, along with our earlier deduction that $x \notin (A - B) - C$, means that $A - (B - C) \nsubseteq (A - B) - C$. $\square$

Summarizing Our Conclusions

At this point we have proved all four of our conjectures, so we really have four theorems. It is always a good idea, in any sort of investigation, to summarize any conclusions that have emerged as a result of the investigation. We want to share with our audience what we believe we have learned as clearly and concisely as possible. Here is what we might write for a conclusion in the situation we have been working in:

> *Our investigation focused on comparing the two sets $(A - B) - C$ and $A - (B - C)$, which were formed out of arbitrary sets A, B, and C. We created Venn diagrams for the two sets and studied these diagrams to develop some conjectures concerning the relationship between the two sets. We were able to prove all of the conjectures we formulated. What did we discover? Although it need not be the case that the two sets $(A - B) - C$ and $A - (B - C)$ are equal, it is always the case that $(A - B) - C$ is a subset of $A - (B - C)$. Moreover, the condition $A \cap C = \varnothing$ is both necessary and sufficient for the equality of the sets $(A - B) - C$ and $A - (B - C)$.*

Chapter 4 Problems

4A. Quantifiers and Arguments

1. Assume that all birds have wings.

    (a) You see a bird. What can you conclude?

    (b) You see a creature with wings. What can you conclude?

(c) You see a creature that is not a bird. What can you conclude?

(d) You see a creature without wings. What can you conclude?

2. What would you need to do in order to refute (i.e., show false) the following claims?

(a) All birds have wings.

(b) Some birds have wings.

(c) No birds have wings.

(d) If a creature has wings, then it is a bird.

4B. Proving an *If... Then* Statement

1. For each of the following claims, apply Proof Strategy 4.8, our standard strategy for proving a conditional statement, to complete the beginning of the claim's proof.

(a) *Claim:* If the function f is differentiable, then f is continuous.
     Proof: Suppose _____. We must show _____.

(b) *Claim:* If the integer k is odd, then the integer $k - 5$ is even.
     Proof: Assume _____. We must show _____.

(c) *Claim:* The number x is nonzero provided that $|x|$ is positive.
     Proof: Suppose _____. We must show _____.

2. Prove that if $1 + 1 = 2$, then $\frac{1}{2} + \frac{1}{2} = 1$.

4C. Proving and Disproving *For All* Statements

1. The conjecture

   For every natural number n, the number $n^2 + n + 17$ is odd

is actually true. Explain why the following "argument" is not acceptable as a proof for this conjecture: *Observe that $1^2 + 1 + 17 = 19$, $2^2 + 2 + 17 = 23$, $3^2 + 3 + 17 = 29$, and $4^2 + 4 + 17 = 37$, which are all odd. From these examples, we may generalize to conclude that $n^2 + n + 17$ will always be odd.*

2. Use Proof Strategy 4.11, our standard strategy for proving a *for all* statement, to complete the first two sentences of the proof of the given claim.

(a) *Claim:* For every prime number p, the number $p + 7$ is not prime.
     Proof: Consider a _____. We must show _____.

(b) *Claim:* For all real numbers a, b, and c, if $a + c = b + c$, then $a = b$.
     Proof: Let _____. We will show _____.

   (c) *Claim:* The intersection of two sets is always a subset of either set.

   Proof: Consider _____. We will show _____.

   (d) *Claim:* The additive inverse of any negative real number is positive.

   Proof: Suppose _____. We must show _____.

   (e) *Claim:* The sum of the additive inverses of two real numbers is equal to the additive inverse of their sum.

   (f) *Proof:* Consider _____. We will show _____.

3. Prove that for all real numbers a, b, and c, $a(b + c) = ca + ab$.

4. Prove that for any nonzero real number a, $a \cdot (1/a) \cdot a = a$.

5. Disprove the given statement.

   (a) For every even integer n, $n/2$ is odd.

   (b) For all sets A and B, $A \cap B \subset A \cup B$.

   (c) For any a, b, $c \in \mathbf{R}$, $a + bc = c + ba$.

   (d) For any set A, $(A - A) - A = A - (A - A)$.

   (e) The product of two irrational numbers is always irrational.

   (f) The product of a rational number and an irrational number is always irrational.

4D. Counterexamples

1. Provide a counterexample to show that the given statement is false.

   (a) $\mathbf{Q} \subseteq \mathbf{Z}$.

   (b) The square of any real number is positive.

   (c) Every set has a proper subset.

2. Find the least natural number n that provides a counterexample to the following conjecture: *For every natural number n, the number $n^2 + n + 17$ is prime.*

4E. Proving Set Inclusion

1. Apply Proof Strategy 4.14, our standard strategy for proving a subset relationship, to complete the beginning of the proof of the given claim.

   (a) *Claim:* $S \subseteq T$.

   Proof: Consider $x \in$ _____. We must show _____.

   (b) *Claim: $S - T \subseteq S \cup T$.*

      Proof: Suppose _____. We must show _____.

2. Prove that for any sets A and B, $A - (A \cap B) \subseteq A - B$.

4F. The *Know/Show* Approach to Developing a Proof

1. Use our standard proof strategies to identify what we know and what we need to show if we want to prove the given statement.

   (a) If the function f is differentiable, then f is continuous.

   (b) The number x is nonzero provided that $|x|$ is positive.

   (c) For every natural number n, the number $n^2 + n + 17$ is odd.

   (d) For all real numbers a, b, and c, if $a + c = b + c$, then $a = b$.

   (e) The additive inverse of any negative real number is positive.

2. Suppose we want to apply the *Know/Show* approach to develop a proof showing that for any sets A, B, and C, $A - (B \cup C) \subseteq B \cup (A - C)$.

   (a) Identify what we know and what we want to show.

   (b) Revise what we know and what we want to show as much as possible based on the definitions of set union and difference.

   (c) Logically, how can we deduce the revised version of what we want to show from the revised version of what we know?

   (d) Write a proof for the claim based on your work in (a), (b), and (c).

3. Use the distributive property and the fact that 0 multiplied by any real number yields 0 to prove that for any real numbers a and b, $ab + a(-b) = 0$.

4G. Asserting Existence

1. Determine whether the given real number has an additive inverse. For each one that does, find the additive inverse.

   (a) 100. (b) 3/8. (c) −11.26.

   (d) $\pi/6$. (e) 0. (f) $4t$. (g) $x^2 + 1$.

2. Determine whether the given real number has a multiplicative inverse. For each one that does, find the multiplicative inverse.

   (a) 100. (b) 3/8. (c) −11.26.

   (d) $\pi/6$. (e) 0. (f) $4t$. (g) $x^2 + 1$.

3. Tell whether the given piece of information allows us to assert the existence of an object. If it does, bring this object into existence and describe any properties the object must satisfy.

    (a) $L \nsubseteq M$. (b) $A \cap B \nsubseteq C$. (c) $X \subseteq Y$. (d) $S \neq \emptyset$.

    (e) $T = \emptyset$. (f) $A \cup B = C$. (g) $E \neq F$. (h) $P \cap Q \neq P$.

4. Prove that for all real numbers a, b, and c, if $a \neq 0$ and $ab = ac$, then $b = c$.

4H. Proving Existence

Prove each of the following.

1. There exists a real number x such that $x^3 = -2x^2$.

2. There is a prime number greater than 100.

3. There exist sets A and B such that $A \neq B$.

4. For every set A, there exists a set B such that $A \neq B$.

5. For any set A, there is a set B such that $B \subseteq A$.

6. Every set with at least two distinct elements has at least three non-empty subsets.

7. A positive integer n is **composite** if $n = ab$ for some integers a and b with $1 < a < n$ and $1 < b < n$.

    (a) Every composite integer has at least three positive factors.

    (b) There is a composite integer having only three positive factors.

4I. Using and Proving Uniqueness

1. Suppose we know that the number t is a solution to a certain algebraic equation and we want to prove that t is the only solution to this equation. According to Proof Strategy 4.28, our standard strategy for proving uniqueness, we should assume _____ and then try to show _____.

2. Prove that, for a given nonzero real number a, the number $1/a$ is the unique multiplicative inverse of a.

3. Use the uniqueness of additive inverses to prove that if $a + b = 0$, then $b = -a$.

4. Use the uniqueness of multiplicative inverses to prove that if $ab = 1$, then $b = 1/a$.

5. For each nonnegative real number r, there is a unique nonnegative real number s such that $s^2 = r$; the number s is called the **principal square**

root of r and is denoted $\sqrt{r}$. Show (in mathematics, *show* is synonymous with *prove*) that the equation $x = \sqrt{6-x}$ has a unique real number solution. (HINT: "Solve" the equation and then "check" candidate solutions.)

6. Show that if a, b, and c, are real numbers with $b \neq 0$, the equation $(a + x) \cdot (1/b) = c$ has a solution and the solution is unique.

7. Prove that there is a unique set A such that for every set B, $A \subseteq B$. (HINT: What is this set A?)

4J. Working with Definitions in Proofs

1. A function f is **continuous** at a real number a if $\lim\limits_{x \to a} f(x) = f(a)$. According to Proof Strategy 4.33, our standard strategy for proving a statement from its definition, if we want to prove that the cosine function is continuous at π, what must we actually show?

2. Use the definitions of *odd* integer and *even* integer given in Section 4.6 to prove each of the following.

   (a) 0 is even.

   (b) −5 is odd.

   (c) The set of even integers is closed under multiplication.

   (d) The set of odd integers is closed under multiplication.

   (e) The sum of an even integer and an odd integer is odd.

   (f) The product of an even integer and an odd integer is even.

3. In this problem, make use of the definitions of *additive inverse* and *multiplicative inverse* given in Section 4.6.

   (a) Prove that the additive inverse of an even integer is even.

   (b) Prove that the additive inverse of an odd integer is odd.

   (c) Show that for each nonzero real number a, $1/(1/a) = a$.

   (d) Let a and b be nonzero real numbers. Prove that $1/(ab) = (1/a) \cdot (1/b)$.

   (e) Show that for each nonzero real number a, $-(1/a) = 1/(-a)$. (HINT: Use the fact that $-a = (-1)a$ along with the result of Problem 3(d) above.)

4. In this problem, make use of the definitions of *subtraction* and *division* given in Section 4.6.

   (a) Provide counterexamples to show that subtraction is neither commutative nor associative.

(b) Provide counterexamples to show that division is neither commutative nor associative.

(c) Let a be a real number. Show that $a \div 1 = a$.

(d) Let a be a nonzero real number. Show that $1 \div a = 1/a$.

(e) Let a be a nonzero real number. Use the result of Problem 3(c) to show that $a \div (1/a) = a \cdot a$.

(f) Let a, b, $c \in \mathbf{R}$ with both b and c nonzero. Show that $a \div (bc) = (a \div b) \div c$.

(g) Let a and b be real numbers with $b \neq 0$. Use the result of Problem 3(e) to show that $-(a \div b) = (-a) \div b$ and $-(a \div b) = a \div (-b)$.

5. It is customary to use a fraction bar to indicate division; that is, we often write a/b in place of $a \div b$. By definition, a real number r is **rational** provided that $r = a/b$ for some integers a and b with $b \neq 0$. Prove each of the following.

   (a) The number 4.812 is rational.

   (b) Every integer is a rational number.

   (c) If $a \neq 0$ is rational, then $1/a$ is rational.

   (d) If a is rational, then $-a$ is rational.

   (e) The set of rational numbers is closed under subtraction. (HINT: Make use of the result of Problem 5d.)

   (f) The set of nonzero rational numbers is closed under division. (HINT: Make use of the result of Problem 5c.)

6. In this problem, make use of the definitions of $<$, $>$, and $\leq$ given in Section 4.6.

   (a) Let a, b, $c \in \mathbf{R}$. Prove that if $a < b$ and $c < 0$, then $ac > bc$.

   (b) Let a and b be real numbers. Use the Law of Trichotomy to prove that if it is not the case that $a \leq b$, then $a > b$.

7. Show that if a, b, c, and d are real numbers with $a \neq c$, the equation $ax + b = cx + d$ has a solution and the solution is unique.

4K. Mathematical Equivalences

1. What two statements must we prove if we want to prove the statement

   A circle's diameter is less than 10 if and only if its radius is less than 5

   using Proof Strategy 4.43, our standard strategy for proving an *if and only if* statement?

2. Suppose you are reading a proof of the statement

   The natural number n is divisible by 3 if and only if the sum of the digits comprising n is divisible by 3.

   If you see the label ($\Leftarrow$) in front of part of the proof, exactly what statement is being proved there?

3. Suppose you are reading a proof of the statement

   The following are equivalent:
   1). *a is a zero of the polynomial p(x);*
   2). *x − a is a factor of the polynomial p(x);*
   3). *(a, 0) is an x-intercept of the graph of p.*

   If you see the label $(3) \Rightarrow (2)$ in front of part of the proof, exactly what statement is being proved there?

4. For any sets A, B, and C, prove each of the following.
   (a) $C \subseteq A \cap B$ if and only if both $C \subseteq A$ and $C \subseteq B$.
   (b) $A \cup B \subseteq C$ if and only if both $A \subseteq C$ and $B \subseteq C$.

5. Let a be a real number. Prove the following are equivalent:
   1). $a < 0$;
   2). a is negative;
   3). $-a$ is positive;
   4). $-a > 0$.

6. Let n be an integer. Prove the following are equivalent:
   1). n is even;
   2). $n + 1$ is odd;
   3). $n/2$ is an integer.

4L. Proving Set Equality

1. Prove that for any sets A and B, $A = B$ if and only if both $A \subseteq B$ and $B \subseteq A$.

2. According to the double containment strategy for proving set equality, if we want to prove $A - B = A - (A \cap B)$, what set containments must we prove?

3. Suppose A and B are sets and you are reading a proof of the statement $A - B = A - (A \cap B)$. If you see the label ($\supseteq$) in front of part of the proof, exactly what statement is being proved there?

4. For any sets A and B, prove that $A - (A - B) = A \cap B$.

5. For any sets A, B, and C, prove each of the following.

   (a) $A \cap (B - C) = (A \cap B) - C$.

   (b) $A - (B \cup C) = (A - B) \cap (A - C)$.

   (c) $(A \cap B) - C = (A - C) \cap (B - C)$.

6. Show that for any sets A and B, the following are equivalent:

   1). $A \subseteq B$;

   2). $A \cap B = A$;

   3). $A \cup B = B$.

4M. Necessary and Sufficient Conditions

1. Tell whether the given condition is necessary but not sufficient, sufficient but not necessary, both necessary and sufficient, or neither necessary nor sufficient for $A \cap B = A$.

   (a) $A = \emptyset$. **(b)** $B = \emptyset$. **(c)** $A \cup B = B$. **(d)** $A \cap B = A \cup B$.

2. Tell whether the given condition is necessary but not sufficient, sufficient but not necessary, both necessary and sufficient, or neither necessary nor sufficient for the natural number n to be even.

   (a) n is divisible by 6. **(b)** $n^2 + 1 > 0$.

   (c) $n + 1$ is even. **(d)** $n + 1$ is odd.

3. Use your answers to (2) to determine whether any of the conditions listed are equivalent to n being even. Then state this equivalence in the form of a theorem in four different ways, as was done at the very end of Section 4.7.

4. Prove that for any sets A, B, and C, if $A \subseteq B$, then $A \cap C \subseteq B \cap C$. Then, show that there exist sets A, B, and C such that $A \cap C \subseteq B \cap C$, but for which $A \not\subseteq B$.

5. Based on Problem (4), is the condition $A \subseteq B$

   (a) sufficient for $A \cap C \subseteq B \cap C$?

   (b) necessary for $A \cap C \subseteq B \cap C$?

4N. Proof by Contradiction

1. Suppose we want to prove the statement

   Every set has at least one subset

using proof by contradiction. Write the first sentence of the proof so that it begins with the phrase *suppose to the contrary.*

2. Suppose we want to prove the given statement using proof by contradiction. Tell what we would assume at the beginning of the proof of the statement.

   (a) The square of any odd integer leaves a remainder of 1 when it is divided by 8.

   (b) If x and y are even integers, then $x + y$ is even.

   (c) If the function f is differentiable, then f is continuous.

3. Use proof by contradiction to prove each of the following.

   (a) The sum of a rational number and an irrational number is always irrational.

   (b) The integer -6 is not odd.

   (c) If the integer n is even, then the integer $n^2 - 1$ is odd.

   (d) For any set A, $A - A = \varnothing$.

   (e) For any set A, $A \cap \varnothing = \varnothing$.

   (f) For any set A, $A \cup \varnothing = A$.

   (g) For any set A, $\varnothing - A = \varnothing$.

   (h) If $A \subseteq B$ and $B \subseteq C$, then $A \subseteq C$.

   (i) If $A \subseteq B$, then $C - B \subseteq C - A$.

4. Prove that for any sets A and B, $A - (A - B) = A \cap B$, using contradiction for the "reverse" containment.

5. Use contradiction to prove that there is no smallest positive real number.

40. Proof by Contraposition

1. Suppose we want to prove the statement

   If the function f is differentiable, then f is continuous

   using proof by contraposition. Our proof would begin as follows:

   Proof (by contraposition): Suppose _____. We must show _____.

2. Use proof by contraposition to show that if $A \subseteq B$, then $C - B \subseteq C - A$.

3. Let n be an integer. Use contraposition to prove that if n is not odd, then $n + 1$ is not even.

4. Let x be a real number. Use contraposition to prove that if x is negative, then $-x$ is positive.

5. Recall that for any set S, the collection of all subsets of S is denoted $\mathcal{P}(S)$. Use contraposition to prove that if $A \subseteq B$, then $\mathcal{P}(A) \subseteq \mathcal{P}(B)$.

6. We can effectively use contraposition to prove that for any sets A and B, $A \cap B \subseteq A$. Do so by assuming A and B are sets and that $x \notin A$.

7. Use contraposition in proving both *if...then* statements needed to establish each of the following.

   (a) $A - B = \emptyset$ if and only if $A \subseteq B$.

   (b) $A \cap B = \emptyset$ if and only if $A - B = A$.

4P. Proving an *Or* Statement

1. Suppose we want to prove the statement

$$A \subseteq B \ \text{ or } \ A - B \neq \emptyset$$

   using Proof Strategy 4.58, our standard strategy for proving an *or* statement. Our proof would begin as follows:

   Proof: Suppose —————. We must show —————.

2. Prove that for any sets A, B, and C, if $A \subseteq B \cup C$, then $A - B \subseteq C$, and conversely.

3. Use proof by contraposition to prove each of the following.

   (a) If $A \subseteq B$ and $B \subseteq C$, then $A \subseteq C$.

   (b) If a and b are both negative, so is $a + b$.

   (c) If a is rational and b is irrational, then $a + b$ is irrational.

   (d) If m is even and n is not even, then $m + n$ is not even.

   (e) For any sets A and B, $A \cup B = \emptyset$ if and only if both $A = \emptyset$ and $B = \emptyset$.

4Q. Proofs Involving Cases

1. Verify that $(Q \vee R) \Rightarrow P$ is logically equivalent to $(Q \Rightarrow P) \wedge (R \Rightarrow P)$. Then explain how this logical equivalence provides justification for the use of cases in attempting to deduce P from the hypothesis $Q \vee R$ (see Proof Strategy 4.61).

2. For any sets A and B, prove each of the following.
   (a) $A = (A \cap B) \cup (A - B)$.
   (b) $(A - B) \cup (B - A) = (A \cup B) - (A \cap B)$.
3. For any sets A, B, and C, prove each of the following.
   (a) $A \cap (B \cup C) = (A \cap B) \cup (A \cap C)$.
   (b) $A \cup (B \cap C) = (A \cup B) \cap (A \cup C)$.
   (c) $A - (B \cap C) = (A - B) \cup (A - C)$.
   (d) $(A \cup B) - C = (A - C) \cup (B - C)$.
   (e) If $A \subseteq B$, then $A \cup C \subseteq B \cup C$.
4. Prove that if a is positive, then $1/a$ is positive.
5. Let a, b, c, $d \in \mathbf{R}$. Prove each of the following.
   (a) If $a < b$ and $c \le d$, then $a + c < b + d$.
   (b) If $a \le b$ and $c \le d$, then $a + c \le b + d$.
6. How does the fact that

$$\textit{If } a < b, \textit{ then } a + c < b + c$$

follow from (5a) above?
7. A proof by contradiction can sometimes lead us to then consider cases. When this occurs, we must reach a contradiction in every one of the cases. Take this approach in proving each of the following.
   (a) For every real number a, $0a = 0$.
   (b) If $ab = 0$, then $a = 0$ or $b = 0$.
   (c) For every real number a, $a^2 \ge 0$.

4R. Miscellaneous Proofs

1. Prove each of the following using whatever methods you deem appropriate. You may make use of other results stated in the problems for this chapter.
   (a) The real number a is positive if and only if $1/a$ is positive.
   (b) The real number a is negative if and only if $1/a$ is negative.
   (c) For any real number a, $a > 1$ if and only if $0 < 1/a < 1$.
2. The **absolute value** $|a|$ of a real number a is defined so that

$$|a| = \begin{cases} a, \text{ if } a \ge 0; \\ -a, \text{ if } a < 0. \end{cases}$$

Prove that, for all real numbers a and b, $|a + b| \le |a| + |b|$. (HINT: Consider cases.)

3. Assume A, B, and C are sets. Complete each of the following proofs.

    (a) *Claim:* If $A - (B \cup C) = (A - B) \cup (A - C)$, then $A \cap (B \cup C)$ $\subseteq A \cap B \cap C$.

        Proof: Suppose $A - (B \cup C) = (A - B) \cup (A - C)$ and suppose to the contrary that $A \cap (B \cup C) \not\subseteq A \cap B \cap C$. Then…

    (b) *Claim:* If $A - (B \cup C) = (A - B) \cup (A - C)$, then $A \cap (B \cup C)$ $\subseteq A \cap B \cap C$.

        Proof: Suppose $A - (B \cup C) = (A - B) \cup (A - C)$ and consider an anonymous x. We must show if $x \in A \cap (B \cup C)$, then $x \in A \cap B \cap C$, and shall do so via contraposition. So…

    (c) *Claim:* $A - (B \cup C) \subseteq (A - B) \cup (A - C)$.

        Proof: Consider an anonymous $x \in A - (B \cup C)$ and suppose to the contrary that $x \notin (A - B) \cup (A - C)$. Then…

    (d) *Claim:* $A - (B \cup C) \subseteq (A - B) \cup (A - C)$.

        Proof: We shall essentially use contraposition. Suppose…

    (e) *Claim:* If $A \cap (B \cup C) \subseteq A \cap B \cap C$, then $A - (B \cup C) = (A - B) \cup (A - C)$.

        Proof: Suppose $A \cap (B \cup C) \subseteq A \cap B \cap C$. We already know [from our proofs in (c) and (d) above] that $A - (B \cup C) \subseteq (A - B) \cup (A - C)$. Thus, it suffices to show that $(A - B) \cup (A - C) \subseteq A - (B \cup C)$. Well, suppose to the contrary that $(A - B) \cup (A - C) \not\subseteq A - (B \cup C)$. Then…

    (f) *Claim:* If $A \cap (B \cup C) \subseteq A \cap B \cap C$, then $A - (B \cup C) = (A - B) \cup (A - C)$.

        Proof: Suppose $A \cap (B \cup C) \subseteq A \cap B \cap C$. We have already proved that $A - (B \cup C) \subseteq (A - B) \cup (A - C)$ in (c) and (d) above, so we need only show that $(A - B) \cup (A - C) \subseteq A - (B \cup C)$. So suppose $x \notin A - (B \cup C)$. We must now show…

    (g) *Claim:* If $A \cap (B \cup C) \subseteq A \cap B \cap C$, then $(A - B) \cup (A - C) \subseteq A - (B \cup C)$.

        Proof: We proceed by contraposition. Suppose…

4S. A Mathematical Research Situation: Investigating the Relationship Between Two Sets

For any sets A, B, and C, we can form the sets $(A \cup B) - C$ and $A \cup (B - C)$. As budding mathematicians, we might then become interested in the relationship between these two sets. For instance, we might ask:

- Are the two sets $(A \cup B) - C$ and $A \cup (B - C)$ always equal?

- If they are not equal, is one always a subset of the other?

- If one of them is not a subset of the other, is there a condition on the sets A, B, and C that *would* guarantee a subset relationship?

By carrying out an investigation of the two sets $(A \cup B) - C$ and $A \cup (B - C)$, we might hope to formulate conjectures that propose answers to these questions, and perhaps others as well. We could then use what we have learned about proof writing to attempt to develop and write proofs for our conjectures.

1. In attempting to answer the questions posed above, we might use Venn diagrams to illustrate the sets $(A \cup B) - C$ and $A \cup (B - C)$. Draw two Venn diagrams, one depicting $(A \cup B) - C$ and the other depicting $A \cup (B - C)$. Then try using the pictures you have drawn to answer the following questions:

   (a) Must the sets $(A \cup B) - C$ and $A \cup (B - C)$ be equal?

   (b) Does it appear that one of the sets $(A \cup B) - C$ or $A \cup (B - C)$ is always a subset of the other?

   (c) Does it appear that one of the sets $(A \cup B) - C$ or $A \cup (B - C)$ is not always a subset of the other? If so, can you find a condition on the sets A, B, and C under which you believe the sets $(A \cup B) - C$ and $A \cup (B - C)$ would be equal?

   Use your answers to these questions to formulate several conjectures concerning the sets $(A \cup B) - C$ and $A \cup (B - C)$ that we could then attempt to prove.

2. For each of the conjectures you formulated in (1), either prove it or give evidence showing why it is not true.

4T. Another Research Situation

Given any sets A, B, and C we may form the sets $A - (B \cup C)$ and $(A - B) \cup (A - C)$. In this investigation, you will explore the relationship between these two sets. In particular, we might ask whether or not these sets must be equal, or, if this does not appear to be the case, whether one of them must be a subset of the other. Conduct the investigation by working through the following.

1. Draw two Venn diagrams, one depicting $A - (B \cup C)$ and one depicting $(A - B) \cup (A - C)$.

   (a) Based on the diagrams, do you believe the sets $A - (B \cup C)$ and $(A - B) \cup (A - C)$ must be equal?

   (b) Based on the diagrams, do you believe that one of the sets $A - (B \cup C)$ and $(A - B) \cup (A - C)$ must be a subset of the other?

   (c) Prove that one of the sets $A - (B \cup C)$ and $(A - B) \cup (A - C)$ must be a subset of the other.

2. Do your Venn diagrams suggest that the sets $A - (B \cup C)$ and $(A - B) \cup (A - C)$ will be equal if $A \cap (B \cup C) \subseteq A \cap B \cap C$? If so, prove that this is the case. If not, provide evidence to support your conclusion.

3. Does it appear from your Venn diagrams that we must have $A \cap (B \cup C) \subseteq A \cap B \cap C$ if $A - (B \cup C) = (A - B) \cup (A - C)$? Formulate a conjecture and then try to write a proof for it.

4. Do your Venn diagrams suggest that $A - (B \cup C) = (A - B) \cup (A - C)$ if A and $B \cup C$ are disjoint? Formulate a conjecture and then try to write a proof for it.

5. According to your Venn diagrams, does it appear that sets A and $B \cup C$ must be disjoint if $A - (B \cup C) = (A - B) \cup (A - C)$? Again, formulate a conjecture and then try to write a proof for it.

Chapter 5

Relations and Functions

Having developed some of the essentials about how proofs are planned and written, we are ready to begin exploring, from a more sophisticated perspective, certain fundamental mathematical ideas with which you are already familiar. In this chapter, we will broaden our understanding of the core concepts of *relation* and *function* to suit a more general mathematical outlook.

5.1 Relations

Questions to guide your reading of this section:

1. How is the notion of *ordered pair* defined? When are two ordered pairs considered to be equal?

2. What is meant by the *Cartesian product* $X \times Y$ of sets X and Y?

3. What is meant by a *relation*? a *relation on $X \times Y$*? a *relation on X*?

4. What is meant by the *domain* of a relation? the *range*?

5. What is an *interval* of real numbers? How is *interval notation* used to describe an interval?

Given a and b, the **ordered pair** (a, b) is formally defined as the set $\{\{a\}, \{a, b\}\}$ whose members are the sets $\{a\}$ and $\{a, b\}$. For the ordered pair (a, b), a is referred to as the **first** or **left coordinate**, and b is referred to as the **second** or **right coordinate**.

Example 5.1 The ordered pair $(2, 5)$ is, by definition, the set $\{\{2\}, \{2, 5\}\}$. The first coordinate of $(2, 5)$ is 2 and the second coordinate is 5. In contrast, the ordered pair $(5, 2)$ is the set $\{\{5\}, \{5, 2\}\}$ and has first coordinate 5 and second coordinate 2. Observe that $(2, 5) \neq (5, 2)$ because, as sets, the ordered pair $(2, 5)$ includes a member, namely the set $\{2\}$, that is not a member of the ordered pair (5, 2). Thus, the order in which coordinates are listed in an ordered pair really does make a difference. We also see how the ordered pair (2, 5) is different from the set {2, 5}, because the order in which elements are listed in a set is immaterial. For instance, whereas (2, 5) ≠ (5, 2), it is true that {2, 5} = {5, 2}.

Practice Problem 1   What ordered pair is represented by the set {{0, −3}, {−3}}?

Ordered pairs can be used to organize information. For instance, to convey the fact that a triangle has three sides and a rectangle has four sides, we might form the ordered pairs (*triangle*, 3) and (*rectangle*, 4). We also point out that the first coordinates of these ordered pairs are *not* numbers (there is no requirement that the coordinates be numbers).

The following theorem confirms our expectation that ordered pairs are the same exactly when their corresponding coordinates are identical. The proof relies heavily on the notion that sets are equal only when they have exactly the same members, but you will have to keep in mind that, formally, the members of an ordered pair are themselves sets.

Theorem 5.2 (*Equality of Ordered Pairs*)   Two ordered pairs (x_1, y_1) and (x_2, y_2) are equal, that is, $(x_1, y_1) = (x_2, y_2)$, if and only if both $x_1 = x_2$ and $y_1 = y_2$.

Proof: (⇒) We will use contraposition. Suppose $x_1 \neq x_2$ or $y_1 \neq y_2$.

Case 1: $x_1 \neq x_2$.
Then $\{x_1\} \neq \{x_2\}$ and, because $x_2 \notin \{x_1\}$, $\{x_1\} \neq \{x_2, y_2\}$. Hence, $\{x_1\} \notin \{\{x_2\}, \{x_2, y_2\}\}$. Of course, $\{x_1\} \in \{\{x_1\}, \{x_1, y_1\}\}$. Therefore, it follows that $\{\{x_1\}, \{x_1, y_1\}\} \neq \{\{x_2\}, \{x_2, y_2\}\}$, which means $(x_1, y_1) \neq (x_2, y_2)$.

Case 2: $y_1 \neq y_2$.
We may assume that $x_1 = x_2$ (otherwise, *Case 1* applies). Let $x = x_1 = x_2$. Because $y_1 \neq y_2$, at least one of y_1 or y_2 is different from x. Without loss of generality, assume $y_1 \neq x$. Then, because $y_1 \notin \{x\}$, it follows that $\{x, y_1\} \neq \{x\}$. Also, because $y_1 \notin \{x, y_2\}$, it follows that $\{x, y_1\} \neq \{x, y_2\}$. Therefore, $\{x, y_1\} \notin \{\{x\}, \{x, y_2\}\}$. But, because $\{x, y_1\} \in \{\{x\}, \{x, y_1\}\}$, we may conclude that $\{\{x\}, \{x, y_1\}\} \neq \{\{x\}, \{x, y_2\}\}$, which means $(x_1, y_1) \neq (x_2, y_2)$.

(⇐) Suppose $x_1 = x_2$ and $y_1 = y_2$. It follows that

$$(x_1, y_1) = \{\{x_1\}, \{x_1, y_1\}\} = \{\{x_2\}, \{x_2, y_2\}\} = (x_2, y_2). \quad \square$$

Note the use of the phrase *without loss of generality* within the proof of Case 2 of our argument for ($\Rightarrow$). This phrase indicates that although there are really two possibilities to consider, $y_1 \neq x$ and $y_2 \neq x$, we understand that the argument for the second possibility $y_2 \neq x$ can be done by simply interchanging y_1 and y_2 in the argument we produced for the first possibility $y_1 \neq x$. Generally, when two or more possibilities must be considered at a certain point in a proof, but the argument supporting each possibility can be obtained by just interchanging names of objects, mathematicians typically incorporate the phrase *without loss of generality* and present the argument for only one of the possibilities.

Having established Theorem 5.2, we will rarely need to work with the formal definition of ordered pair. For example, although technically the ordered pair $(2, 5)$ is defined to be the set $\{\{2\}, \{2, 5\}\}$, it will usually be sufficient for us simply to think of $(2, 5)$ as the ordered pair whose first coordinate is 2 and whose second coordinate is 5.

Cartesian Products

The *xy*-coordinate system we use to plot ordered pairs of real numbers and graphs of equations is also called the *Cartesian coordinate system* after Rene Descartes, the mathematician and philosopher who invented it. More generally, the **Cartesian product** $X \times Y$ of sets X and Y is the set of all ordered pairs whose first coordinates are members of X and whose second coordinates are members of Y; that is, $X \times Y = \{(x, y) \mid x \in X, y \in Y\}$. Thus, the Cartesian product $\mathbf{R} \times \mathbf{R}$ is the set of all ordered pairs of real numbers; geometrically, we view this set as the familiar *xy*-plane.

Example 5.3 $\mathbf{R}^+ \times \mathbf{R}^-$ is the set of all ordered pairs whose first coordinates are positive real numbers and whose second coordinates are negative real numbers. Thus, $(2, -0.5) \in \mathbf{R}^+ \times \mathbf{R}^-$, but $(-2, 0.5) \notin \mathbf{R}^+ \times \mathbf{R}^-$ and $(2, 0) \notin \mathbf{R}^+ \times \mathbf{R}^-$.

Example 5.4 If $X = \{0, 1, 2\}$ and $Y = \{1, 0.5\}$, then

$X \times Y = \{(0, 1), (0, 0.5), (1, 1), (1, 0.5), (2, 1), (2, 0.5)\}$,

$Y \times X = \{(1, 0), (0.5, 0), (1, 1), (0.5, 1), (1, 2), (0.5, 2)\}$,

$X \times X = \{(0, 0), (0, 1), (0, 2), (1, 0), (1, 1), (1, 2), (2, 0), (2, 1), (2, 2)\}$, and

$Y \times Y = \{(1, 1), (1, 0.5), (0.5, 1), (0.5, 0.5)\}$.

Example 5.5 Let S be the set of students majoring in mathematics at a certain college, and let P be the set of mathematics professors at this college. Then, $S \times P$ is the set of all ordered pairs for which the first coordinate is a student majoring in mathematics at this college, and the second coordinate is a mathematics professor at this college.

> **Practice Problem 2**   If $A = \{\pi, -\pi, 0\}$ and $B = \{x \in \mathbf{N} \mid x < 3\}$, describe $A \times B$ and $B \times B$ by listing.

Relations

In mathematics, any set of ordered pairs is called a **relation**. This is because the *relationship* between two variables, say x and y, can be expressed by forming the set of all ordered pairs in which a value of the variable x occupies the first coordinate position and a corresponding value of the variable y occupies the second coordinate position.

Example 5.6   The following table gives the record high temperatures in degrees Fahrenheit recorded in the six New England states according to data collected by the National Climatic Data Center and reported by the U.S. Department of Commerce.

state	ME	NH	VT	MA	CT	RI
record high temperature	105	106	105	107	106	104

The relationship between the variable *state* and the variable *record high temperature* is expressed by the table, but it can easily be converted to a set of ordered pairs in which each New England state is paired with its record high temperature:

$$\{(ME, 105), (NH, 106), (VT, 105), (MA, 107), (CT, 106), (RI, 104)\}.$$

Example 5.7   The set of all ordered pairs of the form (P, V), where P is a former president of the United States and V is a vice president who served with that president, is a relation. The fact that Al Gore served as vice president under Bill Clinton is expressed via the membership of the ordered pair (*Bill Clinton, Al Gore*) in this relation.

Given sets X and Y, any subset of $X \times Y$ is called a **relation on** $X \times Y$. That is, a relation on $X \times Y$ is a set of ordered pairs whose first coordinates are members of X and whose second coordinates are members of Y. A relation on $X \times X$ is usually referred to as a **relation on** X; in other words, a relation on X is really just a set of ordered pairs for which all coordinates are members of X.

Example 5.8   If $X = \{0, 1, 2\}$ and $Y = \{1, 0.5\}$, then $\{(0, 0.5), (1, 1), (1, 0.5)\}$ is a relation on $X \times Y$, $\{(0.5, 0), (0.5, 2)\}$ is a relation on $Y \times X$, $\{(0, 2), (1, 0)\}$ is a relation on X, and $\{(0.5, 0.5), (0.5, 1), (1, 0.5)\}$ is a relation on Y.

Example 5.9 The set $\{(x, x^2) \mid x \in \mathbf{R}\}$ is a relation on $\mathbf{R}$ because all coordinates, both first and second, are real numbers. In a precalculus or calculus course, this relation might be described using the equation $y = x^2$.

Sometimes a letter or a symbol such as "~" (called a *tilde*) is used as the name of a relation. If ~ is a relation, it is common to write $x \sim y$ in place of $(x, y) \in$ ~. In part, this is because we can read $x \sim y$ as x *is related to* y, which is what we intend when (x, y) is a member of the relation ~.

Example 5.7 (continued) Let ~ be the relation consisting of those ordered pairs for which the first coordinate is a former president of the United States and the second coordinate is a vice president who served with that president. Because (*Bill Clinton, Al Gore*) $\in$ ~, we may write *Bill Clinton ~ Al Gore*.

Example 5.9 (continued) If $S = \{(x, x^2) \mid x \in \mathbf{R}\}$, we may write $3S9$ because $(3, 9) \in S$.

Practice Problem 3 Let ∇ be the relation consisting of all ordered pairs of nonzero integers for which either coordinate, when squared, produces the same number.

   (a) Is it true that $-11\nabla 11$?

   (b) Write an equation relating a, b, and c if it is known that $-a\nabla bc$.

   (c) What ordered pairs must be members of ∇ if it is known that $-a\nabla bc$?

We must be careful, however, not to assume that just because x is related to y by a certain relation that it must also be the case that y is related to x.

Example 5.7 (continued) Whereas *Bill Clinton* ~ *Al Gore*, it does not follow that *Al Gore* ~ *Bill Clinton*, as this would signify that Bill Clinton served as a vice president under Al Gore, which is not true. Thus, *Al Gore* $\not\sim$ *Bill Clinton*.

The set of all first coordinates of the ordered pairs in a relation is called the **domain** of the relation. The set of all second coordinates is called the **range**.

Example 5.6 (continued) The relation

$$\{(ME, 105), (NH, 106), (VT, 105), (MA, 107), (CT, 106), (RI, 104)\}$$

in which a New England state is paired with its record high temperature has domain

$$\{ME, NH, VT, MA, CT, RI\},$$

the set consisting of the six New England states, and range

$$\{105, 106, 107, 104\},$$

the set of record high temperatures achieved in these states. Note that we should not refer to, for instance, *NH* as "a domain" of this relation or 106 as "a range." The domain is the *set of all first coordinates*, not a particular first coordinate, and the range is the *set of all second coordinates*, not a particular second coordinate.

Example 5.9 (continued) Because any real number can be the first coordinate of an ordered pair in $S = \{(x, x^2) \mid x \in \mathbf{R}\}$, the domain of S is $\mathbf{R}$.

However, as the square of a real number will never be negative, the range of S includes no negative numbers. Now for each nonnegative real number y, there is at least one real number x for which $x^2 = y$ (for instance, take $x = \sqrt{y}$). Thus, we may conclude that the range of S is $\mathbf{R}^+ \cup \{0\}$.

Example 5.7 (continued) The relation consisting of the set of all ordered pairs (P, V), where P is a former president of the United States and V is a vice president who served with that president, has the set of all former U.S. presidents as its domain and the set of all former U.S. vice presidents as its range.

> **Practice Problem 4** Find the domain and range of the relation $\{(x, y) \in \mathbf{N} \times \mathbf{Z} \mid y = -2x\}$.

The next few examples consider some of the most important relations in all of mathematics.

Example 5.10 For real numbers a and b, we write $a \leq b$ to indicate that a is *less than or equal to* b. Formally, we may view $\leq$ as the relation on $\mathbf{R}$ consisting of those ordered pairs of real numbers for which the first coordinate is less than or equal to the second coordinate. Thus, for any real numbers a and b, $(a, b) \in \leq$ iff $a \leq b$.

For instance, $(-2, \pi) \in \leq$ because the first coordinate -2 is less than or equal to the second coordinate π. Also, $(\pi, \pi) \in \leq$ because $\pi \leq \pi$ is a true statement. But $(4, \pi) \notin \leq$ because $4 \leq \pi$ is not a true statement.

This example also illustrates that the idea of placing the symbol representing a relation between the objects being related is not completely unfamiliar. For example, we are much more used to writing $-2 \leq \pi$ than the equivalent $(-2, \pi) \in \leq$.

Example 5.11 Similarly, $<$ is the relation on $\mathbf{R}$ consisting of all ordered pairs of real numbers for which the first coordinate is *less than* the second coordinate. Thus, for instance, $(-2, \pi) \in <$ because $-2 < \pi$, but $(\pi, \pi) \notin <$ as $\pi \not< \pi$ and $(4, \pi) \notin <$ as $4 \not< \pi$.

Example 5.12 Another important relation on $\mathbf{R}$ is *equality*, denoted $=$. For any real numbers a and b, $(a, b) \in =$ iff $a = b$. Hence, as $\pi = \pi$ we have $(\pi, \pi) \in =$, but as $-2 \neq \pi$ we have $(-2, \pi) \notin =$.

Example 5.13 Given a set A, $\subseteq$ is the relation on $\mathcal{P}(A)$, the power set of A (the set of all subsets of A), where, for any subsets B and C of A, $(B, C) \in \subseteq$ iff $B \subseteq C$. For instance, $(\{1,2\}, \mathbf{N}) \in \subseteq$ because $\{1,2\} \subseteq \mathbf{N}$, but $(\{1,2\}, \{1,3,5\}) \notin \subseteq$ because $\{1, 2\} \not\subseteq \{1, 3, 5\}$.

Intervals and Interval Notation

Because many of the relations we work with in mathematics involve real numbers, this is a good point at which to introduce the notion of an *interval* of real numbers and to review with you notations for intervals that you likely learned in your calculus courses.

A subset I of $\mathbf{R}$ is an **interval** provided that whenever $a, b \in I$ and $a < x < b$, it follows that $x \in I$. In other words, an *interval* of real numbers is a subset of $\mathbf{R}$ having the property that it includes every real number between any two of its members. Intuitively, intervals are "connected" in the sense that they have no "gaps" in them.

Example 5.14 The set $\mathbf{R}^+$ is an interval because between any two positive numbers there are only positive numbers.

But the set $A = \{1, 2, 3\}$ is *not* an interval because it does not include every real number between any two of its members. For instance, the real number 1.5 lies between the numbers 1 and 2, both of which are in A, yet 1.5 is not in A.

Given an interval:

- If there is a real number that is less than or equal to every member of the interval, the largest such real number is called the **left endpoint** of the interval.

- If there is a real number that is greater than or equal to every member of the interval, the smallest such real number is called the **right endpoint** of the interval.

Example 5.15 The interval $A = \{x \in \mathbf{R} \mid 1 \leq x < \pi\}$ has left endpoint 1 because 1 is the largest real number that is less than or equal to every member of A. The interval A has right endpoint π because π is the smallest real number that is greater than or equal to every member of A. Note that an endpoint of an interval may or may not be included as a member of the interval.

Example 5.16 The interval $\{x \in \mathbf{R} \mid x > 4\}$ has left endpoint 4, but no right endpoint. The interval $\{x \in \mathbf{R} \mid x \leq 4\}$ has right endpoint 4, but no left endpoint. And the interval $\mathbf{R}$ has neither a left endpoint nor a right endpoint.

Any interval of real numbers can be described using so-called **interval notation**. We simply write down the interval's left endpoint, or the symbol "$-\infty$" if the interval has no left endpoint, followed by a comma, then write down the interval's right endpoint, or the symbol "∞" if there is no right endpoint, and then use "square" or "round" brackets to indicate, respectively, whether or not an endpoint is included.

Example 5.16 (continued) The interval $\{x \in \mathbf{R} \mid x > 4\}$ can be expressed by writing $(4, \infty)$, and the interval $\{x \in \mathbf{R} \mid x \leq 4\}$ can be expressed by writing $(-\infty, 4]$. The set $\mathbf{R}$ of all real numbers may be expressed as $(-\infty, \infty)$.

Example 5.15 (continued) The interval $\{x \in \mathbf{R} \mid 1 \leq x < \pi\}$ can be expressed as $[1, \pi)$.

When using interval notation, a square bracket is never used next to either of the symbols "∞" or "$-\infty$" because neither of these symbols represents a real number. Rather, "∞" should be interpreted as *unbounded in the positive sense* and "$-\infty$" as *unbounded in the negative sense*.

Practice Problem 5 Express the given interval using interval notation and identify its endpoints (if any).

   (a) $\{t \in \mathbf{R} \mid 0 \geq t > -2\}$.   **(b)** $\{x \in \mathbf{R}^{-} \mid x < -3.7\}$.

The string of symbols (a, b) in isolation is ambiguous because it can refer to either the interval consisting of all real numbers between a and b (including neither a nor b) or the ordered pair having first coordinate a and second coordinate b. Context will dictate which interpretation is correct in a given setting. For example, if the range of a certain relation is $(3, 6)$, we would view $(3, 6)$ as an interval, whereas if $(3, 6)$ is a member of a certain relation, we would view $(3, 6)$ as an ordered pair.

Answers to Practice Problems

1. $(-3, 0)$.

2. $A \times B = \{(\pi, 1), (\pi, 2), (-\pi, 1), (-\pi, 2), (0, 1), (0, 2)\}$ and
    $B \times B = \{(1, 1), (1, 2), (2, 1), (2, 2)\}$.

3. **(a)** yes.   **(b)** $(-a)^2 = (bc)^2$.   **(c)** $(-a,\ bc), (-a,\ -bc), (a,\ bc)$, and
    $(a,\ -bc)$.

4. The domain is **N**, and the range is the set of all negative even
    integers.

5. **(a)** $(-2,\ 0]$, which has left endpoint -2 and right endpoint 0.
    (b) $(-\infty,\ -3.7)$, which has no left endpoint and right endpoint -3.7.

5.2   Equivalence Relations and Partitions

Questions to guide your reading of this section:

1. What does it mean for a relation on a set to be *reflexive? symmetric? transitive?*

2. Under what circumstances is a relation an *equivalence relation?* What is meant by the *equivalence classes* created by an equivalence relation?

3. What is a *partition* of a set? What is meant by the *cells* created from a partition?

4. How are equivalence relations and partitions related to each other?

There are many properties a relation may possess.

Definition 5.17   A relation $\sim$ on a set A is

- **reflexive** if for every $a \in A$, $a \sim a$;
- **symmetric** if whenever $a \sim b$, it follows that $b \sim a$;
- **transitive** if whenever both $a \sim b$ and $b \sim c$, it follows that $a \sim c$.

The choice of terminology for the properties listed in Definition 5.17 makes some sense.

- When you look into a mirror, you *see yourself* via your *reflection*. *Reflexivity* of a relation requires that each member of the set on which the relation is defined be *related to itself*.

- We think of the letter

$$A$$

as being *symmetric* with respect to an imaginary vertical line we can draw straight down through the point at its top. This is because the portion of the letter lying to the left of this line can be *flipped* with the portion of the letter to the right of the line to re-create the original figure. *Symmetry* of a relation tells us that whenever a is related to b, we may *flip* a and b and conclude that b is related to a.

- The prefix *trans-* means *across*. We can think of *transitivity* of a relation as telling us that when there is a "bridge" b linking a to c, meaning a is related to b and b is related to c, we can move *across* the bridge to conclude that a is related to c.

Example 5.18   Let $A = \{1, 2, 3, 4\}$ and consider the relation

$$R_1 = \{(1, 1), (2, 2), (3, 3), (2, 3), (3, 2), (2, 4)\}$$

on A. Recalling that we may write xR_1y iff $(x, y) \in R_1$, we see that R_1 is *not* reflexive because, even though $1R_11$, $2R_12$, and $3R_13$, we do *not* have $4R_14$. Also, as $2R_14$, symmetry would also require that $4R_12$, but this is *not* the case as the ordered pair $(4, 2)$ is *not* in R_1. Thus, R_1 is *not* symmetric. (Even though $2R_13$ and $3R_12$, R_1 is not symmetric because the ability to flip coordinates and still have a member of the relation does not always hold.) Transitivity of R_1 would require, as $3R_12$ and $2R_14$, that $3R_14$. However, it is *not* true that $3R_14$ because the ordered pair $(3, 4)$ is *not* in R_1. Thus, R_1 is *not* transitive.

It is usually relatively straightforward to determine whether a relation is reflexive or symmetric. Determining whether a relation $\sim$ possesses the transitive property can, however, sometimes be challenging because we must consider *all* possible instances for which $a \sim b$ and $b \sim c$; that is, the right coordinate of an ordered pair in $\sim$ is the same as the left coordinate of an ordered pair in $\sim$. If we are able to reach the conclusion that $a \sim c$ in *each* such instance, we conclude that $\sim$ is transitive. But if we can find even one instance where we cannot reach this conclusion, $\sim$ is *not* transitive.

We point out that when looking for situations where $a \sim b$ and $b \sim c$, there is no need to consider the special cases $b = a$ and $b = c$, as we will find no violations of the transitive property in such situations. For example, if $b = a$, the requirement

$$(a \sim b \text{ and } b \sim c) \quad \Rightarrow \quad a \sim c$$

becomes

$$(a \sim a \text{ and } a \sim c) \quad \Rightarrow \quad a \sim c,$$

which is clearly true because the conclusion is actually part of the hypothesis. The situation is similar for the special case $b = c$.

Example 5.19 Define a relation R_2 on $A = \{1, 2, 3, 4\}$ so that

$$1R_21, \ 2R_22, \ 3R_23, \ 4R_24, \ 2R_23, \text{ and } 3R_22.$$

To determine whether R_2 is transitive, we consider all possible instances where the right coordinate of an ordered pair in R_2 is the same as the left coordinate of an ordered pair in R_2, realizing, based on the discussion immediately prior to this example, that we do not need to consider ordered pairs having identical coordinates. In each of these situations, we want to know if the appropriate conclusion can be reached. Knowing that $2R_23$ and $3R_22$, is it also true that $2R_22$? Knowing that $3R_22$ and $2R_23$, is it also true that $3R_23$? Because the answer to each of these questions is "yes," we may conclude that R_2 is transitive.

The relation R_2 is reflexive because *every* member of A is related to itself and is symmetric as interchanging the coordinates of any ordered pair in R_2 yields an ordered pair that is in R_2.

Practice Problem 1 Consider the relation

$$R_3 = \{(1, 1), (2, 2), (3, 3), (2, 3), (3, 2), (2, 4), (3, 4)\}$$

on the set $A = \{1, 2, 3, 4\}$. Show that R_3 is neither reflexive nor symmetric, but is transitive.

Although every relation is a set of ordered pairs, we must remember it is often the case that a relation is not directly specified in this way.

Example 5.20 Let P be a set of people and define the relation $\sim$ on P so that $a \sim b$ iff b has the same father as a. Observe that:

- As any person has the same father as herself/himself, we may conclude that $a \sim a$ for every $a \in P$, thus making $\sim$ reflexive.
- Whenever $a \sim b$, meaning b has the same father as a, we may conclude that a has the same father as b; that is, $b \sim a$, so $\sim$ is symmetric.

- Whenever $a \sim b$ and $b \sim c$, meaning b has the same father as a and c has the same father as b, we may conclude that c has the same father as a (a and c must have the same father, whoever is father to b); that is, $a \sim c$, so $\sim$ is transitive.

Practice Problem 2   Let P be a set of people and define the relation $\sim$ on P so that $a \sim b$ iff b has a grandfather in common with a. Verify that $\sim$ is reflexive and symmetric, but need not be transitive.

Equivalence Relations

A given relation attempts to capture a connection between the first coordinate and the second coordinate of each ordered pair that is a member of the relation. One type of connection a relation may express is a notion of *equivalence*; that is, a notion of one object *being the same as* another object in terms of one or more specified properties or attributes the objects may possess. Here are some examples of "equivalence" that come up in the study of mathematics:

- Numerical or algebraic expressions that have the *same value* are *equal*. For example, we may write $x^2 - 1 = (x - 1)(x + 1)$ because the expressions $x^2 - 1$ and $(x - 1)(x + 1)$ have the same value for each real number x.

- Geometric figures that have the *same shape and the same size* are equivalent in the sense of being *congruent* to one another. For example, the two rectangles shown below are congruent even though they are oriented differently; from the point of view of size and shape, they are equivalent.

- Statement forms that have the *same truth table* are *logically equivalent*. For example, one can easily check that the statement forms $P \vee Q$ and $\sim P \Rightarrow Q$ are logically equivalent.

We are interested in those properties that are shared by all relations that express a notion of equivalence. Because we would expect a to be "equivalent" to itself, b to be "equivalent" to a if a is "equivalent" to b, and a to be "equivalent" to c if a is "equivalent" to b and b is "equivalent" to c, we realize that reflexivity, symmetry, and transitivity must be among these properties. It turns out, in fact, that these three properties completely characterize the abstract notion of "equivalence."

Definition 5.21   A relation that is reflexive, symmetric, and transitive is called an **equivalence relation**. If $\sim$ is an equivalence relation on the set A, then:

- When $x \sim y$, we say that x and y are **equivalent**.
- The set $\{x \in A \mid x \sim a\}$ of all members of A that are equivalent to a specified $a \in A$ is the **equivalence class** of a and is denoted $[a]$.

Thus, to prove a relation is an equivalence relation, we must show it is reflexive, symmetric, and transitive.

Example 5.22   Two geometric figures A and B are **congruent**, denoted $A \cong B$, provided they have the same shape and the same size. Note that:

- Any geometric figure has the same shape and size as itself, thus making congruence reflexive.
- Whenever a geometric figure A has the same shape and size as a geometric figure B, it follows that B has the same shape and size as A, thus making congruence symmetric.
- Whenever a geometric figure A has the same shape and size as a geometric figure B, and B also has the same shape and size as a geometric figure C, it follows that A has the same shape and size as C, thus making congruence transitive.

Hence, if G is a collection of geometric figures, $\cong$ is an equivalence relation on G. If $C \in G$, the equivalence class of C under $\cong$ is $[C] = \{x \in G \mid x \cong C\}$, the set of all figures in G that are congruent to the particular geometric figure C. So if C is a circle of radius 5 inches, then $[C]$ would be the set of all circles of radius 5 inches that are in G.

Example 5.23   Let S be the set of students at a certain college who have declared a major and assume none of these students is pursuing more than one major. Define a relation $\sim$ on S so that $a \sim b$ iff a is pursuing the same major as b. Observe that:

212 Chapter 5 ■ Relations and Functions

- Because any student pursues the same major as himself/herself, $a \sim a$ for every $a \in S$, meaning $\sim$ is reflexive.

- Whenever a pursues the same major as b, it follows that b must be pursuing the same major as a, meaning $\sim$ is symmetric.

- Whenever a pursues the same major as b, and b pursues the same major as c, it follows that a must be pursuing the same major as c, meaning $\sim$ is transitive.

Therefore, $\sim$ is an equivalence relation. If *Nancy* is a declared biology major at the college of interest, then $[Nancy] = \{x \in S \mid x \sim Nancy\}$ is the set of all declared majors at the college pursing the same major as Nancy; in other words, $[Nancy]$ is the set of all declared biology majors at the college. Thus, if *Joe* is also a declared biology major at the college, then $[Joe] = [Nancy]$.

Example 5.12 (continued) Because it is easily seen that equality of real numbers is reflexive, symmetric, and transitive, it follows that = is an equivalence relation on **R**. Because a real number is equal to itself but to no other real number, for each $a \in \mathbf{R}$, the equivalence class of a under equality is $[a] = \{a\}$.

To demonstrate that a relation is *not* an equivalence relation, we need only show that a single one of the properties reflexivity, symmetry, or transitivity fails to hold.

Example 5.10 (continued) In Problem 5E.2 you will show the relation $\leq$ on **R** is not symmetric. Hence, $\leq$ is not an equivalence relation on **R**.

Example 5.19 (continued) Recall that the relation R_2 was defined on the set $A = \{1, 2, 3, 4\}$ so that $R_2 = \{(1, 1), (2, 2), (3, 3), (4, 4), (2, 3), (3, 2)\}$. We previously showed R_2 is reflexive, symmetric, and transitive, so R_2 is an equivalence relation. Note that R_2 is identifying 2 and 3 as being equivalent to each other, but otherwise a member of A is only equivalent to itself. The equivalence classes produced by R_2 are $[1] = \{1\}$, $[2] = \{2, 3\} = [3]$, and $[4] = \{4\}$.

Practice Problem 3 Tell why the relation R_3 described in Practice Problem 1 is not an equivalence relation.

In terms of our *Know/Show* approach to developing proofs, observe that to establish that a relation $\sim$ on a set A is

- *reflexive*: we *know* $a \in A$ and we must *show* $a \sim a$;

- *symmetric*: we *know* $a \sim b$ and we must *show* $b \sim a$;

- *transitive*: we *know* $a \sim b$ and $b \sim c$ and we must *show* $a \sim c$.

In carrying out a proof that an unfamiliar relation is actually an equivalence relation, the relation's definition will be of prime importance in establishing these properties.

Example 5.24 Define a relation $\equiv$ on the set $\mathbf{Z}$ so that for any integers a and b, $a \equiv b$ iff there is an integer k such that $a - b = 4k$. We will prove that $\equiv$ is an equivalence relation. The definition of $\equiv$ will be fundamental to our task. Specifically, to show that $\equiv$ is

- *reflexive*: we *know* a is an integer and we must *show* $a \equiv a$, which means we must *show* there is an integer k such that $a - a = 4k$ (observe that the integer 0 will play the role of k here);

- *symmetric*: we *know* $a \equiv b$, meaning there is an integer k such that $a - b = 4k$, and we must *show* $b \equiv a$, meaning we must *show* there is an integer j such that $b - a = 4j$ (because $b - a$ is the opposite of $a - b$, it appears that the role of j will be taken by the integer $-k$);

- *transitive*: we *know* $a \equiv b$ and $b \equiv c$, meaning there exist integers j and k such that $a - b = 4j$ and $b - c = 4k$, and we must *show* $a \equiv c$, meaning we must *show* there is an integer l such that $a - c = 4l$ [observing that $a - c = (a - b) + (b - c)$ suggests the role of l will be played by $j + k$].

The following proof is constructed based on the plan we have just outlined.

Claim: The relation $\equiv$ is an equivalence relation on $\mathbf{Z}$.
    Proof: We need only show that $\equiv$ is reflexive, symmetric, and transitive.

Subclaim 1: $\equiv$ is reflexive.
Consider $a \in \mathbf{Z}$. Because $a - a = 0 = 4 \cdot 0$ and 0 is an integer, by definition of $\equiv$, we may conclude that $a \equiv a$.

Subclaim 2: $\equiv$ is symmetric.
Suppose $a \equiv b$. Then, by the definition of $\equiv$, there is an integer k such that $a - b = 4k$. It follows that $b - a = -(a - b) = -4k = 4(-k)$, where $-k \in \mathbf{Z}$ as the opposite of an integer is also an integer. Thus, by the definition of $\equiv$, we may conclude that $b \equiv a$.

Subclaim 3: $\equiv$ is transitive.
Suppose $a \equiv b$ and $b \equiv c$. Then, by the definition of $\equiv$, there exist integers j and k such that $a - b = 4j$ and $b - c = 4k$. It follows that

$$a - c = (a - b) + (b - c) = 4j + 4k = 4(j + k),$$

where $j + k \in \mathbf{Z}$ as the sum of integers is also an integer. Thus, by the definition of $\equiv$, we may conclude that $a \equiv c$. $\square$

Note that, as proving a relation is an equivalence relation involves proving three separate properties (reflexivity, symmetry, and transitivity), we have used *subclaims* within our proof in Example 5.24 to indicate where each property is being proved. The proof of each subclaim immediately follows its statement. Such a layout is fairly typical when a proof involves establishing multiple independent results.

Practice Problem 4   Define the relation ~ on the set $\mathbf{Z}$ so that $a \sim b$ iff $a + b$ is even. Prove that ~ is an equivalence relation. You may freely use the fact that the sum and difference of even integers is even.

Partitions

We introduced the notion of *equivalence relation* in attempting to capture the general idea of objects *being the same as one another* with respect to a specified attribute or property. It would appear that another way to encapsulate this idea would be simply to group objects into sets in such a way that *equivalent* objects wind up being in the same set.

Example 5.25   Suppose P is a set consisting of two different triangles T_1 and T_2, four different quadrilaterals Q_1, Q_2, Q_3, and Q_4, and one hexagon H; that is, $P = \{T_1, T_2, Q_1, Q_2, Q_3, Q_4, H\}$. To capture the notion that figures in P having the same number of sides will be considered equivalent, we could sort them into three subsets $\{T_1, T_2\}$, $\{Q_1, Q_2, Q_3, Q_4\}$, and $\{H\}$.

Note that the sorting process illustrated in Example 5.25 puts each member of the original set P into exactly one of the three subsets formed through the consideration of a figure's number of sides.

Definition 5.26   A collection of nonempty subsets of a nonempty set A having the property that each element of A is a member of exactly one of the subsets in the collection is called a **partition** of A. Each of the subsets in a partition of A is called a **cell** of the partition. The cell including an element $a \in A$ is denoted $[a]$.

Example 5.25 (continued)   Thus, $\{\{T_1, T_2\}, \{Q_1, Q_2, Q_3, Q_4\}, \{H\}\}$ is a partition of $P = \{T_1, T_2, Q_1, Q_2, Q_3, Q_4, H\}$. Note that the cell containing T_1 is $[T_1] = \{T_1, T_2\} = [T_2]$.

Practice Problem 5   Explain why $\{\mathbf{Q^+}, \mathbf{R^+} - \mathbf{Q^+}, \{0\}, \mathbf{Q^-}, \mathbf{R^-} - \mathbf{Q^-}\}$ is a partition of the set $\mathbf{R}$ of all real numbers.

Given a nonempty set A and an equivalence relation $\sim$ on A, we may note that:

- For each $a \in A$, the equivalence class $[a] = \{x \in A \mid x \sim a\}$ is nonempty, as it includes a (reflexivity of $\sim$ implies that a is equivalent to itself).

- For each $a \in A$, the equivalence class $[a]$ is a subset of A ($[a]$ consists of those members of A that are equivalent to a).

- Each member of A is included in exactly one equivalence class, as $a \in [x]$ and $a \in [y]$ would tell us that $a \sim x$ and $a \sim y$, from which it follows, using the symmetry and transitivity of $\sim$, that $x \sim y$, which itself implies that $[x] = [y]$.

Thus, the equivalence classes determined by an equivalence relation on a set always form a partition of the set.

Conversely, a partition of a set A naturally induces an equivalence relation on A in such a way that the resulting equivalence classes are precisely the subsets into which A had originally been partitioned. We simply declare that two members of A are equivalent iff they are members of the same cell of the partition.

Example 5.25 (continued)   Define a relation $\sim$ on $P = \{T_1, T_2, Q_1, Q_2, Q_3, Q_4, H\}$ so that $a \sim b$ iff a and b are in the same cell of the partition $\{\{T_1, T_2\}, \{Q_1, Q_2, Q_3, Q_4\}, \{H\}\}$ of P. Note that $\sim$ consists of the ordered pairs (T_1, T_1), (T_1, T_2), (T_2, T_1), (T_2, T_2), (Q_1, Q_1), (Q_1, Q_2), (Q_2, Q_1), (Q_2, Q_2), (Q_1, Q_3), (Q_3, Q_1), (Q_3, Q_3), (Q_1, Q_4), (Q_4, Q_1), (Q_4, Q_4), (Q_2, Q_3), (Q_3, Q_2), (Q_2, Q_4), (Q_4, Q_2), (Q_3, Q_4), (Q_4, Q_3), and (H, H). It is easily verified that $\sim$ is an equivalence relation (do so) with the equivalence classes being precisely the three cells $\{T_1, T_2\}$, $\{Q_1, Q_2, Q_3, Q_4\}$, and $\{H\}$ of the original partition.

Hence, we see that partitions and equivalence relations are simply two alternative methods for imposing a notion of "equivalence" on a set of objects. This observation further validates our choice of including all three of the reflexive, symmetric, and transitive properties when formulating the definition of *equivalence relation*.

Answers to Practice Problems

1. R_3 is neither reflexive nor symmetric, for the same reasons, respectively, that the relation R_1 from Example 5.18 is neither reflexive nor symmetric. To verify that R_3 is transitive, we consider all possible situations where we have two ordered pairs in R_3 for which

(continues)

Answers to Practice Problems (continued)

the right coordinate of one is the same as the left coordinate of the other, excluding ordered pairs having identical coordinates. We now want to know if the appropriate conclusion can be reached in each of these situations: Knowing that $2R_3 3$ and $3R_3 2$, is it also true that $2R_3 2$? Knowing that $2R_3 3$ and $3R_3 4$, is it also true that $2R_3 4$? Knowing that $3R_3 2$ and $2R_3 3$, is it also true that $3R_3 3$? Knowing that $3R_3 2$ and $2R_3 4$, is it also true that $3R_3 4$? Because the answer to each of these questions is "yes," it follows that R_3 is transitive.

2. The relation ~ is reflexive because a person must have a grandfather in common with himself/herself, meaning $a \sim a$ for each $a \in P$. The relation ~ is symmetric because if $a \sim b$, meaning b has a grandfather in common with a, it must follow that a has that same grandfather in common with b, meaning $b \sim a$. The relation ~ need not be transitive because it is possible for $a \sim b$ and $b \sim c$, meaning b has a grandfather in common with each of a and c, without a and c having to possess a common grandfather (because each person has two grandfathers, the common grandfather of a and b can be a person different from the common grandfather of b and c).

3. R_3 is neither reflexive nor symmetric.

4. *Claim:* The relation ~, defined on **Z** so that $a \sim b$ iff $a + b$ is even, is an equivalence relation.

   Proof: We must show ~ is reflexive, symmetric, and transitive.

   Subclaim 1: ~ is reflexive.
   Let $a \in$ **Z**. Because $a + a = 2a$ and a is an integer, we may conclude, by definition, that $a + a$ is even. Hence, $a \sim a$.

   Subclaim 2: ~ is symmetric.
   Assume $a \sim b$. Then $a + b$ is even. But because $b + a = a + b$, we may conclude that $b + a$ is even. Hence, $b \sim a$.

   Subclaim 3: ~ is transitive.
   Assume $a \sim b$ and $b \sim c$, so that $a + b$ and $b + c$ are both even. It follows that $a + c = (a + b) + (b + c) - 2b$ is even, because $2b$ is even as b is an integer and any sum or difference of even integers is even.   □

5. Note that **Q**⁺ is the set of all positive rational numbers, **R**⁺ – **Q**⁺ is the set of all positive irrational numbers, {0} has 0 as its only member, **Q**⁻ is the set of all negative rational numbers, and **R**⁻ – **Q**⁻ is the set of all negative irrational numbers. Each of these five sets is a nonempty subset of **R**. No two of them have members in common as no real number can be both rational and irrational, and the Law of Trichotomy guarantees that each real number has exactly one status: positive, negative, or 0.

5.3   Functions

Questions to guide your reading of this section:

1. What property must a relation have in order to be a *function*? How is this property described using the notions of *input* and *output*?

2. What does the notation $f: X \to Y$ convey?

3. What is meant by a function's *domain*? *range*? What is the distinction between the *codomain* and the *range* of a function?

4. How is *function composition* defined? Is function composition commutative? Is it associative?

5. What is required in order for two functions to be *equal*? How do we usually go about proving two functions are equal?

Functions are the most important type of relation in mathematics.

Definition 5.27   A relation is a **function** if each member of the domain is paired with exactly one member of the range.

Example 5.28   The relation $\{(0, 2), (1, 0), (2, 0)\}$ is a function, as the first coordinate 0 is paired only with a single second coordinate, namely 2; the first coordinate 1 is paired only with a single second coordinate, namely 0; and the first coordinate 2 is paired only with a single second coordinate, namely 0. Observe, as in this example, that it is acceptable for a function to pair the same second coordinate with different first coordinates.

In contrast, the relation $\{(0, 2), (1, 1), (1, 2)\}$ is *not* a function because the first coordinate 1 is paired with two different members of the range, 1 and 2.

Example 5.9 (continued)   The relation $\{(x, x^2) \mid x \in \mathbf{R}\}$, which can be described by the equation $y = x^2$, is a function, as *squaring* any particular real number will produce just one number.

The members of the domain of a relation are sometimes referred to as **inputs** and the members of the range as **outputs**. Using this terminology, a function is simply a relation for which each input has exactly one output. Thus, functions have an apparent advantage over nonfunctional relations in that once we have found the output a function associates with a particular input, there is no possibility that the function would assign another, different, output to that input.

Example 5.29   The circumference C of a circle is a function of the circle's radius r. Specifically, $C = 2\pi r$, as a circle having a given positive real number r as radius (input) will have just one circumference; namely, the number $2\pi r$ (output).

Practice Problem 1   Let ∇ be the relation consisting of all ordered pairs of nonzero integers for which either coordinate, when squared, produces the same number. Provide evidence showing that ∇ is not a function.

In mathematics, we often think of a function as taking an input from a certain set and *sending* it or *mapping* it to an output in another (perhaps different) set. The following definition provides us with a formal method for viewing a function in this way.

Definition 5.30   A relation f on $X \times Y$ is a **function** or **mapping from** X **into** Y, denoted $f : X \to Y$, provided that f is a function and for each $x \in X$ there is some $y \in Y$ such that $(x, y) \in f$.

Example 5.31   Consider the function $\# : \mathcal{P}(\{1, 2, 3\}) \to \{0, 1, 2, 3\}$ that maps a given subset of $\{1, 2, 3\}$ to its size; that is, its number of members. As a set of ordered pairs, we may observe that

$$\# = \{(\varnothing, 0), (\{1\}, 1), (\{2\}, 1), (\{3\}, 1), (\{1, 2\}, 2), (\{1, 3\}, 2), (\{2, 3\}, 2), (\{1, 2, 3\}, 3)\}.$$

To emphasize the view of the function $\#$ as a mapping, we can draw an arrow, $\mapsto$, from each input to its corresponding output.

$\varnothing \mapsto 0$
$\{1\} \mapsto 1$
$\{2\} \mapsto 1$
$\{3\} \mapsto 1$
$\{1, 2\} \mapsto 2$
$\{1, 3\} \mapsto 2$
$\{2, 3\} \mapsto 2$
$\{1, 2, 3\} \mapsto 3.$

Note the distinction between the arrow $\mapsto$ that is used to indicate a function's assignment of a *particular* output to a *particular* input and the arrow $\to$ that is used to indicate that the function is *generally* mapping inputs in one set to outputs in another set.

Example 5.28 (continued)   The function $f = \{(0, 2), (1, 0), (2, 0)\}$ is a function from $\{0, 1, 2\}$ into $\{0, 2\}$, so we may write $f : \{0, 1, 2\} \to \{0, 2\}$. Note, however, that the expression $f : \{0, 1, 2\} \to \{0, 2\}$ by itself does not tell us which ordered pairs are members of f. It simply tells us that 0, 1, and 2 are the first coordinates and that each of these numbers is paired with exactly one second coordinate that itself must be a member of the set $\{0, 2\}$.

When $f : X \to Y$, the set X must be the domain of f, but there is no requirement that the set Y be the range of f, only that Y contain the range as a subset.

Example 5.28 (continued)   The function $f = \{(0, 2), (1, 0), (2, 0)\}$ can be viewed as a function from $\{0, 1, 2\}$ into $\mathbf{R}$, that is, we may write $f : \{0, 1, 2\} \to \mathbf{R}$, as there is no requirement that all of the members of the set we are mapping into be actual outputs of the function (it is only mandated that this set *include* the actual outputs).

Sometimes a generic function $f : X \to Y$ is represented visually by a diagram such as the following:

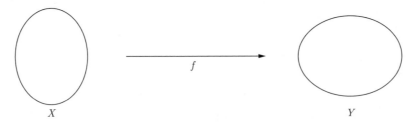

Such a diagram is intended to emphasize the mapping procedure determined by the given function. We can use a similar diagram to indicate that a function f maps a particular input x to a particular output y.

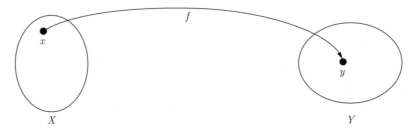

Practice Problem 2   Consider the function $h : \{1, 2, 3, 4\} \to \mathbf{Z}$ defined so that the output corresponding with each input is 2 more than the input.

(a) What output is assigned by h to the input 4?

(b) List all the outputs of h.

(c) Is it true that, for the given function h, $1 \mapsto 3$?

Function Notation

Consider once again the function $f = \{(0, 2), (1, 0), (2, 0)\}$ of Example 5.28. The fact that $(0, 2) \in f$ is usually expressed by writing $f(0) = 2$. Both of the statements $(0, 2) \in f$ and $f(0) = 2$ tell us that the function f assigns the output 2 to the input 0.

Definition 5.32   Given a function f, it is customary to write $f(x) = y$ in place of $(x, y) \in f$; this is what is referred to as **function notation** in algebra and calculus courses. When $f(x) = y$, we say that y is the **value** or **output** of the function f at the **input** x. Thus, for a function f and an input x of f, the expression $f(x)$ represents the output (i.e., value) of f at x.

Example 5.9 (continued)   The function $\{(x, x^2) \mid x \in \mathbf{R}\}$, which we previously described using the equation $y = x^2$, can be defined as the function $S : \mathbf{R} \to \mathbf{R}$ where for each real number x, $S(x) = x^2$. Thus, $S(-3)$ represents the output of the function S at the input -3. Because $(-3)^2 = 9$, we may observe that $S(-3) = 9$.

Example 5.31 (continued)   Because the subset $\{2, 3\}$ of $\{1, 2, 3\}$ has 2 members, $\#(\{2, 3\}) = 2$.

Practice Problem 3   For the function h given in Practice Problem 2, determine $h(3)$.

Domain, Codomain, and Range of a Function

We have already introduced the terms *domain* and *range* in the context of an arbitrary relation. This terminology is used in the same way for relations that are functions, but another term, *codomain*, enters the picture. We now make clear how best to think about domain, codomain, and range in the context of a function.

Definition 5.33   Suppose $f : X \to Y$. The set X is the **domain** of f; that is, the set of inputs of f. The set Y is called the **codomain** of f; the *codomain* can be viewed as specifying the *general type of outputs* a function yields. The set $\{f(x) \mid x \in X\}$ of outputs of f is called the **range** or **image** of f.

The primary reason for introducing the notion of *codomain* is convenience. For instance, whereas it is usually very important to know exactly what inputs a certain function can process, it may not be as important to know precisely what outputs the process can achieve, only the general type of output.

Example 5.34   Recall that whereas $\sqrt{x}$ can be calculated for any real number x, if x is negative, the computed result will be a non-real complex number. Now suppose we plan to work with a function f defined so that

$$f(x) = \sqrt{1 + \sqrt{1 + \sqrt{x}}}$$

and that we are only interested in real number outputs (this would usually be the case in a basic calculus course, for instance). Thus, we could take **R** as our *codomain*. Let us also assume that we are not particularly interested in exactly which real numbers are achieved as outputs, only that whatever outputs f does produce should be real numbers. Even though we are saying that knowing the range of f is not important to us, note that we still need to know the domain of f as we must avoid using inputs that would yield non-real outputs. Hence, we must make sure that we do not take the square root of a negative number. Thus, it would be reasonable to take the domain of f to be $[0, \infty)$. This will guarantee that the outputs will be real numbers. So, we may write $f : [0, \infty) \rightarrow \mathbf{R}$ to indicate that our function f has domain $[0, \infty)$ and codomain **R**. By not requiring that the set to the right of the arrow in this expression be the range of f, that is, the exact set of outputs of f, but just a set that we are sure contains all of the outputs, we are not forced into determining the range of f.

Example 5.35 Consider the functions

$f : \mathbf{R} \rightarrow \mathbf{R}$, where $f(x) = x^2$;

$g : \mathbf{R} \rightarrow [0, \infty)$, where $g(x) = x^2$;

$h : \mathbf{R}^- \rightarrow \mathbf{R}$, where $h(x) = x^2$;

$i : \mathbf{R}^- \rightarrow \mathbf{R}^+$, where $i(x) = x^2$.

The domain of f is **R**, the codomain is **R**, and the range is $[0, \infty)$ (after *squaring* all real numbers, we obtain the set of all nonnegative real numbers).

The domain of g is **R**, the codomain is $[0, \infty)$, and the range is $[0, \infty)$.

The domain of h is $\mathbf{R}^-$, the codomain is **R**, and the range is $\mathbf{R}^+$ (after *squaring* all negative real numbers, we obtain the set of all positive real numbers).

The domain of i is $\mathbf{R}^-$, the codomain is $\mathbf{R}^+$, and the range is $\mathbf{R}^+$.

Practice Problem 4 For the function h given in Practice Problem 2, identify the domain, codomain, and range.

Composition of Functions

The most important and most common method for combining functions is the operation of *function composition*. This is where one function is allowed to act on the output of another function; that is, one function *follows* another.

Example 5.36 If a large hot-air balloon is being inflated at the rate of 10 cubic feet per minute, the function f where $f(t) = 10t$ gives us the volume $f(t)$

of air in the balloon t minutes after the balloon begins to inflate. Assume that once enough air has been pumped into the balloon it will begin to look spherical. Recall that the volume V of a sphere having radius r is given by $V = \frac{4}{3}\pi r^3$. Solving this equation for r yields

$$r = \sqrt[3]{\frac{3V}{4\pi}}, \tag{1}$$

which expresses the radius as a function of the volume.

Now if we want to create a function that gives the radius of the balloon t minutes after the balloon begins to inflate, we first apply the function f to obtain the volume of air in the balloon at time t, and we then use this output of f as the input V to the function described by Equation (1), thus obtaining the balloon's radius r. That is, we let the "radius" function act on the output $10t$ of the function f, as depicted in the following diagram:

$$t \quad \mapsto \quad 10t \quad \mapsto \quad \sqrt[3]{\frac{3(10t)}{4\pi}}$$

Ultimately, this gives us

$$r = \sqrt[3]{\frac{3(10t)}{4\pi}}.$$

The diagram we created in Example 5.36 can be generalized to illustrate the operation of function composition whenever we let a function g "follow" a function f:

$$x \quad \overset{f}{\mapsto} \quad f(x) \quad \overset{g}{\mapsto} \quad g(f(x))$$

The idea is that the function f first operates on one of its allowable inputs x, and then the function g operates on what f just gave us, namely $f(x)$, to produce $g(f(x))$. Note that in order for g to operate on what f has produced, it must be the case that this output of f is an allowable input to g. In the definition of function composition stated below, we ensure that this will happen by taking the codomain of f and the domain of g to be the same set.

Definition 5.37   Given functions $f : A \to B$ and $g : B \to C$, we define the **composite function** $g \circ f : A \to C$ so that $(g \circ f)(x) = g(f(x))$ for each $x \in A$. The operation $\circ$ used here is called **function composition**.

Example 5.36 (continued) In the balloon inflation example, we actually formed the function $g \circ f$ from the functions f and g where

$$f(t) = 10t$$

and

$$g(V) = \sqrt[3]{\frac{3V}{4\pi}}.$$

The composition $g \circ f$ is calculated as

$$(g \circ f)(t) = g(f(t)) = g(10t) = \sqrt[3]{\frac{3(10t)}{4\pi}}.$$

Practice Problem 5 Define $f : \{0, 1, 2\} \to \mathbf{R}$ and $g : \mathbf{R} \to [0, \infty)$ so that $f(x) = x + 3$ and $g(x) = x^2$.

    (a) Fill in the blanks: $g \circ f :$ _____ $\to$ _____.

    (b) Find a formula for $(g \circ f)(x)$.

Equality of Functions

Keeping in mind that a function is a *set* of ordered pairs, and that two sets are equal if and only if they have precisely the same members, it follows that functions f and g are *equal* if and only if they include precisely the same ordered pairs as members. Evidently, this requires that the two functions have the same domain (so that both sets of ordered pairs involve precisely the same first coordinates) and, for each allowable input, the two functions assign the same output. These observations provide us with a standard strategy for proving two functions are equal.

Proof Strategy 5.38 (*Proving Two Functions Are Equal*) To prove two functions f and g are equal, show that f and g have the same domain and show that, for each x in this common domain, $f(x) = g(x)$.

Example 5.39 The function $f : \mathbf{R} \to \mathbf{R}$ defined by $f(x) = x^2$ is *not* equal to the function $g : [0, \infty) \to \mathbf{R}$ defined by $g(x) = x^2$ because the domain $\mathbf{R}$ of f is not the same as the domain $[0, \infty)$ of g. Thus, even if two functions use the same formula to calculate outputs, they will not be equal if their domains do not match.

Example 5.40 Define functions $f : \{-1, 0, 1\} \to \mathbf{R}$ and $g : [-1, 1] \cap \mathbf{Z} \to \mathbf{R}$ so that $f(x) = |x|$ and $g(x) = x^2$. Because the only integers within the interval $[-1, 1]$ are -1, 0, and 1, it follows that $[-1, 1] \cap \mathbf{Z} = \{-1, 0, 1\}$, so that f and g have the same domain. We may also observe that $f(-1) = |-1| = 1 = (-1)^2 = g(-1)$, $f(0) = |0| = 0 = (0)^2 = g(0)$, and $f(1) = |1| = 1 = (1)^2 = g(1)$, so that for every $x \in \{-1, 0, 1\}$, $f(x) = g(x)$. Thus, $f = g$.

> **Practice Problem 6** Consider the functions $f : \mathbf{R} \to \mathbf{R}$ and $g : (-\infty, \infty) \to \mathbf{R}$ defined by $f(x) = (x+1)^2 + 5$ and $g(x) = x^2 + 2(x+3)$. Prove that $f = g$.

The operations of addition and multiplication of real numbers are said to be *commutative* because it does not matter in what order real numbers are added or multiplied. That is, for any real numbers a and b, $a + b = b + a$ and $ab = ba$. The operation of function composition is *not* commutative. That is, in at least some cases, the order in which two functions are composed can result in different composite functions.

Example 5.41 Consider the functions $f, g : \mathbf{R} \to \mathbf{R}$ defined so that $f(x) = x + 1$ and $g(x) = 2x$. Observe that $f \circ g$, $g \circ f : \mathbf{R} \to \mathbf{R}$, but $(f \circ g)(x) = 2x + 1$ whereas $(g \circ f)(x) = 2(x + 1)$. It follows that $f \circ g \neq g \circ f$ because, for instance, $(f \circ g)(0) = 1$ and $(g \circ f)(0) = 2$.

However, function composition does allow for the same sort of "regrouping," $(a + b) + c = a + (b + c)$ and $(ab)c = a(bc)$, allowed for when we are adding or multiplying. That is, the *associative property* holds for function composition.

Theorem 5.42 (*Function Composition Is Associative*) Given functions $f : A \to B$, $g : B \to C$, and $h : C \to D$, it follows that $(h \circ g) \circ f = h \circ (g \circ f)$.

Proof: We must show the functions $(h \circ g) \circ f$ and $h \circ (g \circ f)$ have the same domain and that, for each x in this common domain, $((h \circ g) \circ f)(x) = (h \circ (g \circ f))(x)$.

According to the way function composition is defined, note that $h \circ g : B \to D$. Then, as $f : A \to B$, it follows that $(h \circ g) \circ f : A \to D$. Thus, the domain of $(h \circ g) \circ f$ is A. Also, according to the way function composition is defined, $g \circ f : A \to C$, and as $h : C \to D$, we may conclude that $h \circ (g \circ f) : A \to D$. Thus, the domain of $h \circ (g \circ f)$ is A. So $(h \circ g) \circ f$ and $h \circ (g \circ f)$ have the same domain, the set A.

Now consider any $x \in A$. Observe that, according to the definition of function composition,

$$((h \circ g) \circ f)(x) = (h \circ g)(f(x)) = h(g(f(x)))$$

and

$$(h \circ (g \circ f))(x) = h((g \circ f)(x)) = h(g(f(x))),$$

thus making $((h \circ g) \circ f)(x) = (h \circ (g \circ f))(x)$. $\square$

Answers to Practice Problems

1. Because, for example, (1, 1) and (1, −1) are both members of ∇, as both 1 and −1 produce the same square, 1, the input 1 has more than one output. Thus, ∇ is not a function.

2. **(a)** 6.   **(b)** 3, 4, 5, 6.   **(c)** Yes.

3. 5.

4. The domain is {1, 2, 3, 4}, the codomain is **Z**, and the range is {3, 4, 5, 6}.

5. **(a)** $g \circ f : \{0, 1, 2\} \to [0, \infty)$.   **(b)** $(g \circ f)(x) = g(f(x)) = (x + 3)^2$.

6. Because **R** and $(-\infty, \infty)$ are just two different ways of representing the set of all real numbers, the functions f and g have the same domain. Now, for each real number x, $f(x) = (x + 1)^2 + 5 = (x^2 + 2x + 1) + 5 = x^2 + 2x + 6 = x^2 + 2(x + 3) = g(x)$. It follows that $f = g$.

5.4   One-to-One Functions, Onto Functions, and Bijections

Questions to guide your reading of this section:

1. What property does a *one-to-one function* possess? How do we prove a function is one-to-one?

2. What property does an *onto* function possess? How do we prove a function is onto?

3. When is a function a *bijection*?

4. What is meant by the *identity function* on a set?

A function produces only one output for each input. As mentioned earlier, however, this does not prevent a function from taking the same output at different inputs. A function having the property that distinct inputs always yield distinct outputs is called *one-to-one*.

It is also convenient to have terminology describing the situation in which a function's codomain and range are the same. Such functions are called *onto*.

Definition 5.43 A function is called **one-to-one** or an **injection** if different inputs always produce different outputs. A function is called **onto** or a **surjection** if its range and codomain are equal (as sets). A function that is both one-to-one and onto (i.e., both an injection and a surjection) is called a **bijection**.

The following examples illustrate that a function may be neither one-to-one nor onto, both one-to-one and onto, one-to-one but not onto, or onto but not one-to-one.

Example 5.44 The function $f : \mathbf{R} \to \mathbf{R}$, where $f(x) = x^2$, is neither one-to-one nor onto. It is not one-to-one because distinct inputs can give the same output; for instance, the inputs 3 and −3 produce the same output 9. It is not onto because its codomain includes all negative real numbers, but no such numbers are in the range (set of outputs) because squaring a real number never produces a negative number.

Example 5.45 Consider the function $j : \{0, 1, 2\} \to \mathbf{Z}$ defined so that $j(x) = x + 1$. Taking into account the rule by which outputs of j are obtained from inputs, along with the fact that the only allowable inputs are specified as 0, 1, and 2, we see that $j = \{(0, 1), (1, 2), (2, 3)\}$. Because different inputs to j always yield different outputs, j is one-to-one. Because the codomain $\mathbf{Z}$ includes many members that are not members of the range (for instance, 0), j is not onto.

Example 5.46 Now consider the function $k : \{0, 1, 2\} \to \{1, 2, 3\}$ defined so that $k(x) = x + 1$. Note that $k = \{(0, 1), (1, 2), (2, 3)\}$. Because different inputs always produce different outputs, k is one-to-one. Because the range of k is the same as the codomain $\{1, 2, 3\}$, k is onto. Because k is both one-to-one and onto, k is a bijection.

Example 5.47 Finally, consider the function $l : \{-1, 0, 1\} \to \{0, 1\}$ defined by $l(x) = x^2$. Note that $l = \{(-1, 1), (0, 0), (1, 1)\}$. Because the distinct inputs −1 and 1 both yield the same output 1, l is not one-to-one. However, as the range of l is $\{0, 1\}$, the same set as the codomain of l, the function l is onto.

Practice Problem 1 Provide evidence showing that the function $f : \mathbf{R} \to \mathbf{R}$ defined so that $f(x) = x(x - 1)$ is not one-to-one and not onto.

When a function consists of a relatively small finite set of ordered pairs, as in some of the examples considered earlier, it is possible to investigate whether the function is one-to-one or onto by simply inspecting the ordered pairs. Such an approach is not always feasible, however, so we will develop standard strategies for establishing that a function is injective or surjective.

Proving a Function Is Onto

We begin by stating and illustrating the standard approach to proving a function is onto its codomain when indeed this is the case.

Proof Strategy 5.48 (*Proving a Function Is Onto*)   To prove a function $f : X \to Y$ is onto, consider an anonymous member y of the codomain Y and show that there is a member x of the domain X for which $f(x) = y$.

Thus, relative to our *Know/Show* approach, to prove $f : X \to Y$ is onto: *We know* y is an anonymous member of the codomain Y, and *we want to show* $y = f(x)$ for some member x of the domain X.

   This strategy is based directly on the definition of *onto*. By showing that an arbitrarily chosen member of the codomain is actually a member of the range, we are in effect demonstrating that the codomain is a subset of the range. Because we already know that, for any function, the range is a subset of the codomain, we may then conclude that the range and codomain are identical, thus making the function onto.

Example 5.49   We claim that the function $g : \mathbf{R} \to [0, \infty)$, where $g(x) = x^2$, is onto but not one-to-one.

   First, note that g is not one-to-one for the same reason that the function f of Example 5.44 is not one-to-one.

   Now, to prove that g is onto, we will begin to implement Proof Strategy 5.48 by considering an anonymous y that is a member of the codomain $[0, \infty)$ of g. The strategy then tells us to try to find a member x of the domain $\mathbf{R}$ of g for which $g(x) = y$. Hence, *we know* $g : \mathbf{R} \to [0, \infty)$ is defined so that $g(x) = x^2$ and *we also know* $y \in [0, \infty)$; *we want to show* there exists $x \in \mathbf{R}$ such that $g(x) = y$.

   Thinking about the way in which g is defined, we realize that $g(x) = y$ really requires that $x^2 = y$. One way of satisfying this requirement is to let $x = \sqrt{y}$. Because the principal square root of a nonnegative real number y will be a real number, we can be sure that our choice of x as $\sqrt{y}$ is a member of the domain of g. We are now in a position to write our proof.

Claim: The function $g : \mathbf{R} \to [0, \infty)$, defined by $g(x) = x^2$, is onto.

Proof: Consider $y \in [0, \infty)$. Let $x = \sqrt{y}$ and note that because the principal square root of a nonnegative real number must be a real number, x is in the domain $\mathbf{R}$ of g. Also observe that $g(x) = g(\sqrt{y}) = (\sqrt{y})^2 = y$. $\square$

Note carefully how we applied our standard strategy for proving a function is onto. We started with an anonymous y in the codomain $[0, \infty)$. Then we looked for an input x that would produce y as the output. Because the function g "squares" the input to get the output, it was reasonable to think that $\sqrt{y}$ could serve as the x we were looking for. So we defined x to be $\sqrt{y}$; this was done by letting $x = \sqrt{y}$, and usually some sort of statement defining the input x in terms of the output y we started with will appear in a proof showing a function is onto. Note also that we made sure x really is an allowable input to our function (i.e., a member of the function's domain). Finally, we demonstrated that when the function is applied to the input x we defined, the result is the original codomain member y we began with.

> **Practice Problem 2** Consider the function $g : [-1, 1] \to [0, 1]$ defined so that $g(x) = \dfrac{x+1}{2}$. Prove that g is a surjection.

We now intend to prove that when onto functions are composed, the result is itself an onto function.

Theorem 5.50 (*The Composition of Onto Functions Is Onto*)   If $f : A \to B$ and $g : B \to C$ are both onto, then $g \circ f : A \to C$ is also onto.

This theorem is *abstract* because we only know that the functions f and g satisfy a certain mathematical property, *that of being onto*. We do not have any concrete information about them: We do not know precisely what sets are serving as domain and codomain of these functions (the letters A, B, and C *name* these sets but do not *identify* them), and we do not know exactly how these functions assign outputs to their inputs. Thus, *we know* $f : A \to B$ and $g : B \to C$ are onto, and *we want to show* $g \circ f : A \to C$ is onto.

Beginning with what we want to show, Proof Strategy 5.48 tells us to start with an arbitrary member of the codomain C of the function $g \circ f$; let us call this object c to help us remember from which set it was chosen. We must now find a member of the domain A of $g \circ f$ that this function maps to c; that is, we are looking for an object a for which both $a \in A$ and $(g \circ f)(a) = c$.

At this point, we have a clear idea of what we must do to complete our proof. We now need to examine what we know to see how these hypotheses could be helpful toward achieving our goal. Let us make a revised listing of what we know and what we need to show based on the approach to the proof we have developed thus far.

> *We know (updated):* The functions $f : A \to B$ and $g : B \to C$ are onto, and $c \in C$.

> *We want to show (updated):* There exists $a \in A$ for which $(g \circ f)(a) = c$.

The knowledge that $c \in C$ and that, according to our original hypothesis, $g : B \to C$ is onto allows us to obtain, using the definition of onto, a member b of B such that $g(b) = c$. Again, we revise what we know in order to account for the object b of whose existence we are now aware.

> *We know (further update):* The functions $f : A \to B$ and $g : B \to C$ are onto, $c \in C$, $b \in B$, and $g(b) = c$.
>
> *We want to show:* There exists $a \in A$ for which $(g \circ f)(a) = c$.

Having discovered an object b in B, our original hypothesis that $f : A \to B$ is onto provides us with a member a of A for which $f(a) = b$. Our hope is that this object a is the one we are ultimately looking for. Because $a \in A$, all we would need for our proof to be complete is for $(g \circ f)(a) = c$. But this is indeed the case, as, taking all of our deductions into account,

$$(g \circ f)(a) = g(f(a)) = g(b) = c.$$

We now present a proof for our claim that the composition of onto functions is onto, which is written and structured according to the plan we have developed.

Proof of Theorem 5.50: Suppose $f : A \to B$ and $g : B \to C$ are onto. Consider an anonymous $c \in C$. We must find $a \in A$ such that $(g \circ f)(a) = c$. Because the function g is onto, there exists $b \in B$ such that $g(b) = c$. Then, because the function f is onto, there exists $a \in A$ such that $f(a) = b$. Hence, $(g \circ f)(a) = g(f(a)) = g(b) = c$. $\square$

Pay careful attention to this proof and to its development, as they indicate how Proof Strategy 5.48 may be applied to establish that a function is onto even when we do not know the actual sets serving as the function's domain and co-domain or the precise way in which the function assigns outputs to inputs.

Proving a Function Is One-to-One

By definition, a function is one-to-one if different inputs always produce different outputs. This means that a function f is one-to-one provided that *whenever $x_1 \neq x_2$, it follows that $f(x_1) \neq f(x_2)$*. However, because a conditional statement and its contrapositive are logically equivalent, we may conclude that f is one-to-one provided that *whenever $f(x_1) = f(x_2)$, it follows that $x_1 = x_2$*. The following proof strategy is based on this characterization and is the most often used method for proving a function is one-to-one.

Proof Strategy 5.51 (*Proving a Function Is One-to-One*)   To prove a function f is one-to-one, consider anonymous members x_1 and x_2 of the domain of f, assume that $f(x_1) = f(x_2)$, and show, based on this assumption, that $x_1 = x_2$.

Example 5.52 We claim that the function $h : \mathbf{R}^- \to \mathbf{R}$, where $h(x) = x^2$, is one-to-one, but not onto.

First note that h is not onto for the same reason that the function f of Example 5.44 is not onto.

To prove that h is one-to-one, we begin to implement Proof Strategy 5.51 by considering anonymous members x_1 and x_2 of the domain $\mathbf{R}^-$ of h and assuming that $h(x_1) = h(x_2)$. According to the strategy, our work will be complete when we have shown that $x_1 = x_2$. Thus, *we know* $h : \mathbf{R}^- \to \mathbf{R}$ is defined so that $h(x) = x^2$, x_1 and x_2 are both in $\mathbf{R}^-$, and $h(x_1) = h(x_2)$; and *we want to show* $x_1 = x_2$.

We might observe that thinking about what we want to show at this stage is probably not going to get us very far as there is no "standard" strategy for proving two objects are equal. So, we will need to focus on what we know. In particular, we want to make use of the fact that we know how the function h obtains outputs from its inputs; namely, by squaring the inputs. This tells us that $h(x_1) = h(x_2)$ really means that $x_1^2 = x_2^2$. Our hope is that this equation, together with the other information that we know, will eventually lead us to our desired conclusion, $x_1 = x_2$. In fact, you might think that this conclusion follows immediately from $x_1^2 = x_2^2$, but this is not necessarily the case. To see why, note that the equation $x_1^2 = x_2^2$ can be rewritten as $x_1^2 - x_2^2 = 0$, and factoring allows a further rewrite as $(x_1 - x_2)(x_1 + x_2) = 0$. It now follows that $x_1 = x_2$ or $x_1 = -x_2$. But the latter possibility can be eliminated because the domain of h includes only negative numbers, and it is not possible for $x_1 = -x_2$ if x_1 and x_2 are both negative. Thus, we may now conclude that $x_1 = x_2$ must be true. Having reached this conclusion, Proof Strategy 5.51 now allows us to conclude that the function h is one-to-one. Here is our proof.

Claim: The function $h : \mathbf{R}^- \to \mathbf{R}$, defined so that $h(x) = x^2$, is one-to-one.

Proof: Consider anonymous members x_1 and x_2 of the domain $\mathbf{R}^-$ of h, and suppose $h(x_1) = h(x_2)$. We need only show $x_1 = x_2$. Because $h(x_1) = h(x_2)$ and, by definition, $h(x) = x^2$, it follows that $x_1^2 = x_2^2$, which can be rewritten as $x_1^2 - x_2^2 = 0$, which after factoring becomes $(x_1 - x_2)(x_1 + x_2) = 0$. We may then conclude that either $x_1 = x_2$ or $x_1 = -x_2$. However, because the domain of h is $\mathbf{R}^-$, both x_1 and x_2 are negative, which would be impossible if $x_1 = -x_2$. Having eliminated the possibility that $x_1 = -x_2$, we now have no choice but to accept that $x_1 = x_2$. $\square$

Practice Problem 3 Prove that the function g given in Practice Problem 2 is an injection.

> **Practice Problem 4**   Knowing that the function g from Practice Problem 2 is both an injection and a surjection, we may now conclude that g is what kind of function?

Just as the composition of onto functions is onto, the composition of one-to-one functions is one-to-one.

Theorem 5.53 (*The Composition of One-to-One Functions Is One-to-One*)
If $f : A \to B$ and $g : B \to C$ are both one-to-one, then $g \circ f : A \to C$ is also one-to-one.

In developing a proof for this theorem, *we know* the functions $f : A \to B$ and $g : B \to C$ are one-to-one, and *we want to show* the function $g \circ f : A \to C$ is one-to-one. According to Proof Strategy 5.51, to show $g \circ f : A \to C$ is one-to-one, we should begin with anonymous members of the domain A of $g \circ f$; we will call these objects a_1 and a_2. The strategy then tells us to assume $(g \circ f)$ $(a_1) = (g \circ f)(a_2)$. Our proof will be done once we have shown $a_1 = a_2$. Thus, *we now know* $f : A \to B$ and $g : B \to C$ are one-to-one, both a_1 and a_2 are members of A, and $(g \circ f)(a_1) = (g \circ f)(a_2)$; and *we now want to show* $a_1 = a_2$.

Of course, by definition of composition,

$$(g \circ f)(a_1) = (g \circ f)(a_2)$$

can be rewritten as

$$g(f(a_1)) = g(f(a_2)).$$

This last statement has the form

$$g(something) = g(something \; else),$$

where the *something* is $f(a_1)$ and the *something else* is $f(a_2)$. Recalling that we know by hypothesis that the function g is one-to-one, it must follow, by the definition of one-to-one, that

$$something = something \; else;$$

that is,

$$f(a_1) = f(a_2).$$

Recognizing that we are in virtually the same position as before, except now it is values of the function f that we know to be equal, we can apply the hypothesis that f is one-to-one in order to determine that $a_1 = a_2$, our ultimate conclusion. The proof given below is written up according to the plan we have just outlined.

Proof of Theorem 5.53: Suppose $f : A \to B$ and $g : B \to C$ are one-to-one. Consider anonymous $a_1, a_2 \in A$ and assume $(g \circ f)(a_1) = (g \circ f)(a_2)$. We must show that $a_1 = a_2$. Using the definition of function composition, it follows from $(g \circ f)(a_1) = (g \circ f)(a_2)$ that $g[f(a_1)] = g[f(a_2)]$. Then, because the function g is one-to-one, it follows that $f(a_1) = f(a_2)$. But then, as the function f is also one-to-one, it follows that $a_1 = a_2$. ☐

Proving a Function Is a Bijection

By definition, a function is a bijection iff it is both one-to-one and onto. Thus, proving a function is a bijection usually requires that we do two separate proofs, one showing the function is one-to-one and one showing it is onto.

Proof Strategy 5.54 (*Proving a Function Is a Bijection*) To prove a function is a bijection, prove the function is one-to-one and prove the function is onto.

Example 5.55 We prove the function $i : \mathbf{R}^- \to \mathbf{R}^+$, where $i(x) = x^2$, is a bijection.

Proof: To prove that i is a bijection, it suffices to prove that i is both one-to-one and onto.

The proof that i is one-to-one is identical to the proof that the function h of Example 5.52 is one-to-one.

To show i is onto, consider an anonymous element y of the codomain $\mathbf{R}^+$ of i. Let $x = -\sqrt{y}$ and note that, as $-\sqrt{y}$ is negative, x is a member of the domain $\mathbf{R}^-$ of i. Also observe that $i(x) = i(-\sqrt{y}) = (-\sqrt{y})^2 = y$. ☐

The following theorem follows immediately from the definition of bijection and Theorems 5.50 and 5.53.

Theorem 5.56 (*The Composition of Bijections Is a Bijection*) If $f : A \to B$ and $g : B \to C$ are both bijections, then $g \circ f : A \to C$ is also a bijection.

Identity Functions

For any set A, we define $id_A : A \to A$, the **identity function** on A, so that $id_A(x) = x$ for every $x \in A$. In other words, the identity function on a set maps each member of the set to itself (the output is exactly the input).

Example 5.57   The function $id_{\mathbf{R}}$ is defined so that for every real number x, $id_{\mathbf{R}}(x) = x$. Hence, for instance, $id_{\mathbf{R}}(99) = 99$ and $id_{\mathbf{R}}(-\pi) = -\pi$.

Practice Problem 5   What is the numerical value of $id_{\mathbf{Z}}(t)$ if $2t^2 - 1 = t$?

In Problem 5P.4, you are asked to prove that the identity function on a set is a bijection.

Bijections and Counting

One of the primary uses of bijections involves counting. For finite sets A and B, it appears that we may define a bijection from A to B when A and B have the same number of elements, but not when A and B have different numbers of elements.

Example 5.58   We easily observe that the sets $\{3, 5, 9\}$ and $\{b, c, k\}$ have the same number of elements. Further, the function $f : \{3, 5, 9\} \to \{b, c, k\}$ defined so that $f(3) = b$, $f(5) = c$, and $f(9) = k$ is easily seen to be a bijection.

However, there is no bijection from $\{3, 5\}$ to $\{b, c, k\}$ because there is no way for the range to include all three of b, c, and k as the domain only includes two members; that is, a function from $\{3, 5\}$ to $\{b, c, k\}$ cannot be onto. And there is no bijection from $\{3, 5, 9\}$ to $\{c, k\}$ because there is no way to map the three distinct inputs 3, 5, and 9 to three distinct outputs when the codomain $\{c, k\}$ includes only two members; that is, a function from $\{3, 5, 9\}$ to $\{c, k\}$ cannot be one-to-one.

We say sets A and B have the **same cardinality** if there is a bijection $f : A \to B$. Informally, two sets having the same cardinality are viewed as having the same number of elements.

Practice Problem 6   Demonstrate that the sets $\{1, 2, 3, 4, 5\}$ and $\{a, e, i, o, u\}$ have the same cardinality.

We illustrate the technique of counting via the construction of a bijection in showing that a given nonempty set has just as many subsets that include a particular element as subsets that do not include the element.

Theorem 5.59   Let A be a nonempty set and let $x \in A$. The number of subsets of A that include x is the same as the number of subsets of A that do not include x.

Proof: We will use S_x to denote the collection of all subsets of A that include x and S_{-x} to denote the collection of all subsets of A that do not include x. To show that S_x and S_{-x} have the same cardinality, we need only exhibit a bijection from S_{-x} to S_x. Define $f : S_{-x} \to S_x$ so that $f(B) = B \cup \{x\}$, and note that this mapping makes sense because if B is a subset of A that does not include x, $B \cup \{x\}$ will be a subset of A that does include x (the mapping f simply "throws" x into the subset B).

Claim 1: f is one-to-one.

Assume B_1 and B_2 are subsets of A that do not include x, and that $f(B_1) = f(B_2)$. Thus, $B_1 \cup \{x\} = B_2 \cup \{x\}$. We must show the set equality $B_1 = B_2$, which we shall do via our standard double-containment strategy.

$(\subseteq)$ Let $y \in B_1$. By the definition of union, it follows that $y \in B_1 \cup \{x\}$. So, as $B_1 \cup \{x\} = B_2 \cup \{x\}$, we may conclude that $y \in B_2 \cup \{x\}$. Now, because B_1 does not include x, we determine that $y \neq x$. Thus, knowing $y \neq x$ and $y \in B_2 \cup \{x\}$, it follows by the definition of union that $y \in B_2$.

$(\supseteq)$ This argument is similar to that of $(\subseteq)$ and is left as an exercise for the reader.

Claim 2: f is onto.

Let $B \in S_x$; that is, assume B is a subset of A that does include x. Then $B - \{x\}$ is a subset of A that does not include x, thus making $B - \{x\}$ a member of S_{-x}. We must show $f(B - \{x\}) = B$ to complete our argument. According to the definition of f, $f(B - \{x\}) = (B - \{x\}) \cup \{x\}$. Thus, it suffices to show that $(B - \{x\}) \cup \{x\} = B$.

$(\subseteq)$ Let $y \in (B - \{x\}) \cup \{x\}$. By definition of union, $y \in B - \{x\}$ or $y \in \{x\}$. Thus, either y is a member of B other than x or else $y = x$. But because B is a subset of A that includes x, we may now conclude that $y \in B$.

$(\supseteq)$ Let $y \in B$. Because B is a subset of A that includes x, either y is a member of B different from x or else $y = x$. Hence, either $y \in B - \{x\}$ or $y \in \{x\}$. Thus, by definition of union, $y \in (B - \{x\}) \cup \{x\}$.

Having established that f is both one-to-one and onto, it follows that f is a bijection. Thus, by definition, the sets S_x and S_{-x} have the same cardinality, meaning there are exactly as many subsets of A that include x as subsets of A that do not. $\square$

Example 5.60   There is exactly the same number of subsets of $\{n \in \mathbf{N} \mid n \leq 100\}$ that include 12 as there is of those that do not include 12.

Answers to Practice Problems

1. Note, for instance, that $f(0) = f(1) = 0$, so the two inputs 0 and 1 both yield the same output 0; hence, f is not one-to-one. Also note, for instance, that -1 is a member of the codomain of f but is not a member of the range because the equation $f(x) = -1$, that is, $x(x - 1) = -1$, has no real number solutions (which can be verified by first rewriting the equation as $x^2 - x + 1 = 0$ and then applying the Quadratic Formula, which was stated way back in Chapter 1).

2. *Claim:* The function $g : [-1,\ 1] \to [0,\ 1]$ defined by $g(x) = \dfrac{x+1}{2}$ is onto.

    Proof: Consider $y \in [0,\ 1]$ and let $x = 2y - 1$. Observe that as $0 \le y \le 1$, it follows, by multiplying through by 2, that $0 \le 2y \le 2$ and, therefore, by subtracting 1, that $-1 \le 2y - 1 \le 1$; hence, $x \in [-1,\ 1]$. Also observe that $g(x) = g(2y-1) = \dfrac{(2y-1)+1}{2} = \dfrac{2y}{2} = y.$ □

3. *Claim:* The function $g : [-1,\ 1] \to [0,\ 1]$ defined by $g(x) = \dfrac{x+1}{2}$ is one-to-one.

    Proof: Consider $x_1,\ x_2 \in [-1,\ 1]$ and suppose $g(x_1) = g(x_2)$. Then $\dfrac{x_1+1}{2} = \dfrac{x_2+1}{2}$. Multiplying both sides of this equation by 2 yields $x_1 + 1 = x_2 + 1$. Subtracting 1 from both sides of the equation just obtained yields $x_1 = x_2$. □

4. It is a bijection.

5. Because 1 is the only integer for which the equation $2t^2 - 1 = t$ is true (apply the Quadratic Formula), $id_\mathbb{Z}(t) = id_\mathbb{Z}(1) = 1$.

6. Note that the function $f : \{1, 2, 3, 4, 5\} \to \{a, e, i, o, u\}$ defined so that $f(1) = a$, $f(2) = e$, $f(3) = i$, $f(4) = o$, and $f(5) = u$ is a bijection.

5.5   Inverse Relations and Inverse Functions

Questions to guide your reading of this section:

1. How do we form the *inverse* of a relation?

2. What is the relationship between the domain and range of a relation and the domain and range of the relation's inverse?

3.   Under what condition will the inverse of a function also be a function?

4.   What can be said about the inverse of a bijection?

There are many situations in mathematics in which it is desirable or necessary to *reverse* or *invert* a process.

To solve various types of algebraic equations, we must "undo" what has been done to the unknown we are solving for. For example, to solve the equation $x + 7 = 10$, we must *undo* the addition of 7 by subtracting 7, and to solve the equation $t^3 = 64$, we must apply a cube root to *undo* the third power.

Similarly, if we want to find simpler functions that can be combined under the operation of function composition to obtain a given function, we are *reversing* the function composition process (i.e., engaging in what we might call function *decomposition*). And if we want to find a function having a given function as its derivative, we must *antidifferentiate* the given function.

To better understand what is at the core of these, and many other, examples, we now take up the abstract study of *inverse relations*. Because a relation is really just a set of ordered pairs and because the fundamental purpose of an inverse relation is to *undo* or *reverse* what the relation itself does, we realize an inverse should simply switch the coordinates of the ordered pairs making up the original relation, thus reversing the relation's input–output scheme.

Definition 5.61   For any relation R, the **inverse** of R is defined to be the relation $R^{-1} = \{(y, x) \mid (x, y) \in R\}$. That is, the inverse of a relation is obtained by interchanging the coordinates in all of the ordered pairs that are members of the relation. Note that the superscript -1 appearing in the notation R^{-1} for the inverse of the relation R is not an exponent.

Example 5.62   The inverse of the relation $R = \{(0, 2), (1, 0), (2, 0)\}$ is $R^{-1} = \{(2, 0), (0, 1), (0, 2)\}$.

Example 5.63   Consider the function f defined by $f(x) = 3x - 1$. This function can also be described by the equation $y = 3x - 1$. Interchanging x and y in this equation corresponds with switching the coordinates of the ordered pairs satisfying the equation and, therefore, yields a new equation $x = 3y - 1$ that describes the ordered pairs that comprise f^{-1}. Solving this new equation for y yields $y = \dfrac{x + 1}{3}$ so that $f^{-1}(x) = \dfrac{x + 1}{3}$.

Practice Problem 1   Express the inverse of the function $g : \{-1, 0, 1\} \to$ **R** defined by $g(x) = -|x|$ as a set of ordered pairs.

Because coordinates are interchanged in obtaining an inverse relation, we immediately obtain the following theorem telling us that the roles of domain and range are interchanged as well.

Theorem 5.64 (*Domain and Range of a Relation's Inverse*)   Given a relation R, the domain of R^{-1} is the range of R, and the range of R^{-1} is the domain of R.

The inverse of a relation that is a function need not be a function.

Example 5.62 (continued)   Recall that the inverse of the relation $R = \{(0, 2),$ $(1, 0), (2, 0)\}$ is $R^{-1} = \{(2, 0), (0, 1), (0, 2)\}$. Note that even though R is a function, R^{-1} is not because the input 0 to R^{-1} has two distinct outputs, 1 and 2.

A moment's reflection on Example 5.62 reveals the reason the inverse of the function R is not itself a function: There are two inputs to R, namely 1 and 2, that yield the same output 0. Thus, R is not one-to-one. This motivates our next theorem, whose proof is a straightforward application of the definitions of *inverse*, *function*, and *one-to-one*, along with our standard approach to proving an *if and only if* statement.

Theorem 5.65 (*When Is the Inverse of a Function Also a Function?*)   The inverse f^{-1} of a function f is itself a function if and only if f is one-to-one.

   Proof: ($\Rightarrow$) Our argument will be by contraposition. Suppose f is not one-to-one. Then there exist elements x_1 and x_2 of the domain of f, along with an element y of the range of f, for which $x_1 \neq x_2$, $(x_1, y) \in f$, and $(x_2, y) \in f$. Then, by the way in which f^{-1} is defined, it follows that $(y, x_1) \in f^{-1}$ and $(y, x_2) \in f^{-1}$. Because $x_1 \neq x_2$, we see that there is a member y of the domain of f^{-1} that is paired with two distinct members of the range of f^{-1}, and so f^{-1} is not a function.
   ($\Leftarrow$) Our argument will again be by contraposition. Suppose f^{-1} is not a function. Then there exists a member x of the domain of f^{-1}, along with members y_1 and y_2 of the range of f^{-1}, for which $(x, y_1) \in f^{-1}$, $(x, y_2) \in f^{-1}$, and $y_1 \neq y_2$. It follows, by the way f^{-1} is defined, that $(y_1, x) \in f$ and $(y_2, x) \in f$. Because $y_1 \neq y_2$, we see that there are two distinct inputs to f that yield the same output. Thus, f is not one-to-one. $\square$

Recall that when we are working with a function f, we usually write $f(x) = y$ in place of $(x, y) \in f$. Our next theorem simply uses this function notation in the case where the inverse of a function is itself a function.

Theorem 5.66 (*An Inverse Function Reverses the Original Function's Input–Output Scheme*)   If f is a one-to-one function, $f(x) = y$ if and only if $f^{-1}(y) = x$.

Practice Problem 2 Prove Theorem 5.66.

When a function is a bijection, the inverse is also a bijection but reverses the direction of the mapping.

Theorem 5.67 (*The Inverse of a Bijection Is a Bijection*) If $f : X \to Y$ is a bijection, then $f^{-1} : Y \to X$ is a bijection from Y onto X.

Proof: Suppose $f : X \to Y$ is a bijection. Then f is both one-to-one and onto. Because f is one-to-one, f^{-1} is a function. Because f is onto Y, the range of f is Y, making Y the domain of f^{-1}. Also, by the way an inverse relation is defined, interchanging inputs and outputs, the outputs of f^{-1} must be members of the domain X of f. Hence, $f^{-1} : Y \to X$.

To show f^{-1} is onto X, consider an anonymous $x \in X$. Because X is the domain of f, there exists $y \in Y$ such that $f(x) = y$. Thus, $f^{-1}(y) = x$, making f^{-1} onto X.

To show f^{-1} is one-to-one, consider anonymous members y_1 and y_2 of the domain Y of f^{-1} and assume that $f^{-1}(y_1) = f^{-1}(y_2)$. Because f is onto Y, there exist $x_1, x_2 \in X$ such that $f(x_1) = y_1$ and $f(x_2) = y_2$. It then follows that $f^{-1}(y_1) = x_1$ and $f^{-1}(y_2) = x_2$. Thus, because $f^{-1}(y_1) = f^{-1}(y_2)$, we must have $x_1 = x_2$. But, as a function can only take one output for a particular input, it now follows that $f(x_1) = f(x_2)$; that is, $y_1 = y_2$. So f^{-1} is one-to-one.

Because we have shown that f^{-1} is both one-to-one and onto, we may now conclude that f^{-1} is a bijection. □

The proof for Theorem 5.67 provides an application of the standard strategies for proving a function is one-to-one and proving a function is onto; you can use it to assess your understanding of the implementation of these strategies.

Example 5.55 (continued) Back in Section 5.4, we proved that the function $i : \mathbf{R}^- \to \mathbf{R}^+$ where $i(x) = x^2$ is a bijection. Hence, $i^{-1} : \mathbf{R}^+ \to \mathbf{R}^-$ is also a bijection. Observe that for each $x \in \mathbf{R}^+$, $i^{-1}(x) = -\sqrt{x}$. (Can you explain the reason for the negative sign here?)

Practice Problem 3 Consider the bijection $h : [-4, -1] \to [1, 4]$ defined so that $h(x) = |x|$.

(a) The function h^{-1} is a mapping from what set to what set?

(b) Find a formula for $h(x)$.

In Problem 5Q.4, you are asked to prove the following corollary, which follows from Theorems 5.66 and 5.67.

Corollary 5.68 For a bijection $f : X \to Y$ and a given element $y \in Y$, $f^{-1}(y)$ is the unique input to f that yields y as output.

Answers to Practice Problems

1. $g^{-1} = \{(-1, -1), (0, 0), (-1, 1)\}$.

2. *Theorem:* If f is one-to-one, $f(x) = y$ if and only if $f^{-1}(y) = x$.

   Proof: Suppose f is one-to-one. Observe that

   $$f(x) = y \iff (x, y) \in f \iff (y, x) \in f^{-1} \iff f^{-1}(y) = x.$$

   (Can you justify each of these equivalences?) $\square$

3. **(a)** $h^{-1} : [1, 4] \to [-4, -1]$.   **(b)** $h(x) = -x$.

Chapter 5 Problems

5A. Ordered Pairs

1. Consider the ordered pair $(1 - \sqrt{4}, 1 + \sqrt{4})$.

   (a) What integer serves as this ordered pair's left coordinate?

   (b) What integer serves as this ordered pair's right coordinate?

   (c) By definition, to what set is this ordered pair equal?

2. The set $\{\{0\}, \{-2, 0\}\}$ is really an ordered pair (a, b). Identify the numerical values of a and b.

3. Use the definition of ordered pair to prove $(-1, 1) \neq (1, -1)$.

4. Prove that $(a, b) = (b, a)$ if and only if $a = b$.

5B. Cartesian Products

1. List the members of the set $\{1, \pi\} \times \{-5, 0, \sqrt{2}, e\}$.

2. List the members of $S \times S$ if $S = \{0, 1\}$.

3. Describe the members of the set $\mathbf{N} \times \mathbf{Q}$.

5C. Thinking About Relations

1. Consider the relation $U = \{(x, y) \in \mathbf{R} \times \mathbf{R} \mid x^2 + y^2 = 1\}$ defined on $\mathbf{R}$.

   (a) Suppose that $(0, a) \in U$. What are the possible values of a?

   (b) Find the domain of U.

   (c) Find the range of U.

   (d) If we were to graph U in the xy-plane, what would we get?

2. Consider the relation $S = \{(x, y) \in \mathbf{R} \times \mathbf{R} \mid y = \sin(x)\}$ defined on $\mathbf{R}$.

   (a) If $tS0$, what values could t have?

   (b) Use interval notation to describe the domain of S.

   (c) Use interval notation to describe the range of S.

3. Let R be the relation defined on the set $\mathbf{N} \times \mathbf{Z}$ so that $(x, y) \in R$ iff $x = y^2$.

   (a) If $(25, t) \in R$, what are the possible values of t?

   (b) Explain why there is no ordered pair in R having 10 as its first coordinate.

   (c) Find the domain of R.

   (d) Find the range of R.

4. Consider the relation $\sim$ defined on the set N of New England states where $a \sim b$ iff a and b are distinct states that share a border.

   (a) If $a \sim New\ Hampshire$, what are the possible values of a?

   (b) Is it true that $(Rhode\ Island,\ Vermont) \in \sim$?

   (c) Is it true that $\sim\, \subseteq N \times N$?

   (d) Find the range of $\sim$.

5. Consider the relation $\leq$ defined on $\mathbf{R}$.

   (a) Is $(-3, -2) \in\, \leq$?

   (b) Is $(-2, -3) \in\, \leq$?

   (c) Is $(-3, -3) \in\, \leq$?

   (d) What is the domain of $\leq$?

   (e) What is the range of $\leq$?

6. Consider the relation $\subseteq$ on the power set of $\mathbf{Z}$.

   (a) Is $(\{-1, 5, 6\}, \{-1, 0, 2, 5, 6, 8\}) \in\, \subseteq$?

   (b) Is $(\{-1, 5, 6\}, \{-1, 0, 2, 6, 8\}) \in\, \subseteq$?

   (c) Is $(\{-1, 5, 6\}, \{-1, 6, 5\}) \in\, \subseteq$?

5D. Intervals and Interval Notation

Express the given set using interval notation.

1. $\{t \in \mathbf{R} \mid t > \frac{1}{2}\}$. 2. $\{t \in \mathbf{R} \mid t \geq \frac{1}{2}\}$. 3. $\{t \in \mathbf{R} \mid t < \frac{1}{2}\}$. 4. $\{t \in \mathbf{R} \mid t \leq \frac{1}{2}\}$.

5. $\{t \in \mathbf{R} \mid -\sqrt[3]{5} \leq t < \frac{1}{2}\}$. 6. $\{t \in \mathbf{R} \mid -\sqrt[3]{5} < t < \frac{1}{2}\}$. 7. $\{t \in \mathbf{R} \mid -\sqrt[3]{5} \leq t \leq \frac{1}{2}\}$.

8. $\{t \in \mathbf{R} \mid -\sqrt[3]{5} < t \leq \frac{1}{2}\}$. 9. The set of all negative real numbers.

10. The set of all nonnegative real numbers.

11. The set of all positive real numbers no greater than 5.

5E. Properties of Relations

1. Let $A = \{1, 2, 3, 4\}$ and consider the relation $R = \{(1, 2), (2, 2), (2, 3)\}$ on A. Verify that R is not reflexive, not symmetric, and not transitive.

2. Explain why the relation $\leq$ on $\mathbf{R}$ is reflexive and transitive. Then give an example to illustrate why it is not symmetric.

3. Explain why the relation $<$ on $\mathbf{R}$ is transitive. Then give examples to illustrate why it is neither reflexive nor symmetric.

4. Let A be a set and consider the relation $\subseteq$ on $\mathcal{P}(A)$, the power set of A. Explain why $\subseteq$ is reflexive and transitive. Then give an example of a specific set A for which $\subseteq$ is not symmetric on $\mathcal{P}(A)$. Finally, give an example of another set A for which $\subseteq$ is symmetric on $\mathcal{P}(A)$.

5. Consider the set $A = \{0, 1, 2\}$ along with the following relations on A:

$R_1 = A \times A$
$R_2 = \{(0, 0), (1, 1), (2, 2)\}$
$R_3 = \{(0, 0), (0, 1), (1, 2), (2, 0)\}$
$R_4 = \{(2, 1)\}$
$R_5 = \{(0, 0), (1, 1), (2, 2), (0, 2), (2, 0)\}$
$R_6 = \{(0, 0), (1, 1), (2, 2), (1, 0), (0, 2), (1, 2)\}$
$R_7 = \{(0, 0), (1, 1), (2, 2), (2, 0), (0, 2), (1, 2), (1, 0)\}$
$R_8 = \{(0, 2), (0, 1), (2, 1)\}$.

(a) Determine the domain and the range of each of these relations.

(b) Which of these relations are reflexive?

(c) Which are symmetric?

(d) Which are transitive?

6. Consider the relations in 1, 2, and 4 of Problem 5C.

   (a) Which of these relations are reflexive?

   (b) Which are symmetric?

   (c) Which are transitive?

5F. Equivalence Relations

1. Let E be a set of algebraic expressions involving only the variable x. Define $=$ on E so that $a = b$ iff a and b are defined for the same real number values of x and for each such number the expressions yield the same value. For example, if $x^2 - 1$ and $(x - 1)(x + 1)$ are both members of E, because these expressions yield the same value for any given real number, we may write $x^2 - 1 = (x - 1)(x + 1)$. Explain why the relation $=$ is an equivalence relation on E.

2. Prove that logical equivalence is an equivalence relation on any set of statement forms.

3. Two geometric figures are **similar** if they have the same shape (but not necessarily the same size). For instance, any two circles are similar to each other, but a circle cannot be similar to a square. Prove that similarity is an equivalence relation on any set of geometric figures.

4. Which of the relations in Problem 5E.5 are equivalence relations?

5. Are any of the relations given in 1, 2, and 4 of Problem 5C equivalence relations?

6. Prove the given relation is an equivalence relation. Then describe the equivalence classes determined by the equivalence relation.

   (a) The relation $\equiv$ on $\mathbf{Z}$ where $a \equiv b$ iff $a - b$ is even.

   (b) The relation $\sim$ on $\mathbf{R}$ where $a \sim b$ iff $|a| = |b|$.

   (c) The relation D on the set of polynomial functions where $f \, Dg$ iff f and g have the same derivative.

   (d) The relation R on the set of continuous real-valued functions having domain $[0, 1]$ where $f \, Rg$ iff $\int_0^1 f(x)dx = \int_0^1 g(x)dx$.

   (e) The relation $\sim$ on $\mathbf{R}$ where $a \equiv b$ iff $a - b \in \mathbf{Z}$.

7. A listing of binary digits (0's and 1's) in a specific order is called a **bit string**. The length of a finite bit string is the number of bits (binary digits) listed in the string. For instance, 001 is a bit string of length 3.

   Let B be the set of all bit strings of finite length. Consider two relations, S and D, on B, defined so that for all $x, y \in B$,

   $$xSy \text{ iff } x \text{ and } y \text{ end with the same bit;}$$
   $$xDy \text{ iff } x \text{ and } y \text{ end with different bits.}$$

One of these relations is an equivalence relation and the other is not.

(a) For the relation that is an equivalence relation, prove that this is the case.

(b) For the relation that is not an equivalence relation, give specific evidence showing that at least one of the reflexive, symmetric, or transitive properties fails to hold.

(c) For the relation that is an equivalence relation, list three members of [0100].

8. Define the set F so that $F = \{(a, b) \mid a, b \in \mathbf{Z}, b \neq 0\}$. Then define the relation R on F so that $(a, b)R(c, d)$ iff $ad = bc$.

(a) Prove that R is an equivalence relation.

(b) Give three members of the equivalence class of $(1, 2)$.

(c) What does it appear that this equivalence relation is attempting to model?

9. Prove the given relation $\sim$ defined on $\mathbf{R} \times \mathbf{R}$ is an equivalence relation.

(a) $\sim$ defined so that $(a, b) \sim (c, d)$ iff $a - d = c - b$.

(b) $\sim$ defined so that $(v, w) \sim (x, y)$ iff $v + 2y = x + 2w$.

5G. Partitions

1. For each relation in Problem 5E.5 that is an equivalence relation, form the partition of $A = \{0, 1, 2\}$ it determines.

2. Why must the cells of a partition of a set be pairwise disjoint (i.e., why must each pair of cells be disjoint)?

3. What do we get if we form the union of all the cells of a partition of a set S?

4. Explain why $\{\mathbf{R}^+, \{0\}, \mathbf{R}^-\}$ is a partition of $\mathbf{R}$. Then describe the equivalence relation determined by this partition.

5. Let X and Y be nonempty sets. Suppose $\mathcal{P}_X$ is a partition of X and $\mathcal{P}_Y$ is a partition of Y. Under what circumstances will $\mathcal{P}_X \cup \mathcal{P}_Y$ be a partition of $X \cup Y$?

5H. Order Relations

A relation $\sim$ on a set A is

- **irreflexive** if for every $a \in A$, $a \not\sim a$;
- **antisymmetric** if whenever both $a \sim b$ and $b \sim a$, it follows that $a = b$.

Irreflexivity is actually the opposite extreme from reflexivity in the sense that an irreflexive relation does not allow any member of the set on which the relation is defined to be related to itself. *Antisymmetry* of a relation ~ means that $b \sim a$ cannot follow from $a \sim b$ unless $a = b$.

1. Determine whether or not the given relation is irreflexive and whether or not it is antisymmetric.

   (a) The relation $R_1 = \{(1, 1), (2, 2), (3, 3), (2, 3), (3, 2), (2, 4)\}$ on the set $\{1, 2, 3, 4\}$.

   (b) The relation $R_4 = \{(1, 2), (2, 2), (2, 3)\}$ on the set $\{1, 2, 3, 4\}$.

   (c) The relation $R_5 = \{(1, 2), (2, 3)\}$ on the set $\{1, 2, 3, 4\}$.

   (d) The relation ~ of Example 5.20.

   (e) The relation $=$ on $\mathbf{R}$.

   (f) The relation $\leq$ on $\mathbf{R}$.

   (g) The relation $<$ on $\mathbf{R}$.

   (h) The relation $\subseteq$ on the power set $\mathcal{P}(A)$ of a given set A.

   (i) The relation $\subset$ on the power set $\mathcal{P}(A)$ of a given set A.

   (j) Congruence on a set of geometric figures.

   (k) Similarity on a set of geometric figures.

2. Try to formulate a conjecture concerning the circumstances under which an equivalence relation will be antisymmetric. Then try to prove your conjecture.

Sometimes a relation is used to encode a notion of relative *size* or *magnitude* or, more generally, some sort of *order* on the members of a set. These types of relations are referred to as *order relations* and include *strict orders, preorders, partial orders, total orders,* and *well-orderings.*

A relation that is both transitive and irreflexive is called a **strict order**.

3. Determine which of the relations in Problem (1) are strict orders.

4. Determine which of the relations in Problem 5E.5 are strict orders.

A relation that is reflexive and transitive is called a **preorder**.

5. Which, if any, of the relations given in Problem 5E.5 are preorders?

6. Let S be a nonempty set of statements. Define the relation R on S so that

$$(p, q) \in R \text{ iff the statement } p \Rightarrow q \text{ is true.}$$

Show that R is a preorder. Must R be symmetric? antisymmetric?

A relation that is reflexive, antisymmetric, and transitive is called a **partial order**. Once a partial order has been imposed on a set, the set itself is said to be **partially ordered** and can be referred to as a **partially ordered set**.

7. Determine which of the relations in Problem (1) are partial orders.

8. Determine which of the relations in Problem 5E.5 are partial orders.

Assuming that R is a partial order on the set A:

- Two elements $a, b \in A$ are said to be **comparable** if aRb or bRa (two elements of A that are not comparable are said to be **incomparable**).

- The partial order R is called a **total order** if any two elements of A are comparable.

9. Show that the relation $\leq$ on **R** is a total order.

10. Show that the relation $\subseteq$ on the set $\mathcal{P}(\mathbf{N})$ of all subsets of the counting numbers is not a total order by identifying two members of $\mathcal{P}(\mathbf{N})$ that are incomparable.

Assuming that R is a partial order on the set A:

- An element y in a subset B of A is said to be a **least element** of B if yRx for every $x \in B$.

- A is said to be **well-ordered** by R if every nonempty subset of A has a least element.

11. Use proof by contradiction to show that a subset of a partially ordered set cannot have more than one least element.

12. Show that the set $\{1, 2, 3\}$ is well-ordered by $\leq$.

13. Show that the set **R** is not well-ordered by $\leq$.

5I. Determining When a Relation Is a Function

1. Which of the relations in Problem 5E.5 are functions?

2. Which of the relations in Problem 5C are functions?

3. The relations studied in precalculus and calculus have real number inputs and outputs and therefore may be graphed in the xy-plane.

To determine whether a graph in the xy-plane represents a function, we can use the so-called *Vertical Line Test*:

> *A graph in the xy-plane represents a function if and only if no vertical line intersects the graph at more than one point.*

Explain why the Vertical Line Test is a valid method for determining, in the setting of a relation having real number inputs and outputs, whether or not more than one output has been assigned to a particular input.

5J. Domain, Range, and Codomain

1. Consider the functions

$f : [-3, 3] \to [-3, 3]$, where $f(x) = |x|$;

$g : [-3, 5] \to \mathbf{R}$, where $g(x) = |x|$;

$h : [-1, 0] \to [0, 1]$, where $h(x) = |x|$;

$i : [3, 10] \to \mathbf{R}^+$, where $i(x) = |x|$;

$j : [-2, 2] \to [0, 2]$, where $j(x) = |x|$.

For each of these functions, find the domain, codomain, and range.

2. Find the range of the function f defined in Example 5.34.

5K. Arithmetic Operations on Real-Valued Functions

Given functions f and g that produce real number outputs, we define the **sum function** $f + g$, the **difference function** $f - g$, the **product function** fg, and the **quotient function** f/g so that

$$(f + g)(x) = f(x) + g(x),$$
$$(f - g)(x) = f(x) - g(x),$$
$$(fg)(x) = f(x)g(x),$$
$$(f/g)(x) = f(x)/g(x).$$

The domain of each of $f + g$, $f - g$, and fg is taken to be the intersection of the domains of f and g, whereas the domain of f/g consists of those members of the intersection of the domains of f and g for which the output of g is nonzero (to avoid division by 0).

1. Consider the function $f : \mathbf{R}^+ \to \mathbf{R}$ defined by $f(x) = \ln(x)$ and the function $g : \mathbf{R} - \{2\} \to \mathbf{R}$ defined by $g(x) = 1 + \dfrac{x}{3x - 6}$.

    (a) Find the domains of the functions $f + g$, $f - g$, $g - f$, fg, f/g, and g/f.

    (b) Find formulas for each of the functions listed in (a).

2. Now consider the functions

$$A = \{(0, 2), (1, 2), (2, 1), (3, 0)\} \text{ and } B = \{(1, 3), (2, 0), (3, 2)\}.$$

Write the functions $A + B$, $A - B$, $B - A$, AB, A/B, and B/A as sets of ordered pairs.

5L. Composition of Functions

1. Suppose that the Celsius temperature on a certain spring morning increases linearly from 10° at 7 a.m. to 20° at 11 a.m. Let the function C be defined so that $C(x)$ gives the Celsius temperature at time x when $7 \le x \le 11$. Also, let the function F be defined so that $F(t) = \frac{9}{5}t + 32$ represents the Fahrenheit temperature corresponding with a Celsius temperature of t.

   (a) Find the domain and the range of C.

   (b) Find a formula for $C(x)$.

   (c) Find a formula for $(F \circ C)(x)$.

   (d) For any time x from 7 a.m. to 11 a.m., what does $(F \circ C)(x)$ represent in the context described in this problem?

2. Given functions $f : A \to B$ and $g : X \to Y$, we can form the composite function $g \circ f : D \to Y$, where $D = \{t \in A \,|\, f(t) \in X\}$. Explain why it makes sense to take the set D as the domain of $g \circ f$.

3. Consider the functions $f : \mathbf{R} - \{-2\} \to \mathbf{R}$ and $g : [-3, \infty) \to \mathbf{R}$ defined so that

$$f(x) = \frac{1}{x + 2} \qquad \text{and} \qquad g(x) = \sqrt{x + 3}.$$

   (a) Find the domains of $g \circ f$ and $f \circ g$.

   (b) Find formulas for $(g \circ f)(x)$ and $(f \circ g)(x)$.

   (c) Explain why the functions $g \circ f$ and $f \circ g$ are not equal in two ways:

   - first, by comparing their domains;
   - and then by finding a member a of the intersection of the domains of $g \circ f$ and $f \circ g$ for which $(g \circ f)(a) \ne (f \circ g)(a)$.

4. Consider again the functions f and g of Problem 5K.1. Determine the domains of $f \circ g$ and $g \circ f$. Then find formulas for these composite functions.

5. Consider again the functions A and B of Problem 5K.2. Express the functions $A \circ B$ and $B \circ A$ as sets of ordered pairs.

5M. Characteristic Functions

Given a nonempty set X, for each subset A of X we define the **characteristic function** $\chi_A : X \to \mathbf{R}$ so that

$$\chi_A(x) = \begin{cases} 1, \text{ if } x \in A; \\ 0, \text{ if } x \notin A. \end{cases}$$

1. Given that $X = \mathbf{R}$, determine $\chi_{[0,10]}$.
2. For any nonempty set X, determine $\chi_\varnothing$ and χ_X.
3. Prove each of the following.
    (a) Subsets A and B of a nonempty set X are equal if and only if $\chi_A = \chi_B$.
    (b) For any subset A of a nonempty set X, $\chi_A + \chi_{(X-A)} = \chi_X$.
    (c) For any subsets A and B of a nonempty set X, $\chi_A \chi_B = \chi_{(A \cap B)}$.
    (d) For any subsets A and B of a nonempty set X,
        $\chi_A + \chi_B = \chi_{(A \cup B)} + \chi_{(A \cap B)}$.

5N. The Intersection and Union of Functions

1. Prove that the intersection of two functions is also a function.
2. Give an example to show that the union of two functions need not be a function.

5O. The Equivalence Relation Induced by a Function on Its Domain

Any function determines an equivalence relation on its domain as follows:

Suppose $f : X \to Y$. Define the relation $\sim$ on X so that $a \sim b$ iff $f(a) = f(b)$.

1. Prove that $\sim$ is an equivalence relation.
2. Determine all the members of the equivalence class of 3 under $\sim$ for each of the following functions.
    (a) $f : \mathbf{R} \to \mathbf{R}$ defined by $f(x) = x^2$.
    (b) $g : \mathbf{R} \to \mathbf{R}$ defined by $g(x) = |x|$.
    (c) $h : \mathbf{R} \to \mathbf{R}$ defined by $h(x) = x^3$.
    (d) $i : \mathbf{R} \to \mathbf{R}$ defined by $i(x) = x^2 + 5x$.

3. Do you think the equivalence relations determined on **R** by the functions f and g in (2) above are the same? Explain your conclusion.

5P. One-to-One Functions, Onto Functions, and Bijections

1. Is the function S in Problem 5C.2 one-to-one?

2. Are either of the functions A or B from Problem 5K.2 one-to-one?

3. Determine which of the functions in Problem 5J.1 are injections, which are surjections, and which are bijections. In each case, write proofs that support your conclusions.

4. Prove that the identity function on an arbitrary set A is a bijection.

5. Because the functions studied in precalculus and calculus have real number inputs and outputs, they can be graphed in the xy-plane. To determine whether such a function is one-to-one, we can use the so-called *Horizontal Line Test*:

   A function whose inputs and outputs are real numbers is one-to-one if and only if no horizontal line intersects its graph at more than one point.

   Explain why the Horizontal Line Test is a valid method for determining, in the setting of a function having real number inputs and outputs, whether or not more than one input has been assigned the same output.

6. If possible, find an example of a
   (a) one-to-one function from $\{1, 2, 3\}$ into $\{1, 2, 3, 4\}$;
   (b) one-to-one function from $\{1, 2, 3, 4\}$ into $\{1, 2, 3\}$;
   (c) function from $\{1, 2, 3\}$ onto $\{1, 2, 3, 4\}$;
   (d) function from $\{1, 2, 3, 4\}$ onto $\{1, 2, 3\}$;
   (e) one-to-one function from $\{1, 2, 3\}$ onto $\{1, 2, 3, 4\}$;
   (f) one-to-one function from $\{1, 2, 3, 4\}$ onto $\{1, 2, 3\}$;
   (g) function from $\{1, 2, 3\}$ into $\{1, 2, 3\}$ that is neither one-to-one nor onto;
   (h) one-to-one function from $\{1, 2, 3\}$ into $\{1, 2, 3\}$ that is not onto;
   (i) function from $\{1, 2, 3\}$ onto $\{1, 2, 3\}$ that is not one-to-one;
   (j) one-to-one function from $\{1, 2, 3\}$ onto $\{1, 2, 3\}$ that is different from the identity function on $\{1, 2, 3\}$;
   (k) one-to-one function from **N** into **N** that is not a surjection;

(l) function from **N** onto **N** that is not an injection;

(m) one-to-one function from **N** onto **N** that is different from the identity function on **N**;

(n) function from **N** into **N** that is neither an injection nor a surjection.

7. Consider anonymous functions $f : A \to B$ and $g : B \to C$, along with the composite function $g \circ f : A \to C$ created from them.

   (a) If $g \circ f$ is an injection, it is only necessary that one of the functions f and g also be an injection. Which is it? Write a proof justifying your conclusion. Then find specific functions f and g showing that $g \circ f$ can be injective without both f and g being injective.

   (b) If $g \circ f$ is a surjection, it is only necessary that one of the functions f and g also be a surjection. Which is it? Write a proof justifying your conclusion. Then find specific functions f and g showing that $g \circ f$ can be surjective without both f and g being surjective.

   (c) Give an example showing that $g \circ f$ can be a bijection even if neither f nor g is a bijection.

8. Let $A \subseteq \mathbf{R}$ and let $f : A \to \mathbf{R}$. We say the function f is

   • **increasing** provided that for all $x_1, x_2 \in A$, if $x_1 < x_2$, then $f(x_1) < f(x_2)$;

   • **decreasing** provided that for all $x_1, x_2 \in A$, if $x_1 < x_2$, then $f(x_1) > f(x_2)$.

   In other words, the function f is increasing if, as the inputs become larger, the outputs also become larger, and is decreasing if, as the inputs become larger, the outputs become smaller.

   (a) Prove that if f is increasing, then f is one-to-one.

   (b) Prove that if f is decreasing, then f is one-to-one.

9. Let A be a nonempty finite set having n elements and let $0 \leq k \leq n$. Prove that there are just as many subsets of A having k elements as there are subsets of A having $n - k$ elements. HINT: Construct an appropriate bijection.

5Q. Inverse Relations and Inverse Functions

1. Determine the inverse of each relation in Problem 5E.5.

2. Describe the inverses of the relations given in Problem 5C.

3. For each bijection in Problem 5J.1, find its inverse.

4. Prove Corollary 5.68.

5. Prove that for a bijection $f : X \to Y$, $f \circ f^{-1} = id_Y$ and $f^{-1} \circ f = id_X$.

6. Suppose $f : A \to B$ and $g : B \to C$ are both bijections. Prove that $(g \circ f)^{-1} = f^{-1} \circ g^{-1}$.

7. Use the associativity of function composition to show that if f is a bijection and $f \circ g = h$, then $g = f^{-1} \circ h$.

8. Suppose f is a bijection and $f \circ g = h$. Explain why it may not be true that $g = h \circ f^{-1}$.

5R. Images and Pre-images

Given a function $f : X \to Y$:

- For any set $A \subseteq X$, the (**direct**) **image** of A under f, denoted $f[A]$, is $\{f(x) | x \in A\}$ (in other words, $f[A]$ is the set of outputs that results when f is applied to all the members of A).

- For any set $B \subseteq Y$, the **pre-image** or **inverse image** of B under f, denoted $f^{-1}[B]$, is $\{x | f(x) \in B\}$ (in other words, $f^{-1}[B]$ is the set of all inputs to f that produce outputs in B).

1. For the function $f : \mathbf{R} \to \mathbf{R}$ defined by $f(x) = x^2$, determine each of the following:

   (a) $f[\{-2, 2\}]$.   (b) $f[[-3, 2)]$.   (c) $f^{-1}[\{1\}]$.

   (d) $f^{-1}[[1, 9)]$.   (e) $f^{-1}[[-2, 0]]$.

2. Let $g : \mathbf{R} \to \mathbf{R}$ be defined so that $g(x) = \sin(x)$. Find each of the following:

   (a) $g[[0, \pi/2)]$.   (b) $g^{-1}[[-1, 0]]$.

3. Let $f : X \to Y$. Investigate each of the following to see which of them are true. Then try to prove the statements you think are true and try to find counterexamples to the ones you think are false. If you believe a statement is false, can you prove containment in "one direction" instead of equality?

   (a) For any sets $A, B \subseteq X$, $f[A \cup B] = f[A] \cup f[B]$.

   (b) For any sets $A, B \subseteq X$, $f[A \cap B] = f[A] \cap f[B]$.

   (c) For any sets $A, B \subseteq Y$, $f^{-1}[A \cup B] = f^{-1}[A] \cup f^{-1}[B]$.

   (d) For any sets $A, B \subseteq Y$, $f^{-1}[A \cap B] = f^{-1}[A] \cap f^{-1}[B]$.

4. Let $f : X \to Y$ and suppose that $A \subseteq X$ and $B \subseteq Y$. Prove each of the following.

   (a) $A \subseteq f^{-1}[f[A]]$.

(b) $f[f^{-1}[B]] \subseteq B$.

(c) If f is one-to-one, then $f^{-1}[f[A]] = A$.

(d) $f[f^{-1}[B]] = B$ if and only if $B \subseteq f[X]$.

5S. Limits of Real-Valued Functions

Let f be a function having real number inputs and outputs. The **limit** of f as x approaches a real number a is the real number L, denoted $\lim_{x \to a} f(x) = L$, if for every $\varepsilon > 0$, there exists $\delta > 0$ such that whenever x is in the domain of f and $x \in (a - \delta, a + \delta) - \{a\}$, it follows that $f(x) \in (L - \varepsilon, L + \varepsilon)$. This definition formally encodes the intuitive notion that L is the limit of f as x approaches a provided that the outputs of f are always suitably close (as measured by ε) to L when the inputs of f are suitably close (as measured by δ), but not equal, to a.

1. Let c be a given real number and define $f : \mathbf{R} \to \mathbf{R}$ so that $f(x) = c$. Prove that for any $a \in \mathbf{R}$, $\lim_{x \to a} f(x) = c$.

2. Define $f : \mathbf{R} \to \mathbf{R}$ so that $f(x) = x$. Prove that for any $a \in \mathbf{R}$, $\lim_{x \to a} f(x) = a$.

3. Define $f : \mathbf{R} \to \mathbf{R}$ so that $f(x) = x^2$. Prove that for any $a \in \mathbf{R}$, $\lim_{x \to a} f(x) = a^2$.

Depending on the situation, it is possible that $\lim_{x \to a} f(x)$ does not exist, meaning that for every real number L, $\lim_{x \to a} f(x) \neq L$.

4. Prove that if $\lim_{x \to a} f(x)$ exists, its numerical value is unique.

5. Define $f : \mathbf{R} - \{0\} \to \mathbf{R}$ so that $f(x) = 1/x$. Prove that $\lim_{x \to 0} f(x)$ does not exist.

6. Define $f : \mathbf{R} \to \mathbf{R}$ so that

$$f(x) = \begin{cases} x, \text{ if } x < 3; \\ x^2, \text{ if } x \geq 3. \end{cases}$$

Prove that $\lim_{x \to 3} f(x)$ does not exist.

7. Define $f : \mathbf{R} \to \mathbf{R}$ so that

$$f(x) = \begin{cases} 0, \text{ if } x \in \mathbf{Q}; \\ 1, \text{ if } x \in \mathbf{R} - \mathbf{Q}. \end{cases}$$

Prove that for any $a \in \mathbf{R}$, $\lim_{x \to a} f(x)$ does not exist. You may freely use the fact that between any two distinct real numbers, there exist both rational numbers and irrational numbers.

8. Prove that if $\lim_{x \to a} f(x) = L$ and $\lim_{x \to a} g(x) = M$, then $\lim_{x \to a} [f(x) + g(x)] = L + M$.

Chapter 6

The Natural Numbers, Induction, and Counting

We begin this chapter by setting forth axioms for the natural numbers that will enable us to work with them from a more formal perspective. Considerable attention is paid to the *axiom of induction* and its relationship to the concept of *recursive definition*. We then explore several ideas from elementary number theory and survey some basic, but widely applicable, counting principles.

6.1 Axioms for the Natural Numbers

Questions to guide your reading of this section:

1. What assumptions govern the behavior of the natural numbers?

2. What does the *Principle of Mathematical Induction* tell us?

3. What are *Peano's axioms* for abstract natural numbers? What is meant by a *model* for Peano's axioms?

The notion of *number* is certainly a primitive mathematical concept. For most of us, our very first mathematical experiences involve using the numbers 1, 2, 3, and so on, to count. As we have seen, these counting numbers are given the name *natural numbers*, and the set of all natural numbers is denoted by **N**. The following axiom expresses our formal assumptions about the natural numbers.

Axiom 6.1 (*Existence and Fundamental Properties of the Natural Numbers*) We assume the existence of a subset **N** of the set **R** of real numbers; the members of **N** are called **natural numbers**. We also assume the following properties of the natural numbers:

(N1) The number 1 is a natural number, and, for each natural number n, the number $n + 1$, called the **successor** of n, is also a natural number.

(N2) The natural number 1 is not the successor of any natural number.

(N3) If j and k are natural numbers and $j \neq k$, then $j + 1 \neq k + 1$.

(N4) If A is a subset of **N** and
    (i) $1 \in A$; and,
    (ii) whenever $k \in A$, it follows that $k + 1 \in A$,

then A is the set of all natural numbers; that is, $A = \mathbf{N}$.

Axiom (N1) formally identifies the numbers 1, 2, 3, ... as members of **N**. Axiom (N2) makes clear that there is no natural number "before" 1 (i.e., the sequence of natural numbers has a definite starting point). Axiom (N3) states that distinct natural numbers have distinct successors and is needed in order to establish that the set of natural numbers is infinite (see Problem 6A.2). Axiom (N4), known as the **Principle of Mathematical Induction**, states that there are no other natural numbers besides those appearing in the sequence 1, 2, 3, ... that is generated using Axiom (N1).

The natural numbers 1, 2, 3, ... form the prototype infinite "discrete" set in which one "starting" object is followed by another different object, which is in turn followed by another object different from either of the first two, and so on, without end, creating what we might think of as an infinite sequence. However, any infinite set whose elements have been listed in a sequence possesses structural similarities with the set of natural numbers.

Example 6.2   The set $S = \{0, -2, -4, -6, \ldots\}$ satisfies the following properties:

1. The number 0 is a member of S and, for each member n of S, the number $n - 2$ is also a member of S (in this context, we can think of $n - 2$ as the *successor* of n).

2. The number 0 is not the successor of any member of S.

3. If j and k are members of S with $j \neq k$, then $j - 2 \neq k - 2$.

4. If A is a subset of S and
    (i) $0 \in A$; and,
    (ii) whenever $k \in A$, it follows that $k - 2 \in A$,

then $A = S$.

Note that these four properties closely parallel the assumptions (N1–N4) we recorded in Axiom 6.1 for the set **N**. From the perspective of the two lists of properties, (N1–N4) for **N** and (1–4) for S, there is no significant mathematical distinction to be made. Both sets have a "first" element, 1 in the case of **N** and 0 in the case of S. Both sets use a successor operator, *adding 1* in the case of **N** and *subtracting 2* in the case of *S*, to get from one number in the particular set to the "next" one. For both sets, distinct elements are followed by distinct successors. And both sets have no other members beyond those capable of being generated via the first property in each list.

Because of situations such as the one described in Example 6.2, it is convenient to generalize properties (N1–N4) from Axiom 6.1 so that the fundamental ideas they express can be applied in other settings in which these same ideas arise. Peano's axioms for the abstract natural numbers, given in the next axiom, provide such an abstraction. They describe the specific features of the natural numbers that, on an intuitive level, make them behave the way they do. However, because they are stated in a neutral, abstract setting, they may be applied in any situation in which the members of an infinite set can be put into a sequence with an appropriate *successor operator* telling us how to get from one item in the sequence to the next.

Axiom 6.3   The following statements, taken together, are referred to as **Peano's axioms for abstract natural numbers**:

(P1) There is an abstract natural number * and, for each abstract natural number n, there is an abstract natural number n' called the **successor** of n.

(P2) The abstract natural number * is not the successor of any abstract natural number.

(P3) If j and k are abstract natural numbers and $j \neq k$, then $j' \neq k'$.

(P4) If A is a subset of the set of all abstract natural numbers and

> (i) * ∈ A; and,

> (ii) whenever $k \in A$, it follows that $k' \in A$,

then A is the set of all abstract natural numbers.

Any set that satisfies all of these axioms, assuming appropriate roles have been identified for * and the successor operator, is called a **model for Peano's axioms**.

Example 6.4   The set **N** = {1, 2, 3, ...} of natural numbers is itself a model for Peano's axioms. In this model, the abstract natural number * is the familiar number 1, and the successor n' of a counting number n is the "next" counting number $n + 1$.

Example 6.2 (continued)   The set $S = \{0, -2, -4, -6, ...\}$ is also a model for Peano's axioms. In this model, the role of * is being played by 0, and the successor of a member of S is obtained by subtracting 2 from the given member of S.

Example 6.5   The set $\{0, 1, 2, 3, \ldots\}$ of so-called **whole numbers** is another model for Peano's axioms. Here, 0 plays the role of $*$, and the successor operator is the same as it is in the model $\mathbf{N}$; namely, *adding* 1.

Example 6.6   The set

$$\omega = \{\varnothing, \ \{\varnothing\}, \ \{\varnothing, \{\varnothing\}\}, \ \{\varnothing, \{\varnothing\}, \{\varnothing, \{\varnothing\}\}\}, \ldots\}$$

is yet another model for Peano's axioms (ω is the lowercase version of the Greek letter *omega*). In this case, the empty set $\varnothing$ is playing the role of $*$, and the successor of any set $t \in \omega$ is the set of all predecessors of t; that is, the set of sets in ω generated before t.

Practice Problem   Explain why the set $F = \{5, 25, 125, 625, \ldots\}$ can be regarded as a model for Peano's axioms. Be sure to identify the number playing the role of $*$ and the successor operator.

Answer to Practice Problem
In the set $F = \{5, 25, 125, 625, \ldots\}$, the number 5 is playing the role of $*$, and the successor operator is "multiplying by 5"; thus, the axiom (P1) is satisfied. The axiom (P2) is satisfied because 5 is not the successor of any member of F (in particular, observe that $1 \notin F$). The axiom (P3) is satisfied as distinct members of F have distinct successors (note that if $x, y \in F$ and $5x = 5y$, it follows that $x = y$). Finally, the axiom (P4) is satisfied because there are no other members of F besides 5 and those numbers that can be obtained by multiplying a given member of F by 5. Because all of (P1–P4) are true, we may conclude that F is a model for Peano's axioms.

6.2   Proof by Induction

Questions to guide your reading of this section:

1. How does the Principle of Mathematical Induction lead to the proof method known as *proof by induction*?

2. In a proof by induction, what is meant by the *base claim*, the *inductive claim*, and the *inductive hypothesis*?

3. How can proof by induction be informally modeled by imagining knocking over an infinite sequence of dominos?

4. What is Σ-*notation* (also called *summation notation*) and how is it used to represent a sum?

Suppose that, for each natural number n, $P(n)$ is a statement and that we wish to prove

For every $n \in \mathbf{N}$, $P(n)$ is true.

If we let $A = \{n \in \mathbf{N} \mid P(n) \text{ is true}\}$, proving this claim is equivalent to proving that $A = \mathbf{N}$. But, according to the Principle of Mathematical Induction, (N4) of Axiom 6.1, if we show both

(i) $1 \in A$

and

(ii) *whenever $k \in A$, it follows that $k + 1 \in A$,*

we may conclude that $A = \mathbf{N}$. In our current context, though, $1 \in A$ means the statement $P(1)$ is true, and we may translate (ii) as

whenever $P(k)$ is true, it follows that $P(k + 1)$ is true.

These observations lead to the following strategy.

Proof Strategy 6.7 (*Proof by Induction*) Suppose that for each natural number n, $P(n)$ is a statement and we want to prove

For every $n \in \mathbf{N}$, $P(n)$ is true.

It is enough to prove both of the following instead:

Base Claim: $P(1)$ is true.
Inductive Claim: Whenever $P(k)$ is true, it follows that $P(k + 1)$ is true.

The word *whenever* appearing in the inductive claim indicates the presence of universal quantification. In fact, the inductive claim may be expressed as

For every $k \in \mathbf{N}$, if $P(k)$ is true, then $P(k + 1)$ is true.

Thus, our standard strategies for proving a *for all* statement and an *if ... then* statement tell us that to prove the inductive claim, we should assume $P(k)$ is true for some fixed but anonymous natural number k; this assumption is called the **inductive hypothesis**. We would then try to show that for this same natural number k, the statement $P(k + 1)$ is also true. By keeping k anonymous, the net effect is that we will simultaneously prove each of the infinitely many conditional statements

If $P(1)$ is true, then $P(2)$ is true

If $P(2)$ is true, then $P(3)$ is true

If $P(3)$ is true, then $P(4)$ is true

and so forth. Note especially, though, that our inductive hypothesis is confined to assuming the truth of $P(k)$ for just one anonymous natural number k. In particular, we are not assuming the truth of $P(k)$ for all values of a variable k, as this is precisely the original statement

For every $n \in \mathbf{N}$, $P(n)$ is true

we are trying to use induction to prove, and we never assume what we want to prove!

It is also important to understand that, by itself, the truth of the inductive claim is not enough to guarantee that $P(n)$ is true for even one natural number n (remember that a conditional statement with a false hypothesis is automatically true). It is the combination of the truth of both the inductive claim *and* the base claim that leads to the truth of $P(n)$ for *all* natural numbers n. Formally, this is due to the Axiom of Induction. But the need for establishing both the base and inductive claims can also be seen informally by imagining we have lined up an infinite sequence of dominos, numbered 1, 2, 3, 4, ..., as in the following diagram.

If, for each $n \in \mathbf{N}$, we let $P(n)$ be the statement

The nth domino is knocked over,

note that in order to believe all of the dominos are actually knocked over, it is not enough to know only the truth of the inductive claim

Whenever a domino is knocked over, then so is the one immediately following it.

This statement would still be true if all the dominos are standing upright but are close enough to each other so that *if* a domino were to be knocked over, it would begin a chain reaction that would result in all of the dominos after it also being knocked over. In order for all the dominos to really be knocked over, we would also have to set the chain reaction in motion by knocking over the very first domino, meaning we would also have to establish the truth of the base claim

The first domino is knocked over.

Together, the truth of the base claim and the inductive claim suggest intuitively that all of the dominos will be knocked over.

An Example of a Proof by Induction:
Write-Up and Analysis

Proof by induction is potentially applicable whenever we are trying to prove that a statement is true for all natural numbers.

Example 6.8   We will use induction to prove that

For every $n \in \mathbf{N}$, $n^2 + n$ is even.

Observe that this claim has the form

For every $n \in \mathbf{N}$, $P(n)$,

where $P(n)$ is the statement

$n^2 + n$ *is even.*

Thus, our base claim is

$1^2 + 1$ *is even,*

and our inductive claim is

Whenever $k^2 + k$ is even, it follows that $(k + 1)^2 + (k + 1)$ is even

(the conclusion within this inductive claim is obtained by replacing each occurrence of n in the expression $n^2 + n$ with $k + 1$). Note that in formulating our initial claim, as well as the base and inductive claims, we have suppressed the words *is true* used in the statement of Proof Strategy 6.7. It is taken for granted, whenever we formulate a claim, that we intend to show the claim *is true*, so explicitly writing out these words in a claim is both awkward and redundant.

Practice Problem 1   Suppose we want to use induction to prove that

For each $n \in \mathbf{N}$, $2n \le 2^n$.

Identify the base claim, the inductive claim, and the inductive hypothesis for this situation.

Example 6.8 (continued)   To establish our base claim, we need only supply appropriate evidence that the number $1^2 + 1$ is even. This can be done by calculating the value of $1^2 + 1$ as 2 and then explaining why 2 is even.

We typically apply *Know/Show* to establish an inductive claim in a proof by induction. In this case, *we know $k^2 + k$ is even*, and *we want to show* that $(k + 1)^2 + (k + 1)$ is even. Thus, according to the definition of *even* integer, *we know $k^2 + k = 2j$ for some integer j*, and *we want to show $(k + 1)^2 + (k + 1) = 2m$ for some integer m.*

Generally speaking, in the presence of similar-looking algebraic expressions, it is usually a good idea to attempt to determine any mathematical connections between them. Comparing $k^2 + k$ and $(k + 1)^2 + (k + 1)$, we may note that if we "expand" the latter, the resulting expression will include the former among its terms:

$$(k + 1)^2 + (k + 1) = (k^2 + 2k + 1) + (k + 1) = (k^2 + k) + (2k + 2).$$

Also, in establishing the inductive claim, it is always expected that we will have to make use of the inductive hypothesis (i.e., what we know). Here, we realize that our inductive hypothesis allows for a substitution that, along with some additional algebra, essentially gives us what we wanted to show:

$$(k^2 + k) + (2k + 2) = 2j + (2k + 2) + 2(j + k + 1).$$

Because $j + k + 1$ is the sum of integers, it is itself an integer and may play the role of the integer m we have been looking for.

We are now ready to write up our proof.

Theorem: For every $n \in \mathbf{N}$, $n^2 + n$ is even.
Proof (by induction):
   Base Claim: $1^2 + 1$ is even.

Proof of Base Claim: Note that $1^2 + 1 = 2$ and 2 is even because $2 = 2 \cdot 1$, where 1 is an integer.

Inductive Claim: Whenever $k^2 + k$ is even, it follows that $(k+1)^2 + (k+1)$ is even.

Proof of Inductive Claim: Let k be a fixed but anonymous natural number, and suppose $k^2 + k$ is even. We must show that the number $(k+1)^2 + (k+1)$ is even. Note that

$$(k+1)^2 + (k+1) = (k^2 + 2k + 1) + (k+1)$$
$$= (k^2 + k) + (2k + 2)$$
$$= 2j + (2k + 2)$$
$$= 2(j + k + 1)$$

for some integer j as, by our inductive hypothesis, $k^2 + k$ is even. But as **N** is closed under addition, we may also conclude that $j + k + 1$ is an integer. Thus, by definition, $(k+1)^2 + (k+1)$ is even. ☐

Relative to the written presentation of our proof, note how at the very beginning we indicated our intention to use induction, how we clearly set forth both the base and inductive claims we needed to establish, and how we followed each of these two claims with their proofs. Once we actually proved both the base and inductive claims, our proof was complete. This is the standard format we will use when writing up a proof by induction.

Practice Problem 2 Use induction to prove that for every $n \in \mathbf{N}$, $2n \leq 2^n$.

Using Induction To Prove Formulas for Sums

You may recall from your work in other mathematics courses that when the terms in a sum all have the same form, it is possible to use so-called *Σ-notation* (Σ is the uppercase Greek letter *sigma*), also called *summation notation*, to express the sum.

Definition 6.9 Suppose m and n are integers with $m \leq n$. Also suppose that each of a_m, a_{m+1}, a_{m+2}, ..., and a_n are real numbers. The sum

$$a_m + a_{m+1} + a_{m+2} + \cdots + a_n$$

can be represented using **sigma** or **summation notation** as

$$\sum_{i=m}^{n} a_i.$$

In this case, the letter i is being used as the **index of summation**, the integer m is the **lower index of summation**, and the integer n is the **upper index of summation**.

Example 6.10   The expression $\sum_{i=3}^{6} i^2$ represents the sum $3^2 + 4^2 + 5^2 + 6^2$ using Σ-notation. The idea is that the *index of summation i* starts at the lower index 3 (given below the symbol Σ) and increases by 1 until it reaches the upper index 6 (appearing above the symbol Σ). For each value 3, 4, 5, and 6 of the index i, the expression i^2 appearing after the symbol Σ is evaluated, giving us 3^2, 4^2, 5^2, and 6^2. Then, these quantities are added (think Σ for *sum*). There is nothing special about the letter i being used as the index. For instance, the same sum is represented by the expression $\sum_{k=3}^{6} k^2$.

Induction can be used to justify many "summation"-type formulas.

Theorem 6.11   For any $n \in \mathbf{N}$,

$$\sum_{i=1}^{n} i^2 = \frac{n(n+1)(2n+1)}{6}. \tag{1}$$

Whenever we intend to carry out a proof using induction, our first task is to formulate the base and inductive claims. The base claim states that the property we are trying to show in general is true in the special case where $n = 1$. So, in our proof by induction of Theorem 6.11, our base claim is

$$\textit{Base Claim: } \sum_{i=1}^{1} i^2 = \frac{1(1+1)(2 \cdot 1 + 1)}{6},$$

obtained by replacing each occurrence of n in the general formula (1) with 1. The inductive claim states that if the property is true for a certain natural number k, then it must also be true for the next larger natural number $k + 1$. Thus, the inductive claim for our proof of Theorem 6.11 is

Inductive Claim: Whenever $\displaystyle\sum_{i=1}^{k} i^2 = \dfrac{k(k+1)(2k+1)}{6}$, it follows that

$$\sum_{i=1}^{k+1} i^2 = \frac{(k+1)((k+1)+1)(2(k+1)+1)}{6},$$

where the *then* component of the claim is obtained by replacing each occurrence of n in the general formula (1) with $k+1$.

Having formulated the base and inductive claims, we now write up a proof by induction for Theorem 6.11 that incorporates them. We will discuss the proof further after presenting it.

Proof of Theorem 6.11:   We proceed by induction.

Base Claim: $\displaystyle\sum_{i=1}^{1} i^2 = \dfrac{1(1+1)(2 \cdot 1 + 1)}{6}$.

Proof of Base Claim: Observe that $\displaystyle\sum_{i=1}^{1} i^2 = 1^2 = 1$ and

$$\frac{1(1+1)(2 \cdot 1 + 1)}{6} = \frac{1 \cdot 2 \cdot 3}{6} = \frac{6}{6} = 1.$$

Inductive Claim: Whenever $\displaystyle\sum_{i=1}^{k} i^2 = \dfrac{k(k+1)(2k+1)}{6}$, it follows that

$$\sum_{i=1}^{k+1} i^2 = \frac{(k+1)((k+1)+1)(2(k+1)+1)}{6}.$$

Proof of Inductive Claim: Suppose that $\displaystyle\sum_{i=1}^{k} i^2 = \dfrac{k(k+1)(2k+1)}{6}$ for a particular,

but anonymous, natural number k. Then

$$\sum_{i=1}^{k+1} i^2 = \sum_{i=1}^{k} i^2 + (k+1)^2$$

$$= \frac{k(k+1)(2k+1)}{6} + (k+1)^2 \qquad \text{(by our inductive hypothesis)}$$

$$= \frac{k(k+1)(2k+1)}{6} + \frac{6(k+1)^2}{6}$$

$$= \frac{(k+1)[k(2k+1) + 6(k+1)]}{6}$$

$$= \frac{(k+1)(2k^2 + 7k + 6)}{6}$$

$$= \frac{(k+1)(k+2)(2k+3)}{6}$$

$$= \frac{(k+1)[(k+1)+1][2(k+1)+1]}{6}. \quad \square$$

Pay close attention to the way we presented the proof of the base claim here. It is usually not too difficult to see why the base claim in a proof by induction must be true. But students often have trouble writing up the proof of the base claim in a logically convincing manner. For instance, here is a variation on the write-up that many students try to use:

We start with $\displaystyle\sum_{i=1}^{1} i^2 = \frac{1(1+1)(2 \cdot 1 + 1)}{6}$. *Therefore,* $1^2 = \frac{1 \cdot 2 \cdot 3}{6}$. *So,*

$$1 = \frac{6}{6} \text{ and, finally, } 1 = 1.$$

Such a presentation of the argument is fatally flawed. First of all, it begins with the statement we actually want to prove and then deduces other statements from it. *But we can never assume what we want to prove!* Furthermore, the overall "logic" being used is faulty in another way: Just because we end up deducing a true statement, namely $1 = 1$, does not mean that the statement we started out with must be true. For instance, from the clearly false statement $-1 = 1$ we could square both sides to produce the true statement $1 = 1$. But just because we are able to obtain this true statement does not make the original statement $-1 = 1$ true.

In our write-up of the base claim, we separately calculated the two quantities $\displaystyle\sum_{i=1}^{1} i^2$ and $\dfrac{1(1+1)(2 \cdot 1 + 1)}{6}$, showing that each of them is equal to the same thing; namely, 1. Because two quantities that are both equal to a third quantity must be equal to each other, the desired equality follows immediately.

Now let us consider the development of our argument for the inductive claim. Our *Know/Show* approach gives us the following:

We know: $\displaystyle\sum_{i=1}^{k} i^2 = \frac{k(k+1)(2k+1)}{6}$ for a fixed, but anonymous, natural

number k.

We want to show: For this same natural number k,

$$\sum_{i=1}^{k+1} i^2 = \frac{(k+1)[(k+1)+1][2(k+1)+1]}{6}.$$

As usual, we need to relate what we already know to what we want to show. In this case, we know something about the summation $\sum_{i=1}^{k} i^2$, and we want to show something about the summation $\sum_{i=1}^{k+1} i^2$. But these two summations are clearly related as the latter is the same as the former except that it includes one additional term. Specifically,

$$\sum_{i=1}^{k+1} i^2 = \sum_{i=1}^{k} i^2 + (k+1)^2. \tag{2}$$

If this is not clear to you, remember that we can always write out a summation expressed via Σ-notation in "expanded" form:

$$\sum_{i=1}^{k+1} i^2 = 1^2 + 2^2 + \cdots + k^2 + (k+1)^2 = \sum_{i=1}^{k} i^2 + (k+1)^2.$$

Once we have discovered the relationship between the summations $\sum_{i=1}^{k} i^2$ and $\sum_{i=1}^{k+1} i^2$, we simply substitute the assumed value $\dfrac{k(k+1)(2k+1)}{6}$ of $\sum_{i=1}^{k} i^2$ into Equation (2) to obtain

$$\sum_{i=1}^{k+1} i^2 = \frac{k(k+1)(2k+1)}{6} + (k+1)^2. \tag{3}$$

It is then just a matter of algebraically simplifying the right side of (3) into the expression

$$\frac{(k+1)[(k+1)+1][2(k+1)+1]}{6}.$$

Practice Problem 3   Use induction to prove that for any $n \in \mathbf{N}$,
$$\sum_{i=1}^{n} (-1)^i = \frac{(-1)^n - 1}{2}.$$

The Number of Subsets of an *n* Element Set

Induction is a legitimate proof technique anytime we would like to prove that a statement is true for all members of a set that satisfies Peano's axioms.

For example, the theorem we will prove next makes a statement about all nonnegative integers 0, 1, 2, 3, …. Because we have seen in Example 6.5 that the set {0, 1, 2, 3, …} is a model for Peano's axioms, it is reasonable to attempt a proof for the theorem using induction. In structuring this proof, we note that the base claim should state that the theorem is true for the "first" nonnegative integer 0, and the inductive claim implicitly incorporates universal quantification over the set of all nonnegative integers. Moreover, the inductive claim has been expressed via *if… then* instead of *whenever… it follows that*; this is fine, as either wording can be used to form a conditional statement. The proof also makes use of Theorem 5.59 from Section 5.4, which states that a given nonempty set has exactly the same number of subsets that include a particular element as subsets that do not include the particular element.

Theorem 6.12 For any nonnegative integer n, the number of subsets of a set having n elements is 2^n .

Proof (by induction):

Base Claim: The number of subsets of a set having 0 elements is 2^0.

Proof of Base Claim: The only set having 0 elements is the empty set $\varnothing$. This set has only one subset, namely $\varnothing$ itself. Because $2^0 = 1$, it follows that every set having 0 elements has 2^0 subsets.

Inductive Claim: If every set having k elements has 2^k subsets, then every set having $k + 1$ elements has 2^{k+1} subsets.

Proof of Inductive Claim: Let k be a fixed, but anonymous, nonnegative integer and suppose that every set having k elements has 2^k subsets. We must show that every set having $k + 1$ elements has 2^{k+1} subsets. So consider a set A having $k + 1$ elements. Note that $A \neq \varnothing$ because the value of $k + 1$ is at least 1, so there exists at least one element x in A. Then, the set $A - \{x\}$ has k elements. Hence, by our inductive hypothesis, the set $A - \{x\}$ has 2^k subsets.

Now observe that the subsets of A may be separated into two collections: those that include x as a member and those that do not include x as a member. To count the total number of subsets of A, it suffices to count the number of subsets of A that include x and add this number to the number of subsets of A that do not include x. Because a subset of A cannot both include and not include x, no subset of A will be counted more than once by this procedure.

Note that every subset of A that does not include x is really just a subset of the k-element set $A - \{x\}$. Furthermore, every subset of $A - \{x\}$ is a subset of A. So, as we already know by our inductive hypothesis that there are 2^k subsets of $A - \{x\}$, it follows that there are 2^k subsets of A that do not include x.

But, because we have already proved that there are the same number of subsets of A that do include x as subsets of A that do not include x, we may now conclude that there are 2^k subsets of A that do include x.

Thus, there is a total of $2^k + 2^k = 2 \cdot 2^k = 2^{k+1}$ subsets of A. $\square$

Answers to Practice Problems

1. The base claim is $2 \cdot 1 \le 2^1$, the inductive claim is *whenever* $2k \le 2^k$, *it follows that* $2(k + 1) \le 2^{k+1}$, and the inductive hypothesis is the assumption that $2k \le 2^k$ for some fixed but anonymous natural number k.

2. *Claim:* For each $n \in \mathbf{N}$, $2n \le 2^n$.

   Proof: We use induction.

   Base Claim: $2 \cdot 1 \le 2^1$.

   Proof of Base Claim: Observe that $2 \cdot 1 = 2$ and $2^1 = 2$, making $2 \cdot 1 = 2^1$. Hence, $2 \cdot 1 \le 2^1$.

   Inductive Claim: Whenever $2k \le 2^k$, it follows that $2(k + 1) \le 2^{k+1}$.

   Proof of Inductive Claim: Assume $2k \le 2^k$ for a fixed but anonymous natural number k. Note that

   $$2(k + 1) = 2k + 2 \le 2k + 2k = 2(2k) \le 2 \cdot 2^k = 2^{k+1},$$

   where the first inequality here is justified because, as k is a natural number, $1 \le k$, thus making $2 \le 2k$, and the second inequality is justified because of our inductive hypothesis that $2k \le 2^k$. $\square$

3. *Claim:* For any $n \in \mathbf{N}$, $\displaystyle\sum_{i=1}^{n} (-1)^i = \frac{(-1)^n - 1}{2}$.

   Proof (by induction):

   Base Claim: $\displaystyle\sum_{i=1}^{1} (-1)^i = \frac{(-1)^1 - 1}{2}$.

   Proof of Base Claim: Note that $\displaystyle\sum_{i=1}^{1} (-1)^i = (-1)^1 = -1$ and

   $$\frac{(-1)^1 - 1}{2} = \frac{-1 - 1}{2} = \frac{-2}{2} = -1.$$

(continues)

Answers to Practice Problems (continued)

Inductive Claim: Whenever $\displaystyle\sum_{i=1}^{k}(-1)^i = \frac{(-1)^k - 1}{2}$, it follows that

$$\sum_{i=1}^{k+1}(-1)^i = \frac{(-1)^{k+1} - 1}{2}.$$

Proof of Inductive Claim: Suppose $\displaystyle\sum_{i=1}^{k}(-1)^i = \frac{(-1)^k - 1}{2}$ for a specific, but anonymous, natural number k. Then

$$\sum_{i=1}^{k+1}(-1)^i = \sum_{i=1}^{k}(-1)^i + (-1)^{k+1}$$

$$= \frac{(-1)^k - 1}{2} + (-1)^{k+1} \qquad \text{(using our inductive hypothesis)}$$

$$= \frac{(-1)^k - 1}{2} + \frac{2 \cdot (-1)^{k+1}}{2}$$

$$= \frac{(-1)^k - 1 + 2 \cdot (-1)^{k+1}}{2}$$

$$= \frac{(-1)^k + (-1)^{k+1} + (-1)^{k+1} - 1}{2}$$

$$= \frac{(-1)^k[1 + (-1)] + (-1)^{k+1} - 1}{2}$$

$$= \frac{(-1)^k \cdot 0 + (-1)^{k+1} - 1}{2}$$

$$= \frac{(-1)^{k+1} - 1}{2}. \qquad \square$$

6.3 Recursive Definition and Strong Induction

Questions to guide your reading of this section:

1. What does it mean to say that a sequence has been defined *recursively*?

2. What proof technique is often useful in establishing facts about recursively defined sequences?

3. How does a proof done by *strong* induction differ from a proof done by *ordinary* induction? In what circumstances is it useful to think about pursuing a proof by strong induction?

A **sequence** is a function having domain $\mathbf{N}$, the set of natural numbers. If s is a sequence, we usually write s_n for $s(n)$ and refer to s_n as the **nth term** of the sequence. A sequence s is usually identified with its terms s_1, s_2, s_3, ... in the order corresponding with the subscripts.

Example 6.13 The fifth term of the sequence s defined so that $s_n = 1/n^2$ is $s_5 = 1/5^2 = 1/25$. The entire sequence s may be viewed as $1, \frac{1}{4}, \frac{1}{9}, \frac{1}{16}, \frac{1}{25}, \frac{1}{36}, \dots$.

Practice Problem 1 Define the sequence x so that $x_n = (-1)^{n+1}$. Write out the first five terms of this sequence.

A sequence is said to have been **defined recursively** if one or more of its initial terms is directly specified and later terms are generated from those already specified or generated.

Example 6.14 If we define $s_1 = 1$ and also define, for each integer $n \geq 2$, $s_n = n \cdot s_{n-1}$, we see that a sequence s_1, s_2, s_3, s_4, s_5, ... is being generated. The first term s_1 is specified directly, whereas each successive term is indirectly defined via the immediately preceding term. Specifically, s_2 is defined using s_1 as $s_2 = 2 \cdot s_1 = 2 \cdot 1 = 2$, s_3 is defined using s_2 as $s_3 = 3 \cdot s_2 = 3 \cdot 2 = 6$, s_4 is defined using s_3 as $s_4 = 4 \cdot s_3 = 4 \cdot 6 = 24$, and so forth.

Note that the key feature of recursive definition is the "looping back" to previously defined objects in order to define "new" objects.

Practice Problem 2 Define the sequence a recursively so that $a_1 = 5$ and for each integer $n \geq 2$, $a_n = 2a_{n-1} - 3$. Find the numerical value of a_3.

For each natural number n, the number $n!$, called **n factorial**, is defined to be the product of all the natural numbers 1 up through n; that is, $n! = 1 \cdot 2 \cdot \dots \cdot n$. Thus, for instance, $1! = 1$, $2! = 1 \cdot 2 = 2$, and $3! = 1 \cdot 2 \cdot 3 = 6$. We also define $0!$ to be 1.

Example 6.14 (continued) For our sequence s defined recursively so that $s_1 = 1$ and, otherwise, $s_n = n \cdot s_{n-1}$, we may observe that $s_1 = 1 = 1!$, $s_2 = 2 \cdot s_1 = 2 \cdot 1 = 2!$, $s_3 = 3 \cdot s_2 = 3 \cdot 2! = 3 \cdot 2 \cdot 1 = 3!$, and $s_4 = 4 \cdot s_3 = 4 \cdot 3! = 4 \cdot 3 \cdot 2 \cdot 1 = 4!$. Thus, it seems reasonable to conjecture that

$$\textit{For every } n \in \mathbf{N}, \, s_n = n!. \tag{1}$$

Generally speaking, when a sequence has been defined recursively, it is possible to use induction to prove facts about the sequence. *In the presence of recursion, think induction!*

Example 6.14 (continued)   Because the sequence s was defined recursively, it is natural to consider attempting a proof of (1) by induction. We now present such a proof.

Claim: For every $n \in \mathbf{N}$, $s_n = n!$.
Proof: We will use induction.

   Base Claim: $s_1 = 1!$.

   Proof of Base Claim: According to the way s is defined, $s_1 = 1$. According to the way factorials are defined, $1! = 1$. Thus, $s_1 = 1!$.

   Inductive Claim: Whenever $s_k = k!$, it follows that $s_{k+1} = (k+1)!$.

   Proof of Inductive Claim: Let k be a fixed but anonymous natural number and suppose $s_k = k!$. We must show $s_{k+1} = (k+1)!$. Note that

$$s_{k+1} = (k+1) \cdot s_k = (k+1) \cdot k! = (k+1)(1 \cdot 2 \cdot \ldots \cdot k) = (k+1)!,$$

   where the first equality holds because of the definition of s, the second equality because of our inductive hypothesis, and the third and fourth equalities because of the definition of factorials.   □

Practice Problem 3   Use induction to prove, for the sequence a defined in Practice Problem 2, that $a_n = 3 + 2^n$ for every $n \in \mathbf{N}$.

Example 6.15   Probably the most famous example of a recursively defined sequence is the so-called **Fibonacci sequence** defined so that each of the first two terms is 1 and each subsequent term is the sum of the two immediately preceding terms. If we represent the nth term of the Fibonacci sequence by F_n, we thus have $F_1 = 1$, $F_2 = 1$, $F_3 = F_2 + F_1 = 1 + 1 = 2$, $F_4 = F_3 + F_2 = 2 + 1 = 3$, $F_5 = F_4 + F_3 = 3 + 2 = 5$, and so on. That is, the Fibonacci sequence F is defined recursively via $F_1 = 1$, $F_2 = 1$, and, for each integer $n \geq 3$, $F_n = F_{n-1} + F_{n-2}$, yielding the sequence 1, 1, 2, 3, 5, 8, 13, ... The first two terms, F_1 and F_2, are specified directly, but the equation $F_n = F_{n-1} + F_{n-2}$ tells us that every other term is obtained by adding together the two terms immediately before it.

Practice Problem 4   Find the 10th term of the Fibonacci sequence.

Strong Induction

In establishing the inductive claim

Whenever $P(k)$ is true, it follows that $P(k + 1)$ is true

in a proof by induction, we actually are establishing all of the implications $P(1) \Rightarrow P(2)$, $P(2) \Rightarrow P(3)$, $P(3) \Rightarrow P(4)$, and so forth. By also establishing the base claim

$P(1)$ *is true,*

we are then able to deduce the truth of each of the statements $P(2)$, $P(3)$, $P(4)$, ..., *in this order,* as follows:

- $P(2)$ follows from $P(1)$ and $P(1) \Rightarrow P(2)$;
- then $P(3)$ follows from $P(2)$ and $P(2) \Rightarrow P(3)$;
- then $P(4)$ follows from $P(3)$ and $P(3) \Rightarrow P(4)$;
- then $P(5)$ follows from $P(4)$ and $P(4) \Rightarrow P(5)$;

and so forth. So, in general, by the time we want to use the implication $P(k) \Rightarrow P(k + 1)$ to conclude that $P(k + 1)$ is true, we would really already know that each of $P(1)$, $P(2)$, ..., $P(k)$ is true, not just $P(k)$ alone. Hence, there is no harm in assuming true all of $P(1)$, $P(2)$, ..., $P(k)$ in the inductive hypothesis, rather than only $P(k)$. In fact, sometimes our standard approach to doing a proof by induction is insufficient even when the method of induction seems highly appropriate. In such cases, we would usually try to use so-called *strong induction.*

Proof Strategy 6.16 (*Proof by Strong Induction*)   Suppose that for each natural number n, $P(n)$ is a statement and we want to prove

For every $n \in \mathbf{N}$, $P(n)$ is true.

It is enough to prove both of the following instead:

Base Claim: $P(1)$ is true.

(*Strong*) *Inductive Claim:* Whenever $P(i)$ is true for all positive integers $i \leq k$, it follows that $P(k + 1)$ is true.

Note that the only difference between proof by *strong* induction and proof by what we will sometimes refer to as *ordinary* induction is the formulation of the inductive claim. For ordinary induction, we assume only that the statement $P(k)$ is true and then try to deduce from this single statement alone that $P(k + 1)$ is true. However, for strong induction, we assume that each of the statements $P(1)$, $P(2)$, $P(3)$, ..., $P(k)$ is true [i.e., we assume $P(i)$ is true up through k rather than just for k itself] and then try to deduce from these statements that $P(k + 1)$ is true.

Example 6.17 Except for the first two terms, a particular term of the Fibonacci sequence depends on the *two* terms immediately preceding it. Thus, it is reasonable to expect that to prove a fact about the Fibonacci sequence, ordinary induction may not be enough. Here, we use *strong* induction to prove that the nth term F_n of the Fibonacci sequence is less than 2^n. Note how our proof of the inductive claim uses the assumption that this result holds not only for a specific but anonymous natural number k but also for the natural number $k - 1$ immediately preceding k (this is why our approach is by *strong* induction rather than *ordinary* induction).

Theorem: For every $n \in \mathbf{N}$, $F_n < 2^n$.

Proof (by strong induction):

  Base Claim: $F_1 < 2^1$ and $F_2 < 2^2$.

  Proof of Base Claim: Note that $F_1 = 1$, $2^1 = 2$, and $1 < 2$; hence, $F_1 < 2^1$. Similarly, note that $F_2 = 1$, $2^2 = 4$, and $1 < 4$; hence, $F_2 < 2^2$.

  (Strong) Inductive Claim: Whenever $F_i < 2^i$ for all positive integers $i \le k$, it follows that $F_{k+1} < 2^{k+1}$.

  Proof of Inductive Claim: Assume $F_i < 2^i$ for all positive integers $i \le k$, where k is a fixed but anonymous natural number greater than 1. Thus, in particular, we are assuming that $F_k < 2^k$ and $F_{k-1} < 2^{k-1}$. Hence,

$$F_{k+1} = F_k + F_{k-1} < 2^k + 2^{k-1} = 2^{k-1}(2 + 1) < 2^{k-1} \cdot 4 = 2^{k+1}. \quad \square$$

The reason the base claim establishes the result for both 1 and 2, rather than just 1, is that the recursive portion of the definition of F_n does not apply until $n = 3$ (note that our argument in the proof of the inductive claim relies on knowing the desired inequality holds for at least the first two natural numbers).

Practice Problem 5 Define the sequence s so that $s_1 = 3$, $s_2 = 5$, and, for each $n \ge 3$, $s_n = s_{n-1} + 2s_{n-2} - 2$. Use strong induction to prove that for every $n \in \mathbf{N}$, $s_n = 2^n + 1$.

Answers to Practice Problems

1. The first five terms are $1, -1, 1, -1, 1$.

2. Note that $a_2 = 2a_1 - 3 = 2 \cdot 5 - 3 = 7$ and then $a_3 = 2a_2 - 3 = 2 \cdot 7 - 3 = 11$.

3. Define the sequence a so that $a_1 = 5$ and for each integer $n \geq 2$, $a_n = 2a_{n-1} - 3$.

    Claim: For every $n \in \mathbf{N}$, $a_n = 3 + 2^n$.

    Proof: We proceed via induction.

    Base Claim: $a_1 = 3 + 2^1$.

    Proof of Base Claim: Note that, by definition of the sequence a, $a_1 = 5$. Also, $3 + 2^1 = 3 + 2 = 5$. Hence, $a_1 = 3 + 2^1$.

    Inductive Claim: Whenever $a_k = 3 + 2^k$, it follows that $a_{k+1} = 3 + 2^{k+1}$.

    Proof of Inductive Claim: Let k be a fixed but anonymous natural number and assume $a_k = 3 + 2^k$. Then
    $$a_{k+1} = 2a_k - 3 = 2(3 + 2^k) - 3 = 6 + 2^{k+1} - 3 = 3 + 2^{k+1},$$
    where the first equality holds because of the definition of a and the second equality because of our inductive hypothesis. $\square$

4. Applying the definition of the Fibonacci sequence gives us $1, 1, 2, 3, 5, 8, 13, 21, 34, 55, \ldots$. Thus, $F_{10} = 55$.

5. The sequence s is defined so that $s_1 = 3$, $s_2 = 5$, and, for each $n \geq 3$, $s_n = s_{n-1} + 2s_{n-2} - 2$.

    Claim: For every natural number n, $s_n = 2^n + 1$.

    Proof (by strong induction): Observe that because the definition of s does not provide for a way to relate s_2 to s_1, our base claim must address the special cases of our overall claim for both $n = 1$ and $n = 2$.

    Base Claim: $s_1 = 2^1 + 1$ and $s_2 = 2^2 + 1$.

    Proof of Base Claim: Note that, by the way the sequence s is defined, $s_1 = 3$; because $2^1 + 1 = 2 + 1 = 3$, we may conclude that $s_1 = 2^1 + 1$. Also note that, by definition of the sequence s, $s_2 = 5$; because $2^2 + 1 = 4 + 1 = 5$, we may conclude that $s_2 = 2^2 + 1$.

    (Strong) Inductive Claim: If $s_i = 2^i + 1$ for each positive integer $i \leq k$, then $s_{k+1} = 2^{k+1} + 1$.

    Proof of Inductive Claim: Let k be a fixed but anonymous natural number greater than or equal to 2. Assume that $s_i = 2^i + 1$ for each positive integer $i \leq k$, meaning we are assuming true

    (continues)

Answers to Practice Problems (continued)

all of the statements $s_1 = 2^1 + 1$, $s_2 = 2^2 + 1$, $s_3 = 2^3 + 1$, ...,
and $s_k = 2^k + 1$. Then

$$s_{k+1} = s_k + 2s_{k-1} - 2 = (2^k + 1) + 2(2^{k-1} + 1) - 2$$

$$= 2^k + 1 + 2^k = 2 \cdot 2^k + 1 = 2^{k+1} + 1,$$

where the first equality holds because of the definition of s and
the second because of our strong inductive hypothesis.   □

6.4   Elementary Number Theory

Questions to guide your reading of this section:

   1. By definition, what does it mean for one integer to *divide* another
      integer?

   2. What does the *Division Algorithm* tell us?

   3. When is a positive integer considered *prime*? *composite*? Which positive
      integer is neither prime nor composite?

   4. How many prime numbers are there?

   5. What does the *Fundamental Theorem of Arithmetic* tell us?

   We now survey some basic mathematical results concerning the integers.
Many of these results will be familiar to you from your earlier work in mathematics, but our emphasis here will be on developing proofs that demonstrate
how they are interconnected.
   We say an integer a **divides** an integer b, denoted $a|b$, if there exists an
integer k such that $ak = b$. In this circumstance, we also refer to a as a **divisor**
or **factor** of b, and we call b a **multiple** of a.

Example 6.18   Because $24 = 6 \cdot 4$ and 4 is an integer, we can say that the
integer 6 divides the integer 24; that is, $6|24$. Thus, 6 is a divisor of 24, 6 is
a factor of 24, and 24 is a multiple of 6. However, because there is no integer
that can be multiplied by 6 to get 28, we would say that 6 does not divide 28;
that is, $6 \nmid 28$. Thus, 6 is not a divisor of 28, 6 is not a factor of 28, and 28 is
not a multiple of 6.

   You are no doubt aware that the largest divisor of a positive integer is
itself. Note how the following proof of this result makes use of contradiction
and one of the fundamental properties of inequalities we examined back in

Chapter 4; namely, that for positive numbers s, t, u, and v, when $s > t$ and $u \geq v$, it follows that $su > tv$.

Theorem 6.19 If a and b are positive integers and $a|b$, then $a \leq b$.

Proof: Assume a and b are positive integers and $a|b$. Then, there is a positive integer k such that $ak = b$. If $a > b$, as $k \geq 1$ it follows that $b = ak > b \cdot 1 = b$, a contradiction, as b cannot be greater than itself. So, we must have $a \leq b$. $\square$

The proofs of the next two theorems are left as exercises (see Problems 6I.1, 2). Note that Theorem 6.20 tells us that the relation *divides* is transitive, and Theorem 6.22 documents our ability to "factor out" common factors.

Theorem 6.20 If $a|b$ and $b|c$, then $a|c$.

Example 6.21 Because $6|24$ and $24|48$, we may use Theorem 6.20 to conclude that $6|48$, which is certainly true as 6 multiplied by the integer 8 yields 48.

Theorem 6.22 If $c|a$ and $c|b$, then $c|(ma + nb)$ for any integers m and n.

Example 6.23 Because $6|24$ and $6|48$, we may use Theorem 6.22 to conclude that $6|(-4 \cdot 24 + 5 \cdot 48)$. Of course, this is reasonable because, according to the distributive property,

$$-4 \cdot 24 + 5 \cdot 48 = -4 \cdot (6 \cdot 4) + 5 \cdot (6 \cdot 8) = 6(-4 \cdot 4 + 5 \cdot 8),$$

where, via the closure of $\mathbf{Z}$ under multiplication and addition, $-4 \cdot 4 + 5 \cdot 8$ is an integer.

The Division Algorithm

By third grade, most children have begun to develop some knowledge of the mathematical operation of division. They learn that it is possible to divide one positive integer, called the *dividend*, by another, called the *divisor*, and obtain an integer *quotient* and a nonnegative integer *remainder* that is less than the divisor.

Example 6.24 When 42 is divided by 5, the quotient is 8 and the remainder is 2. A third-grader might express this by saying that 5 "goes into" 42 eight times with a remainder of 2. Note that this remainder 2 is a nonnegative integer that is less than the divisor 5. One way to express our calculation is to note that $42 = 5 \cdot 8 + 2$. That is, by adding the remainder to the product of the divisor and the quotient, we get the original dividend.

It is also possible to divide -42 by 5 and obtain a nonnegative remainder that is less than the divisor 5. When this is done, the quotient is -9 and the remainder is 3 (note that the quotient is not -8 because this would lead to a remainder of -2, and we want our remainders to be nonnegative). We can express the calculation as $-42 = 5(-9) + 3$.

The basis for obtaining quotients and remainders when dividing is the *Division Algorithm*.

Theorem 6.25 (*The Division Algorithm*) Given any integer n and any positive integer m, there exist unique integers q, the **quotient**, and r, the **remainder**, such that $n = mq + r$ and $0 \leq r < m$. In this context, the positive integer m is referred to as the **divisor** and the integer n is referred to as the **dividend**.

Practice Problem 1 According to the Division Algorithm:

(a) When 30 is divided by 7, what is the quotient and what is the remainder?

(b) When -30 is divided by 7, what is the quotient and what is the remainder?

In the following proof of the Division Algorithm, we consider the divisor m to be "fixed" throughout the proof; that is, m is selected (as an anonymous positive integer) at the very beginning of our argument and will not change. To establish that the existence portion of the result holds for an arbitrary dividend n, we first use induction to establish it for nonnegative values of n, then extend to the case where n is negative. Once existence is established, we then go on to prove the uniqueness portion of the result.

Carefully reading through this proof with the goal of fully understanding everything you are reading is an excellent exercise as the proof uses many of the proof strategies you have been learning about in this book (induction, contradiction, use of cases, demonstrating uniqueness, and so on).

Proof of the Division Algorithm (*Theorem 6.25*): Consider a positive integer m. We first establish the existence portion of the result for nonnegative values of the dividend n using induction.

Note that the existence result holds for $n = 0$, because $0 = 0m + 0$, as well as for $n = 1$, because $1 = 0m + 1$ if $m > 1$ and $1 = 1m + 0$ if $m = 1$. Thus, the base claim for the induction is established.

For the inductive claim, assume k is a fixed but anonymous nonnegative integer for which there exist $q, r \in \mathbf{Z}$ such that $k = mq + r$ and $0 \leq r < m$. We must find $q_1, r_1 \in \mathbf{Z}$ such that $k + 1 = mq_1 + r_1$ and $0 \le r_1 < m$. We consider cases based on whether the remainder r obtained when k is divided by m is less than the maximum allowable remainder of $m - 1$ or equal to it.

Case 1: $r < m - 1$.

Then, $k + 1 = mq + (r + 1)$, where $q, r + 1 \in \mathbf{Z}$ and $0 \le r < r + 1 < (m - 1) + 1 = m$.

Case 2: $r = m - 1$.

Then, $k + 1 = mq + r + 1 = mq + (m - 1) + 1 = m(q + 1) + 0$, where $q + 1, 0 \in \mathbf{Z}$ and $0 \le 0 < m$.

Thus, by induction, we may conclude that the existence portion of the Division Algorithm holds for any nonnegative dividend.

We next establish the existence portion of the result for negative values of the dividend n. So consider $n \in \mathbf{Z}^-$. Then $-n \in \mathbf{Z}^+$, so from what we have already proved, there exist $q, r \in \mathbf{Z}$ such that $-n = mq + r$ and $0 \le r < m$. We consider cases based on whether the remainder r when $-n$ is divided by m is 0 or positive.

Case 1: $r = 0$.

Then, $n = -mq - r = m(-q) + 0$, where $-q, 0 \in \mathbf{Z}$ and $0 \le 0 < m$.

Case 2: $r > 0$.

Then, $n = -mq - r = m(-q - 1) + (m - r)$, where $-q - 1, m - r \in \mathbf{Z}$ and $0 < m - r < m$.

Having now fully established the existence result, we move on to uniqueness, assuming $n = mq + r$ and $n = mq_1 + r_1$, where $q, r, q_1, r_1 \in \mathbf{Z}$, $0 \le r < m$, and $0 \le r_1 < m$. Note that if we can show $r_1 = r$, it follows that $q_1 = q$. So assume by way of contradiction that $r_1 \ne r$. Without loss of generality, suppose $r_1 < r$, which means $r - r_1 > 0$. Note that from $0 \le r < m$ and $0 \le r_1 < m$, we may conclude that $-m < r - r_1 < m$. Thus, we have determined that $0 < r - r_1 < m$. Also, from the fact that $mq + r = mq_1 + r_1$, we may conclude that $m(q_1 - q) = r - r_1$. It then follows that $0 < m(q - q_1) < m$ so that $0 < q - q_1 < 1$, which is impossible because $q - q_1 \in \mathbf{Z}$ and there are no integers between 0 and 1.

Practice Problem 2   Recall our discussion, in connection with the proof of Theorem 5.2 back in Section 5.1, of the phrase *without loss of generality*. We used this phrase again near the end of the proof of the Division Algorithm. Specifically, after assuming $r_1 \ne r$, we wrote *without loss of generality, suppose $r_1 < r$*. Justify the use of this phrase in this situation.

We can use the Division Algorithm to establish that a given integer is either even or odd, but not both.

Theorem 6.26   Every integer is either even or odd, and no integer can be both even and odd.

Proof: Let n be an integer. According to the Division Algorithm, there exist unique integers q and r such that $n = 2q + r$ and $0 \leq r < 2$. Thus, $r \in \{0, 1\}$. Observe that if $r = 0$ it follows, by definition, that n is even, whereas if $r = 1$ it follows, again by definition, that n is odd. Hence, n is either even or odd. However, because the quotient and remainder are unique, it follows that it is not possible for n to be both even and odd.   □

The Well-Ordering of the Natural Numbers

A subset of the set $\mathbf{R}$ of real numbers may or may not have a *smallest* member. For example, the set $\{2, \pi, 1/3\}$ has a smallest element, namely $1/3$, but the set $\mathbf{R}^+$ has no smallest element as taking half of any positive number will yield an even smaller positive number.

Given a subset A of the set $\mathbf{R}$, when there exists $l \in A$ such that $l \leq a$ for every $a \in A$, we call l the **least** member of A.

Practice Problem 3   The relation $\leq$ on $\mathbf{R}$ is **antisymmetric**, meaning that whenever $a \leq b$ and $b \leq a$, it must follow that $a = b$. Use this fact to prove that a subset of $\mathbf{R}$ can have at most one least element.

One of the most fundamental properties of the set $\mathbf{N}$ of natural numbers is that every one of its nonempty subsets has a least element.

Theorem 6.27 (*Well-Ordering of* $\mathbf{N}$)   Every nonempty subset of the set $\mathbf{N}$ of natural numbers has a least element. This fact is often expressed by saying that the set $\mathbf{N}$ is well-ordered.

A theorem labeled as a **lemma** is one whose primary purpose is to help prove another more important theorem. To establish the well-ordering of $\mathbf{N}$, we will first use induction to prove the following lemma, which is really just a restricted version of Theorem 6.27. We will then use the lemma to prove the theorem.

Lemma 6.28   For every $n \in \mathbf{N}$, each nonempty subset of $\{1, 2, 3, ..., n\}$ has a least element.

Proof: We use induction.
   Base Claim: Every nonempty subset of $\{1\}$ has a least element.

Proof of Base Claim: The only nonempty subset of {1} is {1}, and the least element of this set is 1.

Inductive Claim: If every nonempty subset of {1, 2, 3, ..., k} has a least element, then every nonempty subset of {1, 2, 3, ..., k, k + 1} has a least element.

Proof of Inductive Claim: Consider a fixed but anonymous natural number k and suppose every nonempty subset of {1, 2, 3, ..., k} has a least element. We must show each nonempty subset of {1, 2, 3, ..., k, k + 1} has a least element. So consider a nonempty subset A of {1, 2, 3, ..., k, k + 1}.

Case 1: k + 1 ∉ A.

Then, A is a nonempty subset of {1, 2, ..., k} and, by hypothesis, has a least element.

Case 2: k + 1 ∈ A.

If k + 1 is the only element of A, then k + 1 is the least element of A so that A has a least element. Otherwise, A ∩ {1, 2, ..., k} is a nonempty subset of {1, 2, ..., k}, and the least element of A ∩ {1, 2, ..., k} is the least element of A. □

Note that the idea used in Case 2 in our proof is that if the set A includes at least one other element besides k + 1, then k + 1 cannot be the least element of A. We now give our proof of Theorem 6.27.

Proof of Theorem 6.27:   Consider any nonempty subset A of **N**. Then there exists a natural number n ∈ A. Let $B = A ∩ \{1, 2, 3, ..., n\}$. As n ∈ B, B is nonempty. As an intersection of sets is always a subset of any of the sets over which the intersection is being taken, $B ⊆ \{1, 2, 3, ..., n\}$. Thus, by the lemma we have proved, B has a least element. As any elements of A larger than n could not be candidates for the least element of A, and B is precisely the set of elements of A no greater than n, it follows that the least element of B is the least element of A. □

Prime Numbers

A positive integer is **prime** if it has exactly two positive integer divisors and is **composite** if it has more than two positive integer divisors. For example, the integer 7 is prime because it has exactly two positive integer divisors, 1 and 7, whereas the integer 8 is composite because it has more than two positive integer divisors, namely, 1, 2, 4, and 8.

Note that the number 1 is neither prime nor composite, the only positive integer divisors of a prime p are 1 and p, and all primes other than 2 are odd. You are asked to prove these results in Problem 6I.

When an integer is composite, it has a "nontrivial" factorization; that is, a factorization into a product of positive integers other than simply 1 multiplied by the ineger itself. (For instance, the composite integer 8 can be expressed as 2 · 4.) Our next theorem formally records this observation.

Theorem 6.29   Let n be an integer greater than 1. If n is not prime, then there are integers a and b with $1 < a < n$, $1 < b < n$, and $n = ab$. It follows that every integer greater than 1 that is not prime is composite.

Proof: Assume n is an integer greater than 1 and that n is not prime. Because $n > 1$ and $n \cdot 1 = n$, n has at least two positive integer divisors. But as n is not prime, we may conclude that n has more than two positive divisors. So there exists a divisor $a > 1$ of n for which $a \neq n$. Hence, there exists an integer $b > 1$ such that $n = ab$ (do you see why b cannot be equal to 1?). Because a and b are positive divisors of the positive integer n, and the largest divisor of a positive integer is that integer itself, it follows that $a \leq n$ and $b \leq n$. Because $a \neq n$ and $a \neq 1$, it follows that $a < n$ and $b < n$ (can you explain why $b \neq n$?).   □

Corollary 6.30   Every integer greater than 1 has a prime divisor.
Proof (by contradiction): Suppose there is a positive integer greater than 1 having no prime divisors. Then, by the well-ordering of $\mathbf{N}$, there is a smallest such positive integer n. Note that n cannot be prime, for if it is, n would be a prime divisor of itself. So, n must be composite. Thus, there exist integers a and b for which $n = ab$, $1 < a < n$, and $1 < b < n$. Now, because a is a positive integer less than n, and n is the least positive integer having no prime divisors, a must have a prime divisor. But, as the relation *divides* is transitive and a is a divisor of n, any divisor of a must be a divisor of n. So, we may conclude that n has a prime divisor.   □

Note how the proof of this corollary uses the well-ordering of $\mathbf{N}$, the fact that every nonempty subset of $\mathbf{N}$ has a least element. Specifically, once we assumed that there is a natural number greater than 1 having no prime divisors, we realized that the set of all such numbers is nonempty, and the well-ordering of $\mathbf{N}$ then allowed us to conclude that there must be a least such number.

Next, we prove that an integer greater than 1 cannot be a divisor of both of two consecutive integers. For instance, we may observe that 4 is a divisor of 12, but not of 13. In fact, 16 is the next multiple of 4 that is larger than 12.

Theorem 6.31   Let a and b be positive integers with $a > 1$. If $a \mid b$, then $a \nmid (b + 1)$.

Proof: Assume a and b are positive integers with $a > 1$ and suppose $a \mid b$. Suppose to the contrary that $a \mid (b + 1)$ also. Then, there exist positive integers c and d such that $ac = b$ and $ad = b + 1$. Then, $1 = (b + 1) - b = ad - ac = a(d - c)$. Because a is positive, it follows that $d - c$ is positive (*why?*). Also, because c and d are integers, so is $d - c$ (*why?*). Thus, $d - c$ is a positive divisor of 1. But any positive divisor of 1 must be no greater than 1, so $d - c \leq 1$. We may now conclude that $d - c = 1$ (*why?*). Thus, $1 = a(d - c) = a \cdot 1 = a$, contradicting the hypothesis that $a > 1$.   □

Practice Problem 4 There are three places in the proof of Theorem 6.31 that we have inserted *"why?"* parenthetically. Provide the missing justification for the conclusions being reached at those points in the proof.

You may wonder whether there is a largest prime number. The answer is no. In fact, it was known to the ancient Greeks that there are infinitely many prime numbers. We give a version of Euclid's proof of this result.

Theorem 6.32 There are infinitely many primes.

Proof: Our approach will be to show that for any finite set of prime numbers, there exists at least one more prime that is not one of the primes we started with. If we can do this, it follows that there must be infinitely many primes.

So, consider any finite set $\{p_1, p_2, ..., p_n\}$, where each of $p_1, p_2, ..., p_n$ is a prime. Now let k be the product of all of these primes; that is, let $k = p_1 \cdot p_2 \cdot ... \cdot p_n$. Recalling that when a positive integer $a > 1$ divides a positive integer b, a cannot divide the next greater integer $b + 1$, we see that as each of $p_1, p_2, ..., p_n$ divides k, it is impossible for any of $p_1, p_2, ..., p_n$ to divide $k + 1$. Yet, as $k + 1$ is clearly greater than 1, it must have a prime divisor (recall that we already proved every integer greater than 1 has a prime divisor). So, there is at least one more prime beyond those we started with. □

Practice Problem 5 Explain why the strategy outlined at the beginning of the proof of Theorem 6.32 can be used to conclude that there are indeed infinitely many primes.

The Fundamental Theorem of Arithmetic

The most important reason for studying primes is that every integer greater than 1 can be expressed as a product of primes, a representation referred to as the integer's **prime factorization**. The Fundamental Theorem of Arithmetic encapsulates this notion, along with the fact that the representation is unique except for the order in which the primes are listed.

Example 6.33 The prime factorization of 60 is $2 \cdot 2 \cdot 3 \cdot 5$, or $2^2 \cdot 3 \cdot 5$. There is no other way to express 60 as a product of primes except to rearrange the factors (for instance, $60 = 3 \cdot 2 \cdot 5 \cdot 2$).

Theorem 6.34 (*The Fundamental Theorem of Arithmetic*) Any integer greater than 1 can be expressed uniquely (up to the order of the factors) as a product of primes.

We note that by a *product of primes*, as used in our statement of the Fundamental Theorem of Arithmetic, we include the possibility of a product consisting of just a single factor. In this way, a prime number such as 7 is considered to be already expressed as a product of primes.

Proof of the Fundamental Theorem of Arithmetic (Theorem 6.34): We prove the existence of prime factorizations for all integers greater than 1 by using strong induction, leaving the uniqueness portion of the proof as an exercise (see Problem 6H.1).

Base Claim: The integer 2 has a prime factorization.

Proof of Base Claim: Because 2 is prime, its prime factorization is itself. Thus, 2 has a prime factorization.

(Strong) Inductive Claim: For any integer k greater than 1, if each of 2, 3, 4, ..., k has a prime factorization, then $k+1$ has a prime factorization.

Proof of Inductive Claim: Consider a fixed but anonymous integer k greater than 1 and assume that each of 2, 3, 4, ..., k has a prime factorization. If the integer $k+1$ is prime, its prime factorization is itself. Otherwise, $k+1$ is composite, so there exist integers a and b such that $k+1 = ab$, $1 < a < k+1$, and $1 < b < k+1$. Thus, by our inductive hypothesis, each of a and b has a prime factorization. Multiplying these prime factorizations together gives us a product of primes that is equal to $k+1$ and, therefore, serves as a prime factorization of $k+1$. □

Observe that, in the proof of the inductive claim within our proof of the Fundamental Theorem of Arithmetic, when we establish that $k+1$ has a prime factorization, we rely on the assumption that all integers greater than 1 but less than $k+1$ have prime factorizations, not just the assumption that k has a prime factorization. This is the reason that the proof is carried out using strong induction rather than ordinary induction.

Example 6.35 Let us illustrate the argument for our inductive claim in the proof of the Fundamental Theorem of Arithmetic with $k = 44$. Observe that the assumption that 44 has a prime factorization does not help us to show that $k+1 = 45$ has a prime factorization. Instead, observing that 45 is composite and can be expressed as, say, $5 \cdot 9$, we rely on the assumption that every integer larger than 1 but less than 45 has a prime factorization in order to obtain prime factorizations of the factors 5 and 9 of 45. Writing these prime factorizations next to each other produces the prime factorization $5 \cdot (3 \cdot 3)$ of 45.

Usually, the prime factorization of an integer n greater than 1 is expressed as the product of positive integer powers of the distinct primes that divide n; this expression is often referred to as the **prime-power factorization** of n.

Example 6.33 (continued) The prime factorization of 60 is usually expressed in prime-power form as $2^2 \cdot 3 \cdot 5$.

Practice Problem 6   Recall that a **perfect square** is an integer n for which there is an integer k such that $n = k^2$. Prove that all the powers in the prime-power factorization of a perfect square that is greater than 1 must be even.

The Fundamental Theorem of Arithmetic has many interesting applications. For instance, we can use it to prove that $\sqrt{2}$ is irrational. In this proof, we also make use of the result of Practice Problem 6 given above.

Theorem 6.36 $\sqrt{2}$ is irrational.

Proof: Suppose to the contrary that $\sqrt{2}$ is rational. Then, there exist positive integers m and n such that $\sqrt{2} = m/n$. Squaring both sides of this equation gives us $2 = m^2/n^2$, from which it follows that $m^2 = 2n^2$. Recall that the Fundamental Theorem of Arithmetic tells us that the prime factorization of an integer is unique. Because $2n^2$ is the product of 2 and a perfect square, the unique prime factorization of $2n^2$ must contain an odd number of factors of 2. However, because m^2 is a perfect square, the unique prime factorization of m^2 must contain an even number of factors of 2. But this is impossible, because $m^2 = 2n^2$, so the one and only prime factorization of this number cannot contain both an odd number and an even number of factors of 2. Thus, we may conclude that $\sqrt{2}$ is irrational. $\square$

You will find some other applications of the Fundamental Theorem of Arithmetic in Problem 6H.

Answers to Practice Problems

1. (a)   The quotient is 4 and the remainder is 2, because $30 = 7 \cdot 4 + 2$.
    (b)   The quotient is -5 and the remainder is 5, because $-30 = 7 \cdot (-5) + 5$. (Remember that the remainder cannot be negative.)

2. We used the phrase *without loss of generality* here because, after assuming $r_1 \neq r$, it was clear that one of the numbers r_1 or r must be less than the other, and for the purposes of our argument, it did not matter which one of them was smaller. Note that interchanging all occurrences of r_1 and r that appear in the proof after making the assumption that $r_1 < r$ would give us the explicit argument for the case in which $r < r_1$.

(continues)

Answers to Practice Problems (continued)

3. *Claim:* A subset of **R** can have at most one least element.
   Proof: Suppose to the contrary that A is a subset of **R** and that A has two least elements, l and m. Then, as l is a least element of A, we must have $l \leq m$, and as m is a least element of A, we must also have $m \leq l$. It then follows, as $\leq$ is antisymmetric, that $l = m$. □

4. Because a is positive, it must be multiplied by a positive number to obtain a positive product, in this case the number 1; this is why $d - c$ is positive. The number $d - c$ is an integer because the set **Z** is closed under subtraction. We may conclude that $d - c = 1$ because 1 is the only positive integer that is less than or equal to 1.

5. The strategy is to show that for any finite set of primes, there exists at least one more prime that is not one of the primes we started with. By establishing this, anytime someone claims to have all the primes included within some finite set, we can contradict this statement by finding another prime not already in that set. In this way, the size of the set of primes continues to expand beyond any finite number.

6. *Claim:* All of the powers in the prime-power factorization of a perfect square that is greater than 1 must be even.
   Proof: Let n be a perfect square that is greater than 1. Thus, $n = k^2$ for some integer k. We may assume k is positive for if it is not we can replace it with its opposite. Note that as $n > 1$ and k is positive, it follows that $k > 1$. So, according to the Fundamental Theorem of Arithmetic, k has a unique prime-power factorization, which we shall represent as $p_1^{j_1} p_2^{j_2} \ldots p_m^{j_m}$, where $p_1, p_2, \ldots, p_m$ are the distinct primes appearing in this prime factorization and $j_1, j_2, \ldots, j_m$ are the respective positive integer powers to which these primes occur in the prime factorization. It follows that $n = k^2 = \left(p_1^{j_1} p_2^{j_2} \ldots p_m^{j_m} \right)^2 = p_1^{2j_1} p_2^{2j_2} \ldots p_m^{2j_m}$, so that the last expression, $p_1^{2j_1} p_2^{2j_2} \ldots p_m^{2j_m}$, is the prime-power factorization of n. Because each of $j_1, j_2, \ldots, j_m$ is an integer, it follows that each of the integers $2j_1, 2j_2, \ldots, 2j_m$ is even, which is what we set out to prove. □

6.5   Some Elementary Counting Methods

Questions to guide your reading of this section:

1. What is our *Fundamental Assumption About Counting*? How does this assumption lead to the *Sum Rule for Counting*?

2. How do *combinatorial proof strategies* generally differ from ordinary proof strategies? What is a *combinatorial proof*?

3. How do we count the number of elements in the Cartesian product of finite sets? How does this lead to the *Product Rule for Counting*?

4. What is meant by a *permutation* of a set? How are permutations counted?

5. What is the distinction between a *k-permutation* of a set having n elements and a *k-combination* of such a set? How is the number of k-permutations of the set calculated? How is the number of k-combinations of the set calculated?

Perhaps the most important use of the natural numbers is in counting. This section considers some of the most basic mathematical methods used for counting. We denote the number of elements in a finite set A by $|A|$; thus, for example, $|\{2, 4, 6\}| = 3$.

Recall that sets are **disjoint** if they have no elements in common. A fundamental aspect of counting is the idea that to find the total number of elements in the union of two disjoint finite sets, we need only add the numbers of elements in each of the sets.

Example 6.37   Suppose that a third-grade class consists of 8 boys and 10 girls. If B is the set of boys in the class and G is the set of girls in the class, then $B \cup G$ is the set of all children in the class. The total number of children in the class is $|B \cup G| = |B| + |G| = 8 + 10 = 18$.

To have a starting point from which to make deductions about how to count the elements of finite sets formed via unions, intersections, differences, and so on, we will take the notion illustrated in Example 6.37 as an axiom.

Axiom 6.38 (*Fundamental Assumption About Counting*)   For any disjoint finite sets A and B, $|A \cup B| = |A| + |B|$.

Counting the Members of Unions, Intersections, and Differences of Finite Sets

Combinatorics is the branch of mathematics concerned with the different arrangements of given objects subject to some given criteria, along with the methods for counting all such arrangements. We will state some of the counting principles that we develop in the form of *combinatorial proof strategies*. In contrast to the proof strategies we have previously introduced in this book, where formal mathematical language and notation were the norm, combinatorial proof strategies tend to be stated fairly informally. This is because most of the applications of these strategies involve counting nonmathematical objects or steps in a process that is not purely mathematical, and the conversion of the situation to formal mathematics is often either awkward or less revealing than simply working directly within the context of the application itself. Axiom 6.38, our Fundamental Assumption About Counting, provides us with our first combinatorial proof strategy.

Combinatorial Proof Strategy 6.39 (*The Sum Rule for Counting*)   When each item to be counted falls into exactly one of finitely many categories, the total number of items is the sum of the numbers of items in each of the categories.

Example 6.37 (continued)   Because the class has 8 boys and 10 girls, the total number of children in the class is $8 + 10 = 18$.

It seems reasonable that when one finite set is a subset of another, we should be able to obtain the number of elements in the set-theoretic difference by simply subtracting.

Example 6.37 (continued)   Let T be the set of all students in the class. We already know that $|T| = 18$. Because the number of boys in the class is $|B| = 8$ and $B \subseteq T$, it follows that $|T - B| = |T| - |B| = 18 - 8 = 10$, the number of girls in the class.

The proof of the following theorem indicates how the intuition expressed in the prior example follows from our Fundamental Assumption About Counting.

Theorem 6.40   If A and B are finite sets with $B \subseteq A$, then $|A - B| = |A| - |B|$.

Proof: Suppose A and B are finite sets with $B \subseteq A$. Observe that $A = (A - B) \cup B$ (*why?*) and that the sets $A - B$ and B are disjoint (*why?*). Thus, according to our Fundamental Assumption About Counting, $|A| = |A - B| + |B|$, from which it follows that $|A - B| = |A| - |B|$.   □

It is important to note, however, that neither Theorem 6.40 nor Axiom 6.38 may apply in more general circumstances.

Example 6.41   Let C be the set of all students in a calculus class, and let P be the set of all students in a physics class. Suppose there are 32 students in the calculus class and 27 students in the physics class. Also suppose there are four students taking both classes. Note that the number of students taking the calculus class who are not taking the physics class, that is, the number of elements of the set $C - P$, is not $|C| - |P| = 32 - 27 = 5$, but rather $|C| - |C \cap P| = 32 - 4 = 28$, as we only want to subtract off the number of students in the physics class who are also in the calculus class.

Similarly, the number of students taking at least one of the two classes, that is, the number of elements of the set $C \cup P$, is not $|C| + |P| = 32 + 27 = 59$, as this calculation "double counts" students taking both classes. The correct number is $|C| + |P| - |C \cap P| = 32 + 27 - 4 = 55$, calculated so that students taking both classes are not included in the count twice.

The next two theorems reflect what was illustrated in Example 6.41. You are asked to prove Theorem 6.42 in Practice Problem 1.

Theorem 6.42   For any finite sets A and B, $|A - B| = |A| - |A \cap B|$.

Theorem 6.43   For any finite sets A and B, $|A \cup B| = |A| + |B| - |A \cap B|$.

Proof: Consider any finite sets A and B. Because $A \cup B = A \cup (B - A)$, where A and $B - A$ are disjoint from one another, we know that $|A \cup B| = |A| + |B - A|$. But, because we also know that $|B - A| = |B| - |A \cap B|$, we may conclude that $|A \cup B| = |A| + |B| - |A \cap B|$. $\square$

Practice Problem 1   Prove Theorem 6.42.

Practice Problem 2   Out of 100 people responding to a survey, 91 own a computer, 58 own a PC, and 41 own a Mac. How many of the people responding to the survey own both a PC and a Mac?

Cartesian Products and the Product Rule

When a counting process consists of multiple independent steps, and each step can be performed in several ways, it is usually appropriate to multiply the numbers of ways the steps can be performed to determine the total number of ways the process can be completed.

Example 6.44   A committee consisting of one chemistry professor and one math professor will study math requirements for the chemistry major. There are two chemistry professors and three math professors. How many possible committees are there?

To answer this question, we view the formation of the committee as a two-step process. Step 1 consists of choosing a chemistry professor, and Step 2 consists of choosing a math professor. These steps are independent from one another as any chemistry professor may serve on the committee along with any math professor. Because there are two chemistry professors, there are two ways to perform Step 1. Then, because there are three math professors, for each way of performing Step 1, there are three ways of performing Step 2. Thus, there are a total of $2 \cdot 3 = 6$ ways to form the committee.

Note that if the two chemistry professors are $c1$ and $c2$, and the three math professors are $m1$, $m2$, and $m3$, each possible committee can be represented by an ordered pair whose first coordinate is a member of the set $C = \{c1, c2\}$ of chemistry professors and whose second coordinate is a member of the set $M = \{m1, m2, m3\}$ of math professors. That is,

$$C \times M = \{(c1, m1), (c1, m2), (c1, m3), (c2, m1), (c2, m2), (c2, m3)\}$$

represents the set of all possible committees. With this model for the committees, we now realize that to count the total number of committees that can be formed, we need only count the number of members of the Cartesian product $C \times M$, which is itself just the product of $|C|$ and $|M|$. The following figure, an example of what is referred to as a **tree diagram**, illustrates this pictorially.

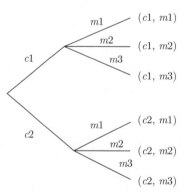

Our work in Example 6.44 suggests the next theorem.

Theorem 6.45   For any finite sets A and B, $|A \times B| = |A| \cdot |B|$.

Proof: Consider any finite sets A and B. If either $|A|$ or $|B|$ is 0, then at least one of A or B is $\varnothing$, in which case $A \times B = \varnothing$, so that both $|A \times B| = 0$ and $|A| \cdot |B| = 0$, from which the desired equation immediately follows.

So, assume now that $|A| = m > 0$ and $|B| = n > 0$, with $A = \{a_1, a_2, ..., a_m\}$ and $B = \{b_1, b_2, ..., b_n\}$. Observe that

$$A \times B = A_1 \cup A_2 \cup ... \cup A_m$$

where, for each $i \in \{1, 2, ..., m\}$,

$$A_i = \{(a_i, b_1), (a_i, b_2), ..., (a_i, b_n)\},$$

the set of members of $A \times B$ having first coordinate a_i. The sets $A_1, A_2, ..., A_m$ are pairwise disjoint (no two of them have a member in common; *do you see why?*) and $|A_1| = |A_2| = \cdots = |A_m| = n$. It follows, using our Fundamental Assumption About Counting, that

$$|A \times B| = |A_1 \cup A_2 \cup ... \cup A_m| = |A_1| + |A_2| + \cdots + |A_m| = mn = |A| \cdot |B|. \quad \square$$

The notions of ordered pair and Cartesian product can be generalized. For each integer $n > 2$, the **ordered n-tuple** $(x_1, x_2, ..., x_n)$ is the ordered pair $((x_1, x_2, ..., x_{n-1}), x_n)$. For each $i \in \{1, 2, ..., n\}$, x_i is the **ith coordinate** of $(x_1, x_2, ..., x_n)$.

Theorem 6.46   Thus, $(5, -2, 0.3)$ is an *ordered 3-tuple* or *ordered triple*. Its first coordinate is 5, its second coordinate is -2, and its third coordinate is 0.3. Formally, $(5, -2, 0.3)$ is the ordered pair $((5, -2), 0.3)$, though we will rarely need to think of it in this way.

In Problem 6J.8, you will prove that ordered n-tuples are equal if and only if their corresponding coordinates are equal.

Given an integer $n \geq 2$ and sets $A_1, A_2, ..., A_n$, the **Cartesian product** $A_1 \times A_2 \times \cdots \times A_n$ is defined to be

$$\{(a_1, a_2, ..., a_n) \mid a_1 \in A_1, a_2 \in A_2, ..., a_n \in A_n\}$$

the set of all ordered n-tuples where, for each $i \in \{1, 2, ..., n\}$, the ith coordinate is a member of A_i.

Example 6.47   The Cartesian product $\mathbf{Z} \times \{0, 1\} \times \mathbf{R} \times \mathbf{R}$ consists of all ordered 4-tuples whose first coordinate is an integer, whose second coordinate is either 0 or 1, and whose third and fourth coordinates are real numbers. Thus, members of $\mathbf{Z} \times \{0, 1\} \times \mathbf{R} \times \mathbf{R}$ include $(-5, 1, \pi, 0.5)$ and $(0, 0, 0, 0)$. However, $(\pi, 1, -5, 0.5)$ is not a member of $\mathbf{Z} \times \{0, 1\} \times \mathbf{R} \times \mathbf{R}$ as π is not an integer.

Example 6.44 (continued)   Suppose now that the committee will also include, in addition to a chemistry professor and a math professor, a student majoring in chemistry. Suppose also that there are five students majoring in chemistry and let $S = \{s1,\ s2,\ s3,\ s4,\ s5\}$ be the set of these students. Then, each possible committee that can be formed can be identified with an ordered triple $(c,\ m,\ s)$, where c is a chemistry professor, m is a math professor, and s is a student majoring in chemistry. Thus, the total number of committees that can be formed is

$$|C \times M \times S| = |C| \cdot |M| \cdot |S| = 2 \cdot 3 \cdot 5 = 30.$$

The following theorem generalizes Theorem 6.45; its proof is left to you as Problem 6J.9.

Theorem 6.48 For any integer $n \geq 2$ and any finite sets $A_1,\ A_2,\ ...,\ A_n$, $|A_1 \times A_2 \times \cdots \times A_n| = |A_1| \cdot |A_2| \cdot \ldots \cdot |A_n|$.

Theorems 6.45 and 6.48 provide the basis for the following counting principle.

Combinatorial Proof Strategy 6.49 (*The Product Rule for Counting*) Suppose the completion of a process requires that n steps be performed. Furthermore, suppose that:

- The first step can be performed in k_1 ways.
- For each way in which the first step can be performed, the second step can be performed in k_2 ways.
- For each way in which the first two steps can be performed, the third step can be performed in k_3 ways.
  $\vdots$
- For each way in which the first $n - 1$ steps can be performed, the nth step can be performed in k_n ways.

Then the total number of ways to complete the process is $k_1 \cdot k_2 \cdot k_3 \cdot \ldots \cdot k_n$.

Example 6.44 (continued) We can apply the Product Rule for Counting to determine the number of committees that can be formed consisting of one of two chemistry professors, one of three math professors, and one of five chemistry students. The process of forming such a committee can be viewed as a three-step process:

> *Step 1:* Choose one of the two chemistry professors to serve on the committee.
>
> *Step 2:* Choose one of the three math professors to serve on the committee.
>
> *Step 3:* Choose one of the five chemistry students to serve on the committee.

Once all three steps have been completed, we will have a committee of the type we want to create. Note that for each of the two ways of choosing the chemistry professor, that is, of performing *Step 1*, there are three ways to choose the math professor, that is, performing *Step 2*. Then, for each way of choosing the chemistry professor and the math professor, that is, of performing *Steps 1* and *2*, there are five ways to choose the chemistry student, that is, performing *Step 3*. Thus, according to the Product Rule for Counting, there are $2 \cdot 3 \cdot 5 = 30$ different committees that can be formed.

Example 6.50   Out of a group of 10 committee members, one member will serve as chair of the committee, another member will serve as vice chair, a third will serve as secretary, and a fourth as treasurer. In how many ways can these assignments be made?

Note that creating an assignment can be thought of as a four-step process: (1) assign a chair; (2) assign a vice chair; (3) assign a secretary; and (4) assign a treasurer. Because there are 10 committee members, there are 10 ways to perform Step 1. Of course, once the chair has been assigned, there are only nine unassigned committee members left. So, for each way of assigning a chair, there are then nine ways of assigning a vice chair. After assigning the chair and vice chair, there are only eight unassigned committee members remaining. So, for each way of assigning a chair and a vice chair, there are eight ways to assign a secretary. Then, once a secretary is assigned, there are only seven unassigned committee members left. So, for each way of assigning a chair, a vice chair, and a secretary, there are seven ways to perform the final step of assigning a treasurer. Thus, in all, there are $10 \cdot 9 \cdot 8 \cdot 7 = 5040$ ways to make the assignments.

> **Practice Problem 3**   A college student plans to take five courses next semester: a math course, a CS course, a psychology course, a music course, and a literature course. The student can choose from 4 different math courses, 3 different CS courses, 6 different psychology courses, 2 different music courses, and 7 different literature courses. In how many different ways can the student choose courses for the next semester?

Permutations

There are times when we need to count the number of ways a set of objects can be arranged in different orders. A **permutation** of a finite set is an arrangement of the members of the set into a particular order (see also Problem 6K).

Example 6.51   Three students, a, b, and c, are waiting to see a professor during office hours. Each of the following represents a permutation of the set of students:

$$abc \quad acb \quad bac \quad bca \quad cab \quad cba.$$

Each permutation represents a different order in which the professor can see the students.

Our next theorem provides us with a method for counting the number of permutations of a finite set. Its proof is an example of a so-called *combinatorial proof*; that is, a proof that relies on combinatorial proof strategies such as the Sum and Product Rules for Counting that we have introduced.

Theorem 6.52   There are $n!$ permutations of an n-element set.

Combinatorial Proof: Consider a set A having n elements. Forming a particular permutation of the elements of A can be thought of as an n-step process:

(1) Choose the element to be listed first.

(2) Choose the element to be listed second.

⋮

(n) Choose the element to be listed last.

Because A has n elements, there are n ways to choose the element to be listed first. Once the element to be listed first has been chosen, there are only $n - 1$ elements left. So, for each way of choosing the first element, there are $n - 1$ ways of choosing the second. Once the first two elements have been chosen, there are only $n - 2$ elements left. So, for each way of choosing the first two elements, there are $n - 2$ ways of choosing the third. This pattern continues until the final step, when there is only one element that has not been listed leaving only one way to choose the element to be listed last. Thus, according to the Product Rule for Counting, the number of permutations of A is $n \cdot (n - 1) \cdot (n - 2) \cdot \ldots \cdot 2 \cdot 1 = n!$.   □

Example 6.51 (continued)   We saw that there were six different orders in which three students could see a professor during office hours. This is correct based on Theorem 6.52 because $3! = 3 \cdot 2 \cdot 1 = 6$.

Practice Problem 4   A home improvement chain has eight stores and eight managers to be assigned, one to each store. In how many ways can this be done?

Sometimes we want to know in how many ways a certain number of objects can be chosen from a set of objects when the order in which elements are chosen is considered relevant. This is just a matter of counting permutations of the set having a certain "size."

A particular ordering of k of the elements of a finite set A having n elements overall is called a **k-permutation** of A. The number of k-permutations of an n-element set is denoted by $P(n, k)$.

Example 6.50 (continued)   Let the 10 committee members be a, b, c, d, e, f, g, h, i, and j. Then $fajc$ and $jfac$ are two different 4-permutations of the set of 10 committee members (remember that order is relevant here).

The next theorem provides us with a formula for $P(n, k)$; you are asked to develop a combinatorial proof of this result in Problem 6J.10.

Theorem 6.53   The number of k-permutations of an n-element set is

$$P(n, k) = n \cdot (n - 1) \cdot (n - 2) \cdot \ldots \cdot (n - k + 1) = \frac{n!}{(n - k)!}.$$

Example 6.50 (continued)   Note that each assignment of a chair, vice chair, secretary, and treasurer is actually a 4-permutation of the set of 10 committee members (the first of the four chosen serves as the chair, the second as the vice chair, the third as the secretary, and the fourth as the treasurer). Thus, there are

$$\frac{10!}{(10 - 4)!} = \frac{10!}{6!} = \frac{10 \cdot 9 \cdot 8 \cdot 7 \cdot 6 \cdot 5 \cdot 4 \cdot 3 \cdot 2 \cdot 1}{6 \cdot 5 \cdot 4 \cdot 3 \cdot 2 \cdot 1} = 10 \cdot 9 \cdot 8 \cdot 7 = 5040$$

possible ways to make the assignments (just as we saw earlier based simply on the Product Rule for Counting).

Because we have discovered that the number of 4-permutations of a 10-element set is 5040, we may write $P(10, 4) = 5040$.

Practice Problem 5   In how many ways can managers be assigned to stores if there are eight managers available but only five stores (so three of the managers will not be assigned to stores)?

Combinations and Binomial Coefficients

Sometimes we just want to count the number of subsets of a set having a certain "size" and do not want to distinguish among the various orders in which the elements of the subset can be arranged.

Example 6.54 How many different ways are there to choose three out of five CDs to take on a trip? The answer to this question is just the number of 3-element subsets of a 5-element set. Note that the order of selection is irrelevant in this situation so the answer is not $P(5, 3)$.

Given a finite set A having n elements, any subset of A having k-elements is called a **k-combination** of A. The number of k-combinations of an n-element set, that is, the number of k-element subsets of an n-element set, is denoted $C(n, k)$.

Theorem 6.55 The number of k-combinations of an n-element set is

$$C(n, k) = \frac{P(n, k)}{k!} = \frac{n!}{k!(n - k)!}.$$

Combinatorial Proof: Consider an n-element set A. Note that each k-permutation of A can be created via a two-step process: first, choose a subset of A having k elements; second, put the elements of this subset into a particular order. The first step can be done in $C(n, k)$ ways, and for each way of performing the first step, the second step can be done in $k!$ ways. So, by the Product Rule for Counting, we may conclude that $P(n, k) = C(n, k) \cdot k!$. Dividing through by $k!$ yields the desired result. □

Example 6.54 (Continued) Because $C(5, 3) = \dfrac{5!}{3!(5 - 3)!} = \dfrac{5!}{3!2!} = \dfrac{5 \cdot 4}{2} = 10,$

we see that there are 10 different ways to choose three CDs from a total of five.

Practice Problem 6 There are 19 experiments of a certain type competing for funding from the National Science Foundation (NSF), but budget limitations will permit only six of them to be funded. In how many ways can the NSF choose the experiments to be funded?

The number $C(n, k)$ is also referred to as a **binomial coefficient** and is denoted $\binom{n}{k}$, which is read as "n choose k" to reflect the fact that it counts the number of ways to choose a k-element subset of an n-element set.

Example 6.56 The binomial coefficient $\binom{5}{3}$ counts the number of 3-element subsets of a 5-element set. Because

$$\binom{5}{3} = \frac{5!}{3!(5-3)!} = \frac{5!}{3!2!} = 10,$$

any set having exactly 5 members has 10 different subsets each having exactly 3 members.

Practice Problem 7   Calculate and interpret the binomial coefficient $\binom{8}{2}$.

In Problem 6L, you are asked to prove a number of important properties of binomial coefficients.

Answers to Practice Problems

1. *Claim:* For any finite sets A and B, $|A - B| = |A| - |A \cap B|$.

    Proof: Let A and B be finite sets. Note that the sets $A - B$ and $A \cap B$ are disjoint, and that $A = (A - B) \cup (A - B)$. Hence, according to the Fundamental Assumption About Counting, we may conclude that $|A| = |A - B| + |A \cap B|$. It follows that $|A - B| = |A| - |A \cap B|$. □

2. Let P be the subset of the 100 respondents who own a PC, and let M be the subset who own a Mac. Then, $|P \cap M| = |P| + |M| - |P \cup M| = 58 + 41 - 91 = 8$, so 8 of the respondents own both a PC and a Mac.

3. Choosing the courses can be thought of as a five-step process: (1) choose a math course, (2) choose a computer science course, (3) choose a psychology course, (4) choose a music course, and (5) choose a literature course. Because there are four math courses to choose from, Step 1 can be completed in four ways. Then, for each way of completing Step 1, because there are three computer science courses to choose from, Step 2 can be completed in three ways. Continuing in this way, we see that the Product Rule for Counting applies and leads us to a total of $4 \cdot 3 \cdot 6 \cdot 2 \cdot 7 = 1008$ ways to choose the courses.

4. The number of ways to assign the managers to the stores is $8!$ because, if we imagine the stores already being listed in a fixed order, each assignment of managers to stores corresponds with a different way of ordering the eight managers (the first manager in a particular ordering is assigned to the first listed store, the second manager to the second listed store, and so forth).

(continues)

Answers to Practice Problems (continued)

5. The number of ways to assign five of the eight managers to the five stores is $P(8, 5) = \dfrac{8!}{(8-5)!} = 8 \cdot 7 \cdot 6 \cdot 5 \cdot 4 = 6720$, the number of 5-permutations of an 8-element set.

6. The number of ways the NSF can choose 6 of the 19 experiments to fund is $C(19, 6) = \dfrac{19!}{6!(19-6)!} = 27,132$.

7. The binomial coefficient $\binom{8}{2}$ counts the number of 2-element subsets of an 8-element set; that is, the number of ways to choose two items from a set of eight items when the order of selection is not relevant. Note that

$$\binom{8}{2} = \frac{8!}{2!(8-2)!} = 28.$$

Chapter 6 Problems

6A. Peano's Axioms

1. The set $\{2, 4, 6, \ldots\}$ is a model for Peano's axioms.

    (a) What number is playing the role of $*$?

    (b) What number is the successor of 16?

    (c) In general, how is the successor of a member of this set obtained?

2. This problem deals with the set $\{*, *', *'', \ldots\}$ of *abstract natural numbers* described in Axiom 6.3. For each nonnegative integer n, we shall denote by $*^n$ the abstract natural number previously expressed by writing $*$ with n primes on it. Thus, $*^0 = *$, $*^1 = *'$, $*^2 = *''$, and so forth.

Intuitively, the set of abstract natural numbers will be infinite if we can show that each "new" abstract natural number produced via (P1) of Axiom 6.3 is distinct from all "previously constructed" abstract natural numbers. This amounts to establishing the following:

Claim: For each nonnegative integer n, $*^{n+1} \notin \{*^0, *^1, *^2, \ldots, *^n\}$.

Use induction to prove this claim.

6B. More Intuition Regarding Induction

1. Imagine a ladder that has infinitely many steps. Suppose you want to convince a friend of yours that you can reach any step on this ladder. To do this via induction, you would have to prove both of the following claims:

   Base Claim: You can reach the first step on the ladder.

   Inductive Claim: Whenever you are able to reach the kth step on the ladder, it follows that you can reach the $(k + 1)$ step on the ladder.

Explain why it is not clear that you are able to reach any step on the ladder if you only establish the inductive claim. HINT: Imagine the ladder is of the type used in a fire escape, with its first (bottom) step some distance above the ground, and you are on the ground wanting to climb up it.

2. There is a flaw in the following argument that attempts to prove that for any natural number n, all the crayons in any box of n crayons are the same color (obviously not true unless $n = 1$). Find the flaw.

   All the crayons in any box having only one crayon are certainly the same color. Now let k be a fixed but anonymous natural number and assume that all the crayons in any box of k crayons are the same color. Consider a box of $k + 1$ crayons and let $c_1, c_2, \ldots, c_k, c_{k+1}$ be the crayons in this box. If we put the crayons $c_1, c_2, \ldots, c_k$ into a box, we would have a box of k crayons, so our assumption tells us that they all are the same color. Similarly, if we put the crayons $c_2, \ldots, c_k, c_{k+1}$ into a box, we would obtain another box of k crayons, so our assumption again tells us that they all are the same color. Then, because the boxes containing $c_1, c_2, \ldots, c_k$ and $c_2, \ldots, c_k, c_{k+1}$ would have crayons $c_2, \ldots, c_k$ in common, it follows that all of the crayons $c_1, c_2, \ldots, c_k, c_{k+1}$ are the same color. Thus, the Principle of Mathematical Induction allows us to conclude that, for any natural number n, all the crayons in any box of n crayons are the same color.

6C. Proving Summation Formulas by Induction

Use induction to prove each of the following.

1. For any $n \in \mathbf{N}, \displaystyle\sum_{i=1}^{n} i = \frac{n(n + 1)}{2}$. (Note that this gives a formula for the sum of the first n natural numbers. Legend has it that the great mathematician Karl Friedrich Gauss discovered this result when he was 8 years old.)

2. For any $n \in \mathbf{N}$, $\displaystyle\sum_{i=1}^{n} i^3 = \frac{n^2(n+1)^2}{4}$.

3. For any nonnegative integer n, $\displaystyle\sum_{k=0}^{n} 2^k = 2^{n+1} - 1$.

4. For any $n \in \mathbf{N}$, $\displaystyle\sum_{k=1}^{n} (2k - 1) = n^2$.

5. For any $n \in \mathbf{N}$, $\displaystyle\sum_{j=1}^{n} \frac{1}{j(j+1)} = \frac{n}{n+1}$.

6D. Induction and Problem Solving

1. Let n be any natural number and imagine you have been given 3^n coins, all identical except for one that is heavier than the others. Use induction to show that you can use a pan balance to identify the heavier coin in n weighings.

2. A checkerboard is an 8×8 grid of identical squares. Let n be any natural number and imagine a generalized checkerboard having dimensions $2^n \times 2^n$ (checkerboards usually have the squares colored red and black, but the colors of the squares are irrelevant here). Remove any one square from this generalized checkerboard. Use induction to show that what remains can be tiled with L-shaped blocks made up of three squares like the one shown below with the squares in the L-shaped block being the same size as the squares making up the checkerboard. NOTE: By "tiled" we mean that the L-shaped blocks must be placed on the board so that each one covers up exactly three squares of the original board, and they cannot be placed so as to overlap one another; also, what remains of the board after the removal of the one square must ultimately be entirely covered by the L-shaped blocks.

3. (*Towers of Hanoi Problem*) Let n be any natural number. In front of you are three pegs. Stacked on one of the pegs, in order of size from the largest on the bottom to the smallest on the top, are n different-sized disks. The other two pegs are empty. Your goal is to restack the disks onto one of the other pegs. You can only move one disk at a time, and you can never put a larger disk on top of a smaller disk. The third peg is there to help you accomplish this task. Each move of a single disk is

counted as a "step." Determine how many steps are required to transfer the entire stack of n disks (your answer will be a formula involving n). Then use induction to verify your formula.

6E. Recursively Defined Sequences

1. Define $A_0 = \{0\}$ and for each $n \in \mathbf{N}$, define $A_n = A_{n-1} \cup \{n\}$. Use induction to prove that for every nonnegative integer n, $A_n = \{0, 1, 2, ...,n\}$.

2. Let P be the amount of money you deposit into a bank account today. Suppose that you do not plan to make any additional deposits into the account and that you will not withdraw any of the money. Furthermore, assume the account earns interest at an annual rate r (expressed as a decimal) compounded monthly. Let $B(n)$ be the account balance after n months. Then $B(0) = P$, and for each $n \in \mathbf{N}$, $B(n) = B(n-1) + \frac{r}{12} B(n-1)$, as the balance after n months is obtained by adding the interest earned in the nth month to the previous balance. Use induction to prove that, for each nonnegative integer n,

    $$B(n) = P\left(1 + \frac{r}{12}\right)^n.$$

3. Let a_0 and r be fixed real numbers with $r \neq 0$ and $r \neq 1$, and suppose that for each $n \in \mathbf{N}$, $a_n = ra_{n-1}$. This produces a special type of sequence $a_0, a_1, a_2, a_3, \ldots$ of real numbers called a **geometric sequence**.

    (a) Use induction to prove that, for every nonnegative integer n, $a_n = a_0 r^n$.

    (b) Use induction and the result from (a) to prove that, for every nonnegative integer n, $\displaystyle\sum_{i=0}^{n} a_0 r^i = \frac{a_0(1 - r^{n+1})}{1 - r}$.

    (c) Recall from calculus that for a fixed number r with $-1 < r < 1$, $r^n \to 0$ as $n \to \infty$. Use this fact, along with the result from (b), to show that for any constants a and r with $-1 < r < 1$, $\displaystyle\sum_{i=0}^{\infty} ar^i = \frac{a}{1 - r}$.

    (This is the formula for the sum of a geometric series having first term a and ratio r.)

4. It is possible to recursively define a sequence of functions. Define the function f_1 so that $f_1(x) = -\frac{1}{x}$. Then, for each integer $n \geq 2$, define the function f_n to be the derivative of f_{n-1}. Write out the first five terms of the sequence f.

5. An interval is **closed** if it includes all its endpoints and **open** if it does not include any of its endpoints. So, for instance, the interval $[0, 1]$ is closed, the interval $(0, 1)$ is open, and the interval $(0, 1]$ is neither open nor closed.

Define C_1 to be the closed interval $[0, 1]$. For each integer $n \geq 2$, define C_n to be the union of the closed intervals obtained from C_{n-1} by removing the open "middle third" of all the closed intervals comprising C_{n-1}. Thus, for example,

$$C_2 = [0, 1] - (1/3, 2/3)$$

$$= [0, 1/3] \cup [2/3, 1]$$

and

$$C_3 = ([0, 1/3] - (1/9, 2/9)) \cup ([2/3, 1] - (7/9, 8/9))$$

$$= [0, 1/9] \cup [2/9, 1/3] \cup [2/3, 7/9] \cup [8/9, 1].$$

(a) Express C_4 as a union of closed intervals.

The set $C_1 \cap C_2 \cap C_3 \cap \ldots = \bigcap_{n=1}^{\infty} C_n$ formed by intersecting all of the sets in the sequence of sets C_1, C_2, C_3, ... is called the **Cantor set**. One thing that makes the Cantor set unusual is that it is an infinite set whose "length" can be regarded as 0.

(b) Find infinitely many numbers that are in the Cantor set.

(c) Show that it is reasonable to regard the "length" of the Cantor set as being 0 by finding the sum of the lengths of all the intervals deleted from $[0, 1]$ in constructing all of the sets C_1, C_2, C_3, ... (The sum you are being asked to find is actually the sum of a geometric series so you can make use of the formula given previously in Problem 3c.)

6. Suppose that $a_1 = 5$, $a_2 = 13$, and for each natural number $n \geq 3$, $a_n = 5a_{n-1} - 6a_{n-2}$. Use strong induction to prove that for every $n \in \mathbf{N}$, $a_n = 2^n + 3^n$.

6F. The Fibonacci Sequence

Let F_n be the nth term of the Fibonacci sequence (see Example 6.15).

1. The famous mathematician Fibonacci, also known as Leonardo of Pisa, stated the following problem in the year 1202.

Start with two baby rabbits, one male and one female. These baby rabbits become adult rabbits the next month. The month after that they produce a pair (one male, one female) of baby rabbits. Now in each subsequent month, each pair of baby rabbits reaches adulthood and each pair of adult rabbits produces a new pair (one male, one female) of baby rabbits. How many pairs of rabbits will there be after n months?

Explain why the answer to this problem is F_n.

2. Note that it appears every third term of the Fibonacci sequence is even. Use induction to prove that, for every $n \in \mathbf{N}$, F_{3n} is even.

3. Use induction to prove that, for each $n \in \mathbf{N}$, $\sum_{k=1}^{n} F_k = F_{n+2} - 1$.

4. Take any line segment and divide it into two unequal segments in such a way that the ratio of the shorter segment formed to the longer segment formed is the same as the ratio of the longer segment to the original whole. This common ratio is known as the **golden ratio** and is denoted γ (the lowercase Greek letter "gamma"). Show that $\gamma = \dfrac{\sqrt{5} - 1}{2}$.

5. Use strong induction to prove that, for each integer $n \geq 3$, $F_n > (1/\gamma)^{n-2}$. You may find it useful to first show that the reciprocal of the golden ratio γ is $\dfrac{1 + \sqrt{5}}{2}$.

6. Use strong induction to prove that, for every $n \in \mathbf{N}$, $F_n = \dfrac{1}{\sqrt{5}} ((1/\gamma)^n - (-\gamma)^n)$. This provides an *explicit*, as opposed to *recursive*, formula for the terms of the Fibonacci sequence.

6G. The Sieve of Eratosthenes

1. Prove that every composite integer n has a prime divisor less than or equal to $\sqrt{n}$.

2. Thus, for each prime p less than or equal to $\sqrt{n}$, if we eliminate all multiples of p greater than p but less than or equal to n from a list of the positive integers $2, \ldots, n$, we will be left with all of the prime numbers less than or equal to n. This method for determining primes is called the *Sieve of Eratosthenes*. Use this method to find all primes less than 100.

6H. The Fundamental Theorem of Arithmetic and Prime Factorizations

1. Recall that the *Fundamental Theorem of Arithmetic* tells us any integer greater than 1 can be expressed uniquely (up to the order of the factors) as a product of primes. Back in Section 6.4 we proved the existence portion of this result. In this problem you will eventually establish the uniqueness portion.

(a) Prove that if a and b are positive integers having 1 as their greatest common divisor, then there exist integers s and t such that $sa + tb = 1$. HINT: By the well-ordering of **N** there must be a smallest positive integer d for which there exist integers s and t such that $sa + tb = d$. Use the Division Algorithm to show that $d|a$ and $d|b$, thus making d a common positive divisor of a and b. Conclude, since 1 is the greatest common divisor of a and b, that $d = 1$.

(b) Prove that if a and b are positive integers having 1 as their greatest common divisor and $a|bc$ for some positive integer c, then $a|c$. HINT: Make use of the result from (a).

(c) Let p be a prime and a_1, a_2, ..., a_n be positive integers. Use induction to prove that if $p|a_1 a_2 \dots a_n$, then $p|a_i$ for some $i \in \{1, 2, ..., n\}$. HINT: Make use of the result from (b).

(d) Use the result from (c) to prove that if p is a prime and k is a positive integer, the only positive divisors of p^k are 1, p, ..., p^k.

(e) Use the results from (c) and (d) to prove that it is not possible for an integer greater than 1 to be expressed as a product of primes in two different ways (different in more than just the ordering of the factors). HINT: Try contradiction.

2. Explain how the prime factorization of an integer n greater than 1 can be used to determine whether a particular prime p divides n.

3. Explain how the prime factorization of an integer n greater than 1 can be used to determine the largest power of a prime p that divides n.

4. Given two integers m and n both greater than 1, how can we use the integers' prime factorizations to determine whether $m|n$?

5. Prove $\sqrt[3]{5}$ is irrational. HINT: Find a way to adapt the argument used in the text to show $\sqrt{2}$ is irrational.

6I. More Results from Elementary Number Theory

1. Prove that if $a|b$ and $b|c$, then $a|c$.

2. Prove that if $c|a$ and $c|b$, then $c|(ma + nb)$ for any integers m and n.

3. Let a, b, and m be given integers with $m > 1$. Prove $m|(a - b)$ if and only if a and b both yield the same remainder (as determined from the Division Algorithm) when divided by m.

4. Use factoring and the Division Algorithm to prove that $n^2 + n$ is even for any integer n.

5. Verify that for any integer n, $3|(n^3 - n)$ in two ways:

    (a) first, by factoring $n^3 - n$ and applying the Division Algorithm;

    (b) then, by using induction.

6. Show that the integer 1 is neither prime nor composite.

7. Show that the only positive integer divisors of a prime p are 1 and p.

8. Show that all primes other than 2 are odd.

9. Prove there are no other sets of three consecutive odd primes other than 3, 5, 7. HINT: Try applying the Division Algorithm with divisor 3.

10. Prove that every integer greater than 11 can be written as the sum of two composite integers.

11. Prove that for any positive integer n, the n consecutive integers

$$(n + 1)! + 2, \ (n + 1)! + 3, \ ..., \ (n + 1)! + (n + 1)$$

are all composite. (Thus, although there are infinitely many primes, for any positive integer n, we can find a set of n consecutive integers none of which are prime.)

6J. Counting Problems

1. Find the number of positive integers that divide 10,000.

2. If a valid license plate consists of two letters followed by three single-digit numbers, how many valid license plates are there?

3. A pizza parlor has five meat and six vegetable toppings that can be added to a basic cheese pizza. The pizzas come in small, medium, and large sizes.

    (a) How many different kinds of pizzas can be ordered?

    (b) How many different kinds of pizzas with at least one vegetable topping can be ordered?

4. How many three-digit positive integers with distinct digits have all odd or all even digits?

5. A binary string is a listing of 0's and/or 1's in some particular order. How many binary strings of length 8 contain exactly four 1's?

6. Twelve runners compete in a race for which prizes are awarded for first-place, second-place, and third-place finishers. In how many different ways could the prizes be assigned?

7. A large company has a total of 20,000 employees and wishes to pilot a new health insurance plan with a random sample of 500 of them. Use a computer algebra system to find the number of different such random samples that could be obtained.

8. Use induction to prove that ordered n-tuples are equal if and only if their corresponding coordinates are equal. HINT: The base claim is established by Theorem 5.2. So you need only use the formal definition of ordered n-tuple to show that if ordered k-tuples are equal precisely when their corresponding coordinates are equal, it follows that ordered $(k + 1)$-tuples are equal precisely when their corresponding coordinates are equal.

9. Use induction to prove that for any integer $n \geq 2$ and any finite sets A_1, A_2, ..., A_n, $|A_1 \times A_2 \times \cdots \times A_n| = |A_1| \cdot |A_2| \cdot \ldots \cdot |A_n|$. HINT: The base claim has already been established in Theorem 6.45, so you really just need to formulate and prove the appropriate inductive claim.

10. Give a combinatorial proof showing that the number of k-permutations of an n-element set is $n \cdot (n - 1) \cdot (n - 2) \cdot \ldots \cdot (n - k + 1) = \dfrac{n!}{(n - k)!}$.

6K. Permutations

The notion that a *permutation* of a set is an ordering of the elements of the set is actually relatively informal. Formally, a **permutation** of a set S is a bijection from S onto itself. Note that this more formal definition of *permutation* allows us to consider permutations of infinite sets as well as finite sets.

1. The identity function on a set is always a permutation of that set. Give a different example of a formalized version of a permutation of $\{1, 2, 3\}$. Then explain why the formal definition of *permutation of a set* captures the intuitive notion that a permutation of a set is an arrangement of the members of the set in a particular order.

2. Give an example of a permutation of $\mathbf{N}$ different from the identity function on $\mathbf{N}$.

3. Give an example of a permutation of $\mathbf{R}$ different from the identity function on $\mathbf{R}$.

4. Explain why the composition of permutations is a permutation.

5. Explain why the inverse of a permutation is a permutation.

6L. Binomial Coefficients

While it is possible to use the "factorial formula" for computing $\dbinom{n}{k}$ to establish various properties of binomial coefficients, it is usually more instructive to attempt to develop combinatorial proofs of such properties. In this spirit, your work in this problem should not use the factorial formula. Instead, your interpretations and explanations must be based on the fact that $\dbinom{n}{k}$ represents the number of k-element subsets of an n-element set.

Sample Problem.   Interpret $\binom{n}{1}$ and explain why $\binom{n}{1} = n$.

Sample Solution.   The binomial coefficient $\binom{n}{1}$ represents the number of 1-element subsets of an n-element set. Because an n-element set has exactly n singleton subsets, $\binom{n}{1} = n$.

1. Interpret $\binom{n}{n}$ and explain why $\binom{n}{n} = 1$.

2. Interpret $\binom{n}{0}$ and explain why $\binom{n}{0} = 1$.

3. Interpret $\binom{n}{n-1}$ and explain why $\binom{n}{n-1} = n$.

4. Interpret $\binom{n}{k}$ and $\binom{n}{n-k}$, and then explain why $\binom{n}{k} = \binom{n}{n-k}$.

5. Interpret $\binom{n+1}{k+1}$, $\binom{n}{k+1}$, and $\binom{n}{k}$, and then explain why
    $$\binom{n+1}{k+1} = \binom{n}{k} + \binom{n}{k+1}.$$

    HINT: From a set containing $n+1$ elements, consider one of the elements to be "special," the others not. A subset of the $(n+1)$-element set will either include this "special" element or it will not. How many of each kind of subset are there?

6. Consider $\binom{n}{0} + \binom{n}{1} + \binom{n}{2} + \binom{n}{3} + \cdots + \binom{n}{n-1} + \binom{n}{n}$. Give as simple a description as you can of what this sum represents. Do you know a simple formula for this quantity?

6M. The Binomial Theorem

The Binomial Theorem states that for any nonnegative integer n,

$$(x + y)^n = \sum_{k=0}^{n} \binom{n}{k} x^k y^{n-k}.$$

1. Verify the special cases of the Binomial Theorem for the powers $n = 2$ and $n = 3$ by expanding and simplifying $(x + y)^2$ and $(x + y)^3$.

2. The Binomial Theorem is actually helpful in expanding any natural number power of a multitermed expression. Find a way to use the Binomial Theorem to simplify $(a - b + \sqrt{2})^4$.

3. When multiplying two or more algebraic expressions together, the distributive property tells us that the result will be the sum of all products obtained by choosing a single term from each factor and multiplying these terms together. Here are some examples:

$$(a + b)(c + d) = ac + ad + bc + bd,$$
$$(a + b)(c + d + e) = ac + ad + ae + bc + bd + be,$$
$$(a + b)(c + d)(e + f) = ace + acf + ade + adf + bce + bcf + bde + bdf.$$

Use this method for multiplying algebraic expressions to write a combinatorial proof of the Binomial Theorem. HINT: You will need to use the interpretation of the binomial coefficient $\binom{n}{k}$ as the number of k-element subsets of an n-element set, but you will have to make clear exactly what this binomial coefficient is counting in your argument.

4. It is also possible to prove the Binomial Theorem by induction. Do so. HINT: In proving the inductive claim, use the fact that $\binom{n}{k-1} + \binom{n}{k} = \binom{n+1}{k}$, which is essentially the result of Problem 6L.5.

6N. Pascal's Triangle

Shown below is the top portion of a triangular array of numbers known as *Pascal's Triangle*.

$$
\begin{array}{c}
1 \\
1 \quad 1 \\
1 \quad 2 \quad 1 \\
1 \quad 3 \quad 3 \quad 1 \\
1 \quad 4 \quad 6 \quad 4 \quad 1 \\
1 \quad 5 \quad 10 \quad 10 \quad 5 \quad 1 \\
1 \quad 6 \quad 15 \quad 20 \quad 15 \quad 6 \quad 1 \\
\vdots
\end{array}
$$

As indicated, Pascal's Triangle really never ends. We can always attach more rows if we understand the process by which one row generates the next one.

1. Examine the portion of Pascal's Triangle displayed and determine how, mathematically, each row generates the next one. Then give the next three rows of Pascal's Triangle.

2. Think of each row in Pascal's Triangle as corresponding with a value of n, beginning at the top with $n = 0$ and continuing downward with $n = 1$, $n = 2$, and so forth. For instance, the row corresponding with $n = 2$ is

1   2   1.

Note that we may also subdivide Pascal's Triangle into diagonals running parallel to the left side of the triangle. Because the triangle really never ends, neither do these diagonals. Think of each of these diagonals as corresponding with a value of k, beginning at the left with $k = 0$ and continuing to the right (and downward) with $k = 1$, $k = 2$, and so forth. For instance, the diagonal corresponding with $k = 0$ consists of infinitely many 1's (this is the left side of the triangle), and the diagonal corresponding with $k = 2$ begins (starting at the upper right and working toward the lower left)

1   3   6   10   15   ....

(a) For each of the following pairs of values of n and k, identify the number in the nth row and along the kth diagonal of Pascal's Triangle: $n = 4$ and $k = 3$; $n = 5$ and $k = 2$; $n = 7$ and $k = 3$; $n = 8$ and $k = 4$.

(b) For each pair of values of n and k given in (a) determine the numerical value of $\binom{n}{k}$.

(c) Based on your answers to (a) and (b), formulate a conjecture about the number located in the nth row and along the kth diagonal of Pascal's Triangle.

(d) Look at your expansions of $(x + y)^2$ and $(x + y)^3$ obtained in Problem 6M.1. What do you notice about the coefficients of the resulting terms in these expansions in relation to Pascal's Triangle? Can you make a more general conjecture?

(e) If you conjectured, back in (c), that the number located in the nth row and along the kth diagonal of Pascal's Triangle is $\binom{n}{k}$, then you are correct. Use this fact to indicate how the process used to get from one row of Pascal's Triangle to the next helps to explain why $\binom{n + 1}{k + 1} = \binom{n}{k} + \binom{n}{k + 1}$.

3. This problem asks you to delve into the history of Pascal's Triangle.

(a) Where and when did Blaise Pascal live?

(b) Discuss two appearances of the triangular arrangement of binomial coefficients within the mathematics of non-European cultures before the time of Pascal.

(c) Though Pascal cannot be credited with inventing the triangle that bears his name, what did he accomplish in his work *Triangle Arithmetique* that makes it reasonable to refer to the triangle as Pascal's Triangle?

4. A **fractal** is a geometric figure that exhibits some degree of self-similarity under appropriate magnification. That is, as we begin to look more closely at a fractal image, we begin to see the entire image replicate itself on a more microscopic level. Pascal's Triangle, if viewed the right way, can reveal a fractal image that is called **Sierpinski's Triangle**. This problem provides one way of beginning to create this image.

Instead of creating Pascal's Triangle itself, imagine a version in which each odd number is replaced by 1 and each even number is replaced by 0 (so the triangular array being created consists of 1's and 0's only); we will call this a "binary" version of Pascal's Triangle. You can create it by using the equations $0 + 0 = 0$, $0 + 1 = 1$, $1 + 0 = 1$, and $1 + 1 = 0$, which reflect, respectively, the facts that the sum of two even integers is even, the sum of an even and an odd is odd, the sum of an odd and an even is odd, and the sum of two odds is even.

Use a piece of graph paper to construct a binary version of Pascal's Triangle to at least its first 20 rows, writing in pencil. Then erase all of the 0's from your figure. You should begin to see Sierpinski's Triangle appear (the more rows of Pascal's Triangle that you construct, the more apparent the fractal image).

Chapter 7

Further Mathematical Explorations

The text culminates by considering three different mathematical topic areas in which you can apply the problem-solving, reasoning, and proof-writing skills developed in the earlier chapters. Each section is independent of the others, and the problems are embedded within the narrative so as to promote active reading and investigation. In attempting to solve a given problem in a certain section, you will sometimes need to make use of the results from problems stated earlier in the section.

7.1 Exploring Graph Theory

Loosely speaking, a *graph* is a finite set of points, called *vertices* (singular: *vertex*), some of which may be joined to one another by *edges*. The idea is that an edge between two vertices indicates some sort of connection or relationship between whatever those vertices represent.

In the graph displayed in Figure 7.1, there are exactly seven vertices M, A, B, W, S, H, and P, representing the seven New England cities Manchester, Augusta, Boston, Worcester, Springfield, Hartford, and Providence, respectively. There are also exactly seven edges in this graph; each edge indicates that two of the cities are connected via an interstate highway. The only vertices "on" an edge are the "endpoints" of the edge.

More formally, a **graph** consists of a nonempty finite set of **vertices**, together with a collection of two-element subsets of the set of vertices, the **edges** of the graph. For example, the graph in Figure 7.1 has $\{M, A, B, W, S, H, P\}$ as its set of vertices and $\{\{M, B\}, \{A, B\}, \{S, W\}, \{W, B\}, \{B, P\}, \{S, H\}, \{H, W\}\}$ as its collection of edges. Typically, though, a graph is represented by a diagram rather than set-theoretically, and an edge, say $\{M, B\}$, is usually denoted either MB or BM.

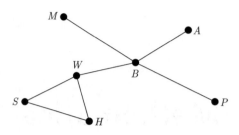

Figure 7.1

Given an edge vw in a graph, we refer to v and w as the **endpoints** of the edge, and we say that the edge **joins** the vertices v and w. Two vertices in a graph are **adjacent** if there is an edge joining them. An endpoint of an edge is said to be **incident** to the edge, and an edge is said to be **incident** to each of its endpoints.

Problem 7.1.1 Consider the graph depicted in Figure 7.1.

(a) What are the endpoints of the edge incident to P?

(b) Is there an edge joining S and B?

(c) Which vertices are adjacent to B?

(d) Which edges are incident to W?

The only items of interest in a graph are the vertices and the adjacency relationships described by the edges. Thus, in a diagram representing a graph, the following are not relevant:

- the specific positioning of a vertex;
- the length of an edge;
- whether an edge is straight or curved;
- whether edges cross one another or not.

Hence, the "same" graph may be drawn in infinitely many ways.

The graph in Figure 7.1 models part of a transportation network. Graphs are also used to model communication networks, power grids, game schedules for sports teams, information storage and retrieval in computer databases, and many other situations.

Problem 7.1.2 Draw a graph representing a computer network consisting of five computers C_1, C_2, C_3, C_4, and C_5, with C_1 directly linked to each of the others, but no other direct connections.

The **complete graph** on n vertices is the graph consisting of n vertices and an edge joining each pair of distinct vertices.

Problem 7.1.3 Draw the complete graphs on 1, 2, 3, and 4 vertices.

Problem 7.1.4 Find a formula for e_n, the number of edges in the complete graph on n vertices.

As we have defined the notion of *graph*, an edge always joins two distinct vertices, and two distinct vertices can be joined by at most one edge. Sometimes, though, we want to allow an edge to join a vertex to itself. At other times we would like to allow for more than one edge to join specified vertices. We will refer to the generalization of a graph that allows for these possibilities as a **multigraph**. Figure 7.2 shows a multigraph in which the vertices are a, b, c, and d, and the edges are $e1$, $e2$, $e3$, and $e4$ (we have separately named the edges to avoid ambiguity; for instance, it would not be clear which of two edges is being referred to by our usual edge notation ba).

An edge in a multigraph having only one endpoint is called a **loop**. Two distinct edges having the same endpoints are called **parallel edges**. Note that a *graph* is just a multigraph having no loops or parallel edges. The terms *endpoint, joins, adjacent,* and *incident* are used in the context of a multigraph in essentially the same ways that they are used in the context of a graph.

Problem 7.1.5 Answer these questions about the multigraph in Figure 7.2.

(a) Are the vertices a and b adjacent to each other?

(b) Are the vertices a and c adjacent to each other?

(c) Which vertex is not adjacent to any vertices?

(d) What is the endpoint of the loop in this multigraph?

(e) What are the endpoints of the parallel edges in this multigraph?

(f) What vertices are incident to edge $e2$? to edge $e1$?

(g) What edges are incident to the vertex b?

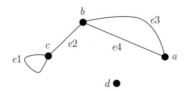

Figure 7.2

Problem 7.1.6 (*The Handshake Problem*)   Four married couples, including the host and hostess, attend a party. Several handshakes take place, but no one shakes hands with himself or herself, with his or her spouse, or with anyone else more than once. Suppose the hostess asks each of the other people how many persons they shook hands with, and each person gives her a different answer. Determine the number of people the hostess shook hands with and the number of people the host shook hands with. HINT: Model the situation with a graph in which the vertices represent the eight people who attended the party, and an edge between two vertices indicates those people shook hands. Consider different possibilities based on the scenario described, eliminating those that do not conform to the given requirements.

The **degree** of a vertex in a (multi)graph is the number of edges incident to the vertex, with the provision that any loops incident to the vertex are counted twice.

Problem 7.1.7   Determine the degree of each vertex in the graph of Figure 7.1 and the multigraph of Figure 7.2.

Problem 7.1.8   Prove that in any (multi)graph, the sum of the degrees of the vertices is twice the number of edges.

Problem 7.1.9   Draw a multigraph having *all* of the following properties:

- there are five vertices v, w, x, y, and z, no more;
- the sum of the degrees of the vertices is 12;
- v and y are adjacent;
- y is adjacent to itself;
- the degree of z is 1 and the degree of v is 4;
- an edge called e is incident to x and is parallel to another edge f that is incident to w;
- there is a loop having endpoint v.

A graph in which the vertices and edges are unnamed is referred to as an **unlabeled graph**. Each of the graphs G_1, G_2, G_3, G_4, and G_5 shown in Figure 7.3 is an unlabeled graph.

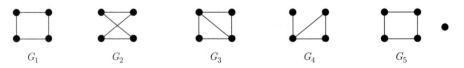

G_1 G_2 G_3 G_4 G_5

Figure 7.3

Note that the graphs G_1, G_2, G_3, and G_4 each have four vertices, whereas G_5 has five vertices. Also, each of the graphs G_1, G_2, G_4, and G_5 has four edges, whereas G_3 has five edges. In the graph G_2, two edges are drawn so that they "cross," but this crossing point is not a vertex.

We are interested in situations in which two graphs, though they may be drawn so as to look different from one another, are actually the same in terms of their essential structure.

Problem 7.1.10   Consider the graphs G_1, G_2, G_3, G_4, and G_5 displayed in Figure 7.3. Argue why, fundamentally, G_1 and G_2 are "the same" structurally, whereas G_3, G_4, and G_5 exhibit "different" structures from each other and from G_1 and G_2.

Graphs possessing exactly the same underlying structure are said to be *isomorphic*. Intuitively, we view two graphs as being isomorphic if they can be labeled or relabeled so that they have exactly the same vertex sets and exactly the same edge sets.

To facilitate more efficient communication, we shall sometimes use $V(G)$ to refer to the set of vertices of a graph G and $E(G)$ to refer to the set of edges of G. Formally, two graphs G and H are **isomorphic** if there is a bijection $f:V(G){\rightarrow}V(H)$ between their vertex sets that preserves adjacency of vertices, meaning that for any vertices $v, w \in V(G)$, $vw \in E(G)$ if and only if $f(v)f(w) \in E(H)$. A bijection with this property is called an **isomorphism**.

Problem 7.1.11   Create an isomorphism between the graphs G_1 and G_2 from Figure 7.3. Verify that all the required properties hold.

Problem 7.1.12   Let $f:V(G){\rightarrow}V(H)$ be an isomorphism of the graphs G and H. Prove that, for each vertex v of G, the degree of v in G is equal to the degree of $f(v)$ in H.

Problem 7.1.13   Use the conclusion of Problem 7.1.12 to explain why the graphs G_1, G_3, G_4, and G_5 from Figure 7.3 are all non-isomorphic to each other.

Problem 7.1.14   How many non-isomorphic graphs are there each having exactly

(a)   three vertices? (Sketch them.)

(b)   four vertices? (Sketch them.)

In 1736, the great mathematician Leonard Euler solved the following problem, known as the *Königsberg Bridge Problem*. A river flowing through the old city of Königsberg separated it into four land masses, labeled A, B, C, and D

in Figure 7.4. The diagram also displays the seven bridges that connected these land masses. The problem is to determine whether it is possible to start at one of the land masses and then walk over each of the seven bridges exactly once, returning to the original starting point.

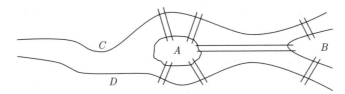

Figure 7.4

Euler modeled the problem via the multigraph shown below, with each land mass represented by a vertex and each bridge represented by an edge.

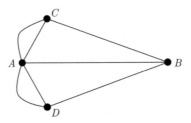

Euler's use of this model initiated the study of graph theory. Problem 7.1.22 provides the basis for solving the Königsberg Bridge Problem.

A **path** in a (multi)graph from a vertex v to a vertex w is a finite sequence of alternating vertices and edges that begins with the vertex v and ends with the vertex w, and where each edge in the sequence is incident to the vertices listed before it and after it. In describing a path in a multigraph that is not a graph, it is important to include the edges being traversed because there may be more than one edge between two vertices. For instance, in the multigraph of Figure 7.2, referring to a path abc is ambiguous; we should write either $ae3be2c$ or $ae4be2c$ to make clear which path we are referring to. In the case of a multigraph that is actually a graph, we can omit the edges when describing a path because there is at most one edge between two distinct vertices. Thus, considering the graph from Figure 7.1, the path $BWSH$ is unambiguous.

If the beginning vertex of a path is the same as the ending vertex, the path is said to be **closed**. Thus, for example, $BWSHWB$ is a closed path in the graph from Figure 7.1.

The **length** of a path is the number of edges in the path (repeated edges are counted in the length according to the number of times they appear). For instance, the lengths of the paths $BWSH$ and $BWSHWB$ in the graph from Figure 7.1 are, respectively, 3 and 5.

A path is **simple** if it has no repeated vertices (hence, no repeated edges either). The path *BWSH* in the graph from Figure 7.1 is simple, whereas the path *BWSHWB* is not. Note that any single vertex of a graph may be viewed as a simple path of length 0.

Problem 7.1.15   Consider again the graph displayed in Figure 7.1.

(a) Give an example of a path of length 2 in this graph.

(b) Give an example of a closed path of length 3 in this graph.

(c) Give an example of a path of length 5 in this graph that is not simple.

(d) Give an example of a simple path in this graph that is at least as long as any other simple path in the graph.

Problem 7.1.16   Consider again the multigraph you created in Problem 7.1.9.

(a) If possible, find a path from v to x.

(b) If possible, find a path from y to z of length 4.

(c) Find a closed path in this graph.

(d) Find a simple path from y to z.

Problem 7.1.17   Explain why, in any (multi)graph, a path from v to w must "contain" a simple path from v to w. (Thus, a given path may always be "cut back" to form a simple path.) Then, illustrate your reasoning by finding a simple path having the same beginning vertex and same ending vertex as the path you gave as your answer to Problem 7.1.15(c).

A **cycle** in a (multi)graph is a closed path with no repeated edges and in which the only repeated vertex is the one appearing first and last.

Problem 7.1.18   Find a cycle in the graph shown in Figure 7.1.

A path that includes exactly once every edge of a (multi)graph and that begins and ends with different vertices is called an **Euler path**. A path that includes exactly once every edge of a (multi)graph and that begins and ends with the same vertex is called an **Euler circuit**.

Problem 7.1.19   Determine whether the given (multi)graph contains either an Euler path or an Euler circuit:

(a) the graph from Figure 7.1.

(b) the multigraph from Figure 7.2.

(c) the graph G_1 from Figure 7.3.

(d) the multigraph consisting of four vertices a, b, c, and d and five edges $e1$, $e2$, $e3$, $e4$, and $e5$, and for which $e1$ and $e3$ are parallel edges with endpoints a and b, $e3$ has endpoints b and c, $e4$ has endpoints c and d, and $e5$ has endpoints b and d.

A (multi)graph is **connected** if there is a path between any two distinct vertices; otherwise it is **disconnected**. Thus, the graph from Figure 7.1 is connected, whereas the multigraph from Figure 7.2 is disconnected.

Problem 7.1.20   Find two non-isomorphic connected graphs each of which has exactly three vertices of degree 1, exactly two vertices of degree 2, exactly one vertex of degree 3, and no other vertices.

Problem 7.1.21   Prove that if one of two isomorphic graphs is connected, so is the other.

Problem 7.1.22   Show that a connected (multi)graph contains an Euler circuit if and only if each of its vertices has even degree. Then explain how this result solves the Königsberg Bridge Problem.

A connected graph having no cycles is called a **tree**. The graph in Figure 7.5 is a tree. However, as the graph from Figure 7.1 contains a cycle, it is not a tree. A graph having exactly four vertices and exactly two edges cannot be a tree because, although it will not contain a cycle, it is necessarily disconnected.

Figure 7.5

Problem 7.1.23   Explain how the tree shown in Figure 7.5 provides a model for a single-elimination playoff series that begins with eight teams.

Problem 7.1.24   Illustrate how the organization of documents into files and subfiles in a computer can be modeled by a tree.

Problem 7.1.25   Prove each of the following:

(a)   There is exactly one simple path between any two given vertices of a tree.

(b)   A tree having more than one vertex must have at least two vertices of degree 1. HINT: Consider a simple path of maximal length.

(c)   The number of edges in a tree is always one less than the number of vertices. HINT: Try induction.

(d)   If an edge is removed from a tree, but in doing so no vertices are removed, the resulting graph must be disconnected.

(e)   If an edge is inserted into a tree, and in doing so no additional vertices are inserted, the resulting graph must contain a cycle.

Problem 7.1.26   Draw all unlabeled trees that are non-isomorphic to each other and that contain no more than six vertices.

A **spanning tree** for a graph G is a tree having precisely the same vertices as G and whose edges are all among the edges of G. Observe that a graph has a spanning tree if and only if the graph is connected.

Problem 7.1.27   A series of pipelines will be built among seven refineries owned by a particular oil company. As the pipelines are expensive to build and maintain, only a minimum number of pipelines will be built to link all the refineries. Also, the nature of the terrain in certain areas would make it impossible to build pipelines directly between some of the refineries. In the graph in Figure 7.6, the vertices represent the refineries, and the edges indicate between which refineries pipelines could be constructed.

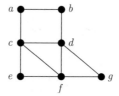

Figure 7.6

Find a spanning tree for this graph (there is more than one), and observe that any spanning tree indicates a way to link all the refineries with the minimum number of pipelines.

A **planar graph** is a graph that can be drawn (or redrawn) in a plane so that none of its edges cross. A drawing of a planar graph in which none of the edges cross is called a **plane graph**. One important application of planar graphs is in the design of electric circuits, as it is usually required that the connections in such a circuit not cross one another.

To provide further motivation for the study of planar graphs, consider the following problem, referred to as the *Three Houses–Three Utilities Problem.* Three new houses a, b, and c must each be connected to each of three utilities: gas g, water w, and electricity e. The graph in Figure 7.7 models this situation.

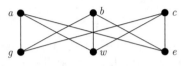

Figure 7.7

The utility companies would prefer to make the connections so that none of the utility lines cross (that way a problem with a gas line, for instance, will be less likely to impact a water line). Is this possible? In other words, is the graph in Figure 7.7 planar? You will determine the answer to this question in Problem 7.1.32(d) below, but in order to do so we first need to learn a bit more about planar graphs.

Given a plane graph, the parts of the plane that remain after the vertices and edges are removed are called the **regions** of the graph. One of these regions, called the **exterior** region, extends indefinitely (outward). The set of regions of a plane graph G is denoted by $R(G)$.

Problem 7.1.28   Verify that each of the graphs G_1, G_4, and G_5 from Figure 7.3 has two regions. Also verify that the graph G_3 has three regions.

Problem 7.1.29   *Euler's Formula* states that for any connected plane graph G, $|V(G)| - |E(G)| + |R(G)| = 2$. In this problem, we will help you to see why this formula holds.

Let G be a connected plane graph. Note that a diagram of G may be constructed as follows:

Step 1: Begin with one of the vertices of G.

Step 2: As long as there are still vertices of G that need to be inserted into the diagram, bring them in one at a time. Also, each time a vertex is incorporated into the diagram, also incorporate an edge that is incident to this vertex and to a vertex that was previously included (such an edge must exist as G is connected).

Step 3: After all of the vertices of G have been included in the diagram, there still may be some additional edges that have not yet been incorporated. Incorporate these into the diagram one at a time until the entire graph G has been created.

Observe that at any stage in this construction of G, we will have created a partial representation of G that is itself a plane graph. Let v, e, and r be, respectively, the number of vertices, number of edges, and number of regions in the part of G we have constructed so far at any point in the construction process. To show $|V(G)| - |E(G)| + |R(G)| = 2$, we need only show that the quantity $v - e + r$ has the value 2 in Step 1 of the construction and that this value is maintained throughout the entire construction.

   (a) Explain why $v - e + r = 2$ immediately after applying Step 1.

   (b) Explain why the value of $v - e + r$ cannot change at any stage during the application of Step 2.

   (c) Explain why the value of $v - e + r$ cannot change at any stage during the application of Step 3.

The **boundary** of a region in a plane graph is the closed path that includes all edges bordering the region the minimum number of times to allow for the creation of a closed path and no other edges. The **degree** of a region in a plane graph is the length of the region's boundary. Thus, the exterior region of the graph from Figure 7.1 has boundary *SWBMBABPBWHS* and degree 11.

Problem 7.1.30   Find the boundary and degree of each region in the plane graph from Figure 7.6.

Problem 7.1.31   Do all of the following:

   (a) Prove that the sum of the degrees of the regions of any connected plane graph G is $2|E(G)|$.

   (b) Prove that if G is a connected plane graph having at least two edges, $|E(G)| \leq 3|V(G)| - 6$. HINT: Make use of the result from (a); show that, under the given hypotheses, each region must have degree at least 3; and apply Euler's Formula.

   (c) Verify that the conclusion of (b) holds for the graph in Figure 7.1.

   (d) Use (b) to show that the complete graph on five vertices is nonplanar.

A graph is **bipartite** if its set of vertices can be partitioned into two disjoint nonempty sets in such a way that every edge of the graph has one of its endpoints in one of the sets and the other endpoint in the other set.

Problem 7.1.32   Do all of the following:

   (a) Show that the graph for the Three Houses–Three Utilities Problem given in Figure 7.7 is bipartite.

(b) Show that for any connected bipartite plane graph G having at least two edges, $|E(G)| \leq 2|V(G)| - 4$. HINT: Show that G cannot contain any cycles of length 3; argue that it then follows that $4|R(G)| \leq 2|E(G)|$; and apply Euler's Formula.

(c) Verify the result from (b) holds for a bipartite graph in which there are five vertices a, b, c, d, and e in all, with each of the vertices in $\{a, b\}$ adjacent to each of the vertices in $\{c, d, e\}$.

(d) Use the result from (b) to solve the Three Houses–Three Utilities Problem.

In the middle of the 19th century, mathematicians became interested in the problem of coloring maps of states or countries in such a way that no two states or countries sharing a border would receive the same color. These mathematicians were not limiting themselves to actual maps of existing states or countries, but wondered about the minimum number of colors required to color any conceivable such map so that regions bordering one another would be assigned different colors. To *border* one another, it was agreed that two regions would have to share a boundary of some positive length. Thus, for instance, in a pie cut in the traditional way into, say, eight slices, each slice would only border two other slices; the point at the center of the pie where all the slices "meet" would not be considered a border.

Graphs can be used to model maps of states or countries. Each state or country is represented by a vertex, and two vertices are joined by an edge if the states or countries represented by the vertices border each other. The graph in Figure 7.8 models the part of a map of the United States including Nevada and the states bordering Nevada.

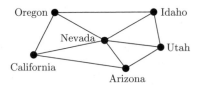

Figure 7.8

Problem 7.1.33 Explain why at least four colors are needed to color a map of the continental United States so that states sharing a border receive different colors. (Think about the states bordering Nevada as modeled by the graph in Figure 7.8.)

A **coloring** of a graph is an assignment of colors to the vertices of the graph. A **proper coloring** of a graph is a coloring of the graph for which no two vertices joined by an edge receive the same color. The **chromatic number** of a graph G, denoted $\chi(G)$, is the minimum number of colors required to obtain a proper coloring of G (the symbol χ is the lowercase Greek letter "chi").

Problem 7.1.34   Determine the chromatic number for the following graphs:

(a) the graph from Figure 7.1.

(b) the graphs in Figure 7.3.

(c) the graph in Figure 7.7.

Problem 7.1.35   Show that any planar graph has a vertex of degree no more than 5. HINT: First show the result holds for a planar graph having only one edge. You can then assume, for the remainder of your proof, that G is a planar graph having at least two edges. If G is connected, use the result of Problem 7.1.31(b) to show that the average of the degrees of the vertices of G is less than 6 (the desired conclusion follows; why?). If G is disconnected, try a proof by contradiction and make use of the fact that you have already obtained the desired conclusion for any connected "component" of G.

As far back as the 1800s, when graph coloring originated with the study of map colorings, mathematicians conjectured that at most four colors were needed to properly color any planar graph. However, a proof of this result was not obtained until 1976. This proof, by Kenneth Appel and Wolfgang Haken, is famous for being the first instance of a mathematical proof that relies on computations provided by a computer (over 1000 hours worth of computations!). The proof is quite difficult, but it is possible to prove, with considerably less effort, that five colors are enough to properly color any planar graph.

Problem 7.1.36 (*The Five-Color Theorem*)   Show that for any planar graph G, $\chi(G) \le 5$. HINT: Try induction. Establish the base claim for any planar graph having fewer than six vertices. To establish the inductive claim, fix $k \ge 5$ and assume that any planar graph having k vertices can be properly colored with no more than five colors. Let G be a planar graph having $k + 1$ vertices that has been drawn as a plane graph. You must show G can be properly colored using at most five colors. To do this, apply the result of Problem 7.1.35. You will then need to consider cases (and draw appropriate pictures to guide your thinking).

7.2   Exploring Groups

Algebra is the study of operations on sets. A **binary operation** $*$ on a set S is a function $*: S \times S \to S$. We usually write $a * b$ for the output $*(a, b)$ of $*$ corresponding with the input (a, b). For instance, addition is a binary operation on $\mathbf{Z}$, and we write $3 + 5$ rather than $+(3, 5)$. The requirement that the computed result of a binary operation on a set must always be a member of that same set is called **closure** (a set must be **closed** under any binary operation defined on the set).

324 Chapter 7 ■ Further Mathematical Explorations

Problem 7.2.1 Determine whether the given "operation" is a binary operation on the given set:

(a) addition on $\mathbf{Z}^-$.

(b) subtraction on $\mathbf{Z}^+$.

(c) multiplication on $\mathbf{R} - \{0\}$.

(d) division on $\mathbf{R}$.

(e) $*$ on $\mathbf{R} - \{-1\}$, where $a * b = a + b + ab$.

Algebraists are interested in the properties a binary operation on a set must possess in order to solve equations involving the operation. For instance, consider the problem of solving the equation $7 + x = 12$. What properties of the addition of integers are needed to solve this equation? The following sequence of implications outlines the "standard" solution process:

$$7 + x = 12$$

$$\Downarrow$$

$$(-7) + (7 + x) = (-7) + 12$$

$$\Downarrow$$

$$[(-7) + 7] + x = (-7) + 12$$

$$\Downarrow$$

$$0 + x = (-7) + 12$$

$$\Downarrow$$

$$x = (-7) + 12$$

$$\Downarrow$$

$$x = 5$$

Observe that, along with the closure of $\mathbf{Z}$ under addition, the solution process makes use of the following properties: the associativity of addition; the existence of an additive identity 0; and the existence of an additive inverse -7 of 7. These properties also allow for the reversibility of the steps in the solution process, which confirms that 5 is a solution to the equation $7 + x = 12$.

Problem 7.2.2 Using only the binary operation of multiplication of rational numbers (no division allowed), together with properties of this operation, show how to solve the equation $3x = 10$ in a manner analogous to that used previously to solve $7 + x = 12$. Compare the properties of multiplication of rational numbers you had to use with the properties of addition of integers used above. What do you notice?

An array of real numbers arranged into two rows and two columns, such as $\begin{bmatrix} 2 & -1 \\ 3 & 6 \end{bmatrix}$, is called a 2×2 **real matrix**. The numbers in the matrix are called the **entries** of the matrix and may be referenced via row and column position. For instance, in the matrix above, -1 is the entry in the first row and second column. Matrices (plural of **matrix**) are **equal** if their corresponding entries are equal. For example, in order for $\begin{bmatrix} 2 & -1 \\ 3 & 6 \end{bmatrix} = \begin{bmatrix} 2 & x \\ 3 & 6 \end{bmatrix}$, we would need $x = -1$.

A binary operation of **matrix multiplication** can be defined on the set of 2×2 real matrices so that, for

$$A = \begin{bmatrix} a_{11} & a_{12} \\ a_{21} & a_{22} \end{bmatrix} \quad \text{and} \quad B = \begin{bmatrix} b_{11} & b_{12} \\ b_{21} & b_{22} \end{bmatrix},$$

the **matrix product** AB is

$$\begin{bmatrix} a_{11}b_{11} + a_{12}b_{21} & a_{11}b_{12} + a_{12}b_{22} \\ a_{21}b_{11} + a_{22}b_{21} & a_{21}b_{12} + a_{22}b_{22} \end{bmatrix}.$$

For example,

$$\begin{bmatrix} 4 & 1/2 \\ 3 & -2 \end{bmatrix}\begin{bmatrix} 7 & 2 \\ 1 & -1 \end{bmatrix} = \begin{bmatrix} 4 \cdot 7 + (1/2) \cdot 1 & 4 \cdot 2 + (1/2)(-1) \\ 3 \cdot 7 + (-2) \cdot 1 & 3 \cdot 2 + (-2)(-1) \end{bmatrix}$$
$$= \begin{bmatrix} 57/2 & 15/2 \\ 19 & 8 \end{bmatrix}.$$

Problem 7.2.3 Consider the equation

$$\begin{bmatrix} 2 & -1 \\ 3 & 6 \end{bmatrix}X = \begin{bmatrix} 4 & 13 \\ -9 & -18 \end{bmatrix}$$

where a solution

$$X = \begin{bmatrix} x_{11} & x_{12} \\ x_{21} & x_{22} \end{bmatrix}$$

within the set of 2×2 real matrices is to be found. How does this equation compare with the equations $7 + x = 12$ and $3x = 10$ already considered? Might we be able to mimic the process used to solve the earlier equations in order to solve this new matrix equation? What properties of matrix multiplication would be needed to carry out a similar solution process?

Problem 7.2.4 Show that the operation of matrix multiplication, as defined on the set of 2×2 real matrices, is associative.

The matrix $I = \begin{bmatrix} 1 & 0 \\ 0 & 1 \end{bmatrix}$ is called the 2×2 **identity matrix**.

Problem 7.2.5 Show that for any 2×2 real matrix A, $IA = A$ and $AI = A$.

A 2×2 real matrix A is **invertible** if there is a 2×2 real matrix B with the property that $AB = BA = I$. This matrix B is called the **inverse** of A and is usually denoted A^{-1}.

Problem 7.2.6 Show that

(a) the matrix $\begin{bmatrix} 2 & -1 \\ 3 & 6 \end{bmatrix}$ is invertible by finding its inverse;

(b) the matrix $\begin{bmatrix} 4 & -1 \\ -8 & 2 \end{bmatrix}$ is not invertible;

(c) the matrix I is invertible (what is its inverse?).

Problem 7.2.7 Use the results of Problems 7.2.4, 7.2.5, and 7.2.6(a) to solve the equation

$$\begin{bmatrix} 2 & -1 \\ 3 & 6 \end{bmatrix} X = \begin{bmatrix} 4 & 13 \\ -9 & -18 \end{bmatrix},$$

where

$$X = \begin{bmatrix} x_{11} & x_{12} \\ x_{21} & x_{22} \end{bmatrix}$$

is an unknown 2×2 real matrix. Then determine whether the solution to that equation is also a solution to the equation

$$X \begin{bmatrix} 2 & -1 \\ 3 & 6 \end{bmatrix} = \begin{bmatrix} 4 & 13 \\ -9 & -18 \end{bmatrix}.$$

If it is not, find a solution. What do your conclusions here tell us about the operation of matrix multiplication?

Problem 7.2.8   Assume

$$X = \begin{bmatrix} x_{11} & x_{12} \\ x_{21} & x_{22} \end{bmatrix}$$

is an unknown 2×2 real matrix.

(a) Determine whether either of the following matrix equations has any solutions. If it does, find all solutions. (Note what you showed in Problem 7.2.6b.)

$$\begin{bmatrix} 4 & -1 \\ -8 & 2 \end{bmatrix} X = \begin{bmatrix} 3 & 5 \\ 1 & 4 \end{bmatrix} \quad \text{and} \quad \begin{bmatrix} 4 & -1 \\ -8 & 2 \end{bmatrix} X = \begin{bmatrix} 1 & 3 \\ -2 & -6 \end{bmatrix}$$

(b) Based on your work in (a) and in Problem 7.2.7, try to formulate conjectures concerning the existence and uniqueness of solutions to an arbitrary matrix equation $AX = B$. Try to prove your conjectures.

Problem 7.2.9   Consider the set $\mathfrak{I}$ of increasing functions from $[0, 1]$ onto $[0, 1]$ (see Problem 5P.8 for the definition of **increasing** function).

(a) Show that for any $f \in \mathfrak{I}$, $f(0) = 0$ and $f(1) = 1$.

(b) Give three specific examples of functions in $\mathfrak{I}$.

(c) Prove that every function in $\mathfrak{I}$ is one-to-one (thus, the functions in $\mathfrak{I}$ are bijections).

(d) Prove that function composition is a binary operation on $\mathfrak{I}$.

(e) Consider the equation $f \circ x = g$, where f and g are given members of $\mathfrak{I}$, and x is an unknown member of $\mathfrak{I}$. Does this equation have a solution in $\mathfrak{I}$? If so, is the solution unique? Does the equation-solving process we have applied previously work here? What do we need to know about function composition and the members of $\mathfrak{I}$ in order to get it to work?

Given a binary operation $*$ on a set S, along with $a, b \in S$, the equations $a * x = b$ and $x * a = b$ are called **linear equations** in the unknown x. Based on the examples we have investigated, we introduce the notion of a *group* as an abstract algebraic structure consisting of a set and a binary operation on that set that satisfies just enough properties so that all linear equations involving the operation can be solved. Specifically, a **group** $(G, *)$ is a set G together with a binary operation $*$ on G for which

- $*$ is associative;

- there exists $e \in G$ such that for all $a \in G$, $e * a = a * e = a$ (e is called the **identity** of the group);

- for each element $a \in G$, there is an element $a^{-1} \in G$ with the property that $a * a^{-1} = a^{-1} * a = e$ (a^{-1} is called the **inverse** of a in the group).

Note that the binary operation in a group is not required to be commutative, for commutativity is not required in the solution of linear equations (recall that neither matrix multiplication nor function composition is commutative). A group is called **abelian** if its binary operation is commutative.

Problem 7.2.10   Verify each of the following.

(a) $(\mathbf{Z}, +)$ is an abelian group, but $(\mathbf{Z}^+, +)$ is not a group, and neither is $(\mathbf{Z}^+ \cup \{0\}, +)$. (When a group's binary operation is addition, it is customary to write $-a$ rather than a^{-1} for the inverse of a; use this convention here.)

(b) With $\cdot$ used to represent multiplication, $(\mathbf{R}, \cdot)$ is not a group, but both $(\mathbf{R}^+, \cdot)$ and $(\mathbf{R} - \{0\}, \cdot)$ are abelian groups.

(c) The set of all invertible 2×2 real matrices, taken under matrix multiplication, is a group.

(d) The set of all increasing functions from $[0, 1]$ onto $[0, 1]$, taken under function composition, is a group.

(e) Define $* : \mathbf{Z} \times \mathbf{Z} \to \mathbf{Z}$ by $a * b = a + b - 5$. Then $(\mathbf{Z}, *)$ is an abelian group.

(f) The set $\mathbf{R} - \{-1\}$, with the binary operation $*$ defined so that $a * b = a + b + ab$, is an abelian group.

Problem 7.2.11   Let $(G, *)$ be a group with identity e. Prove each of the following.

(a) The identity in $(G, *)$ is unique (i.e., if $f \in G$ has the property that $f * a = a * f = a$ for all $a \in G$, it follows that $f = e$).

(b) Each element in G has a unique inverse (i.e., given $a, b, c \in G$ with $a * b = b * a = e$ and $a * c = c * a = e$, it follows that $b = c$).

(c) Left and right cancellation laws hold (i.e., if either $a * b = a * c$ or $b * a = c * a$, it follows that $b = c$).

(d) Each linear equation in $(G, *)$ has a unique solution (specifically, if a and b are given elements of G, the equation $a * x = b$ has the unique solution $x = a^{-1} * b$, and the equation $y * a = b$ has the unique solution $y = b * a^{-1}$).

(e) Given $a, b \in G$, $(a * b)^{-1} = b^{-1} * a^{-1}$.

A binary operation defined on a (small) finite set is often represented by means of a **Cayley table**. The elements of the set on which the operation is defined are listed in the same order along the **left border** and across the **top border**, with the operation symbol in the upper left corner. The region to the right of the left edge and below the top row is referred to as the **body** of the table and displays the results of computations. Specifically, the value of $p * q$ is found in the body of the table where the row beginning with p at the left edge meets the column headed at the top by q. For example, the Cayley table shown below defines a binary operation $*$ on the set $\{a, b\}$ for which $a * a = a$, $a * b = b$, $b * a = a$, and $b * b = a$.

$*$	a	b
a	a	b
b	a	a

Problem 7.2.12   Does the Cayley table shown above exhibit a group structure? What about the Cayley table shown below?

$*$	a	b
a	a	b
b	b	a

Problem 7.2.13   Explain why, in order for a Cayley table to yield a group structure:

(a) One row in the body of the table must match, element by element, the row along the top border of the table, and one column in the body of the table must match, element by element, the column along the left border of the table.

(b) Each element of the set on which the Cayley table is defining a binary operation must appear exactly once in each row and in each column of the body of the table.

Thus, we may conclude that a Cayley table yields a group structure if and only if both of the above conditions (a) and (b) are satisfied and the binary operation defined by the table is associative.

Problem 7.2.14 Use a Cayley table to exhibit the only possible group structure on

  (a) a set $\{e\}$ consisting of one element e (necessarily the identity);

  (b) a set $\{e, a\}$ consisting of two elements, with e being the identity;

  (c) a set $\{e, a, b\}$ consisting of three elements, with e being the identity.

Consider the groups determined by the following Cayley tables:

$\circ$	I	L	A	R
I	I	L	A	R
L	L	A	R	I
A	A	R	I	L
R	R	I	L	A

$\cdot$	1	i	-1	$-i$
1	1	i	-1	$-i$
i	i	-1	$-i$	1
-1	-1	$-i$	1	i
$-i$	$-i$	1	i	-1

In the table on the left, I represents maintaining one's position, L turning to the left, A turning around, and R turning right, and the binary operation $\circ$ is function composition. In the table on the right, i is the imaginary unit, defined so that $i^2 = -1$, and the binary operation $\cdot$ is multiplication.

Note that the tables are structurally identical; that is, they are the same except for the names of the elements in the sets on which the operations have been defined and the symbols used to designate the operations. Replacing the names of the elements and the operation symbol in one table with the corresponding names in the other table would produce the other table. This notion is called *isomorphism*; the two groups here are *isomorphic*.

Replacing the names of the elements of one group with those of another is achieved via a bijection between the groups' underlying sets; thus, two groups can only be isomorphic if the sets on which they have been defined have the same number of elements (i.e., same cardinality; see Example 5.58 and Theorem 5.59). But renaming the elements is not, by itself, enough for an isomorphism. The bijection must also preserve all computations performed via the groups' respective binary operations. Preservation of computations also forces a correspondence between the groups' identities, as well as correspondences between inverses of elements in the respective groups, because the identity and inverses are all defined via computations involving the groups' binary operations.

Problem 7.2.15 Define a bijection $f : \{I, L, A, R\} \to \{1, i, -1, -i\}$ so that $f(I) = 1, f(L) = i, f(A) = -1,$ and $f(R) = -i$. Observe that, under this bijection, the computation $L \circ A = R$ corresponds with $f(L) \cdot f(A) = f(R)$, that is, $i \cdot -1 = -i,$

meaning that $f(L \circ A) = f(L) \cdot f(A)$. Verify that, for any $a,\ b \in \{I,\ L,\ A,\ R\}$, $f(a \circ b) = f(a) \cdot f(b)$.

Formally, groups $(G,\ *_G)$ and $(H,\ *_H)$ are **isomorphic** if there exists a bijection $f\colon G \to H$ such that for all $a,\ b \in G$, $f(a\ *_G\ b) = f(a)\ *_H\ f(b)$. The function f is then called an **isomorphism**. In other words, an isomorphism is a bijection between groups that preserves computations.

Problem 7.2.16   Let $(G,\ *_G)$ and $(H,\ *_H)$ be groups with identities e_G and e_H, respectively, and let $f\colon G \to H$ be an isomorphism.

(a) Prove that $f(e_G) = e_H$.

(b) Prove that for any $a \in G$, $f(a^{-1}) = (f(a))^{-1}$.

Problem 7.2.17   Prove that the function $f\colon (\mathbf{R}, +) \to (\mathbf{R}^+, \cdot)$ defined by $f(x) = 2^x$ is an isomorphism [hence, the groups $(\mathbf{R},\ +)$ and $(\mathbf{R}^+,\ \cdot)$ are isomorphic].

We will use the notation $n\mathbf{Z}$ to represent the set $\{0,\ \pm n,\ \pm 2n,\ \pm 3n,\ \dots\}$ of all integer multiples of a given positive integer n. Thus, for instance, $2\mathbf{Z}$ is the set of all even integers.

Problem 7.2.18   Let n be a positive integer. Prove that $(\mathbf{Z},\ +)$ is isomorphic to $(n\mathbf{Z},\ +)$.

Problem 7.2.19   A function between isomorphic groups need not be an isomorphism. Consider the functions $f,\ g : (\mathbf{R},\ +) \to (\mathbf{R},\ +)$, where $f(x) = -x$ and $g(x) = x + 1$. Only one of f and g is an isomorphism. Which one? Defend your conclusions.

Problem 7.2.20   In Chapter 5, we proved that the inverse of a bijection is a bijection and that the composition of bijections, when defined, is a bijection (see Theorems 5.56 and 5.67). You may freely use these results here.

(a) Prove that if $f\colon(G,\ *_G) \to (H,\ *_H)$ is an isomorphism, so is $f^{-1}\colon(H,\ *_H) \to (G,\ *_G)$.

(b) Prove that if $f\colon(G,\ *_G) \to (H,\ *_H)$ and $g\colon(H,\ *_H) \to (K,\ *_K)$ are both isomorphisms, so is $g \circ f\colon(G,\ *_G) \to (K,\ *_K)$.

(c) An isomorphism of a group with itself is called an **automorphism**. The collection of all automorphisms of a group G is denoted $\text{Aut}(G)$. Show that $(\text{Aut}(G),\ \circ)$ is a group.

Problem 7.2.21   An **algebraic property** of groups is a property that, if possessed by a group, is possessed by all groups isomorphic to that group.

Show that the property that every element of a group is its own inverse is an algebraic property.

Problem 7.2.22   In this problem, you will explore the possible group structures on a set having four elements. Let the four distinct elements be e, a, b, and c.

  (a)   Develop the four different Cayley tables that give potential group structures on the four-element set $\{e, a, b, c\}$ if e serves as the identity element and we make use of the results of Problem 7.2.13 (do not worry about associativity of the binary operations defined by these tables; you will deal with this later).

  (b)   Three of the four potential groups you described via Cayley tables in (a) are isomorphic to each other and to the group $(\{1, i, -1, -i\}, \cdot)$ defined earlier via a Cayley table. Which ones? [Because $(\{1, i, -1, -i\}, \cdot)$ is a group, we now know that three of your tables from (a) also represent groups. Hence, the binary operations defined by these three tables must be associative.]

  (c)   Explain why one of the four tables you created in (a) is not isomorphic to any of the other three. Then, verify that the binary operation described by this table is associative so that, together with your work from (a), we can be sure the table does actually represent a group. This group is known as the **Klein 4-group**.

In working with particular examples of groups, for instance $(\mathbf{Z}, +)$, $(\mathbf{R}^+, \cdot)$, or $(\mathrm{Aut}(G), \circ)$, it is important to make clear both the set *and* the binary operation. But in stating abstract theorems that apply to many or all groups, it is customary to refer to a group G rather than $(G, *)$ and to use multiplicative notation ab in place of $a * b$. We even refer to ab as a **product**, even though the group's operation may not be actual multiplication. The primary exception is when the operation is known to be some form of addition or is expressed using the symbol $+$; in such circumstances, we use additive notation for computations ($a + b$ rather than ab) and inverses ($-a$ rather than a^{-1}).

Given a group G, a subset H of G is **closed under the binary operation of G** if whenever $a, b \in H$, it follows that $ab \in H$. For example, relative to the group $(\mathbf{Z}, +)$, the subset of all even integers is closed, whereas the subset of all odd integers is not.

If a subset H of a group G is closed under the binary operation of G, and H together with its inherited binary operation forms a group, we call H a **subgroup** of G.

Problem 7.2.23   Show that $(2\mathbf{Z}, +)$ is a subgroup of $(\mathbf{Z}, +)$, but that $(\mathbf{Z}^+, +)$ is not a subgroup of $(\mathbf{Z}, +)$ [even though $\mathbf{Z}^+$ is a subset of $\mathbf{Z}$ that is closed under $+$].

Problem 7.2.24 Show that if H is a subgroup of a group G, then

(a) the identity of H is the identity of G;

(b) for each $a \in H$, the inverse of a in the subgroup H is the same as the inverse of a in the group G.

Problem 7.2.25 Let G be a group and suppose $H \subseteq G$. Then, H is a subgroup of G if and only if *all* of the following are true:

- H is closed under the binary operation of G.

- the identity e of G is a member of H.

- for each element $a \in H$, we have $a^{-1} \in H$.

Problem 7.2.26 Use the result of Problem 7.2.25 to prove the following results about an arbitrary group G.

(a) G is a subgroup of G.

(b) If e is the identity of G, then $\{e\}$ is a subgroup of G.

(c) Any intersection of subgroups of G is also a subgroup of G.

(d) $\{x \in G \mid xg = gx \text{ for every } g \in G\}$ is a subgroup of G.

Given an element a of a group G whose binary operation is being expressed multiplicatively, we define **integer powers of a** as follows:

- a^1 represents a.

- a^{-1} represents the inverse of a.

- For each $n \in \{2, 3, 4, ...\}$, a^n represents the product (via the group's binary operation) of n factors of a (e.g., $a^3 = aaa$).

- For each $n \in \{-2, -3, -4, ...\}$, a^n represents the product of $-n$ factors of a^{-1} (e.g., $a^{-3} = a^{-1}a^{-1}a^{-1}$).

- a^0 represents the identity element.

Problem 7.2.27 Determine all integer powers of all elements of the Klein 4-group.

Problem 7.2.28 Show that if a subgroup H of a group G includes a particular element $a \in G$, then $a^n \in H$ for every integer n. Then show that $\{a^n \mid n \in \mathbf{Z}\}$ is itself a subgroup of G.

Given a group G and an element $a \in G$, the subgroup $\{a^n \mid n \in \mathbf{Z}\}$ of G is called the **cyclic subgroup of G generated by** a and is denoted by $<a>$.

Problem 7.2.29   Find all the cyclic subgroups of the given group.

(a) $(\{1,\ i,\ -1,\ -i\},\ \cdot)$.

(b) $(\mathbf{Z},\ +)$.

(c) The Klein 4-group.

A group G is **cyclic** provided that $G = <a>$ for some $a \in G$. When $G = <a>$, we refer to a as a **generator** of G.

Problem 7.2.30   Determine which of the groups from Problem 7.2.29 are cyclic. For each that is cyclic, determine each generator of the group.

A bijection of a set onto itself is called a **permutation** of the set (see Problem 6K).

Problem 7.2.31   Cut out a square piece of paper, and imagine the square is located in an xy-coordinate system, centered at the origin. Label the corners of the square with the numbers 1, 2, 3, and 4 to reflect the quadrants in which the corners lie (1 for upper right, 2 for upper left, 3 for lower left, and 4 for lower right). Also label the corners on the reverse side so that each corner receives the same number as on the front (i.e., a single corner of the square is marked 1 on both sides, a single corner is marked 2 on both sides, and so forth). You will consider motions of the square that bring the square back onto itself and during which the square maintains its shape.

(a) If we are only interested in the final position of the square relative to its initial position, convince yourself that there is a total of eight such motions (the numbering of the square's corners will help you to sort these out). We will designate them by

- I (identity, or counterclockwise rotation through 360°);
- R_{90} (counterclockwise rotation through 90°);
- R_{180} (counterclockwise rotation through 180°);
- R_{270} (counterclockwise rotation through 270°);
- F_x (flip across the x-axis);
- F_y (flip across the y-axis);
- F_+ (flip across the diagonal $y = x$ having positive slope);
- F_- (flip across the diagonal $y = -x$ having negative slope).

These eight motions are referred to as the **symmetries of the square**. Note that each motion is a permutation of the square.

(b) We can use our numbering of the corners of the square to represent the square's symmetries. For example, we can represent R_{90} as $\begin{pmatrix} 1 & 2 & 3 & 4 \\ 2 & 3 & 4 & 1 \end{pmatrix}$, by which we are indicating that each number in the top row is mapped by R_{90} to the number directly below it in the bottom row. Physically, this agrees with the counterclockwise rotation through 90°, as the corner labeled 1 moves to the position of the corner originally labeled 2, and so forth. With this understanding of the notation, use your square to verify the following:

$$I = \begin{pmatrix} 1 & 2 & 3 & 4 \\ 1 & 2 & 3 & 4 \end{pmatrix}, \quad R_{180} = \begin{pmatrix} 1 & 2 & 3 & 4 \\ 3 & 4 & 1 & 2 \end{pmatrix}, \quad R_{270} = \begin{pmatrix} 1 & 2 & 3 & 4 \\ 4 & 1 & 2 & 3 \end{pmatrix}, \quad F_x = \begin{pmatrix} 1 & 2 & 3 & 4 \\ 4 & 3 & 2 & 1 \end{pmatrix},$$

$$F_y = \begin{pmatrix} 1 & 2 & 3 & 4 \\ 2 & 1 & 4 & 3 \end{pmatrix}, \quad F_+ = \begin{pmatrix} 1 & 2 & 3 & 4 \\ 1 & 4 & 3 & 2 \end{pmatrix}, \quad \text{and } F_- = \begin{pmatrix} 1 & 2 & 3 & 4 \\ 3 & 2 & 1 & 4 \end{pmatrix}.$$

(c) Verify the calculations in the following Cayley table, which shows that the set of symmetries of a square is closed under function composition, thus making composition a binary operation on the set of symmetries of the square. You may want to use the representations of the symmetries from (b) in making your calculations. Also, keep in mind that in computing, for instance, $F_x \circ R_{90}$, we must first apply the symmetry R_{90} appearing on the right of $\circ$, then apply the symmetry F_x appearing on the left.

$\circ$	I	R_{90}	R_{180}	R_{270}	F_x	F_y	F_+	F_-
I	I	R_{90}	R_{180}	R_{270}	F_x	F_y	F_+	F_-
R_{90}	R_{90}	R_{180}	R_{270}	I	F_+	F_-	F_y	F_x
R_{180}	R_{180}	R_{270}	I	R_{90}	F_y	F_x	F_-	F_+
R_{270}	R_{270}	I	R_{90}	R_{180}	F_-	F_+	F_x	F_y
F_x	F_x	F_-	F_y	F_+	I	R_{180}	R_{270}	R_{90}
F_y	F_y	F_+	F_x	F_-	R_{180}	I	R_{90}	R_{270}
F_+	F_+	F_x	F_-	F_y	R_{90}	R_{270}	I	R_{180}
F_-	F_-	F_y	F_+	F_x	R_{270}	R_{90}	R_{180}	I

(d) Use what you know about function composition and Cayley tables to explain why the set of symmetries of the square, taken under function composition, forms a group.

Problem 7.2.32 Prove that the set of all permutations of a given set S is a group under the operation of function composition.

When applied to permutations, function composition is usually called **permutation multiplication**, and we write gf in place of $g \circ f$. It is evident that the groups of all permutations of two sets having the same number of elements will be isomorphic. Thus, for each positive integer n, we define the **symmetric group on n letters** to be the group of all permutations of the set $\{1, 2, ..., n\}$. This group is denoted by S_n.

Problem 7.2.33 Consider the permutations $f = \begin{pmatrix} 1 & 2 & 3 & 4 \\ 3 & 1 & 2 & 4 \end{pmatrix}$ and $g = \begin{pmatrix} 1 & 2 & 3 & 4 \\ 2 & 3 & 4 & 1 \end{pmatrix}$ in S_4. Compute gf and fg.

Problem 7.2.34 Explain why S_n has $n!$ elements.

Problem 7.2.35 Note that the group of symmetries of a square is a subgroup of S_4. We can also generalize the group of symmetries of the square. For each integer $n > 2$, let D_n be the set of all symmetries of a regular n-gon (i.e., n-sided polygon for which all sides have the same length and all angles have the same measure).

(a) Show that D_n has $2n$ elements by identifying this many symmetries of a regular n-gon.

(b) Prove that D_n is a subgroup of S_n (D_n is called the **nth dihedral group**).

We have observed that each row and each column in a Cayley table describing a group gives a permutation of the elements of the set on which the group structure is being defined. In fact, every group is isomorphic to a group of permutations (this result is known as *Cayley's Theorem*).

Problem 7.2.36 Let G be a group.

(a) Prove that, for each $g \in G$, the function $\lambda_g : G \to G$ defined by $\lambda_g(x) = gx$ is a permutation of G (λ is the lowercase Greek letter "lambda").

(b) Prove that $\Lambda = \{\lambda_g \mid g \in G\}$ is a subgroup of the group S_G of all permutations of G (Λ is the uppercase Greek letter "lambda").

(c) Prove that G is isomorphic to Λ.

7.3 Exploring Set Cardinality

It is natural to compare the "sizes" of finite sets. For instance, the finite set {2, 5, 8, 11} is "bigger" than the finite set {1, 5, 25} because the former has four elements whereas the latter has only three. When it comes to infinite sets, though, a discussion of "size" is more subtle and less intuitive. Typical descriptors of "the infinite" as something that "goes on forever," "has no end," or "is beyond the finite" really only serve to mark out the infinite as that which is "not finite." And yet it appears there is a case to be made for distinguishing among the relative sizes of infinite sets.

Problem 7.3.1 Argue that the infinite set {0, 1, 2, 3, ...} is "bigger" than the infinite set {1, 2, 3, ...}.

Your response to Problem 7.3.1 was no doubt based on the notion of set inclusion, the essential observation being that {0, 1, 2, 3, ...} includes all of the elements of {1, 2, 3, ...} along with an "extra" element, 0, which suggests it is possible to regard {0, 1, 2, 3, ...} as "bigger" than {1, 2, 3, ...}. You might argue, somewhat generally, that an infinite set A should be considered "bigger" than an infinite set B if $B \subset A$; that is, if B is a subset of A and there is at least one element in A that is not in B.

The primary difficulty with such an approach, however, is that it cannot be applied to the comparison of the sizes of *arbitrary* infinite sets, only those for which one is a subset of the other. For instance, the approach could not be used to compare {1, 2, 3, ...} and {−1, −2, −3, ...}.

Problem 7.3.2 Argue that the sets {1, 2, 3, ...} and {−1, −2, −3, ...} have the same size, but build your argument on more than just the fact that both sets are infinite.

We imagine your response to Problem 7.3.2 was based on the observation that the members of $\mathbf{Z}^- = \{-1, -2, -3, ...\}$ are precisely the opposites of the members of $\mathbf{N} = \{1, 2, 3, ...\}$. By pairing the positive integers with their negative counterparts

$$\begin{array}{cccc} 1 & 2 & 3 & 4 \quad ... \\ \updownarrow & \updownarrow & \updownarrow & \updownarrow \\ -1 & -2 & -3 & -4 \quad ... \end{array}$$

it would then seem reasonable to regard the sets {1, 2, 3, ...} and {−1, −2, −3, ...} as having the same size.

Problem 7.3.3 Verify that the function $f: \mathbf{N} \to \mathbf{Z}^-$ defined so that $f(n) = -n$, which describes the pairing of each positive integer with its opposite, is a bijection.

Your work in Problems 7.3.2 and 7.3.3 in comparing the sizes of the sets **N** and $\mathbf{Z}^-$ suggests a possible general approach to comparing the sizes of infinite sets. The approach is based on whether it is possible to pair the members of one of the sets with the members of the other so as to create a bijection from one of the sets onto the other. Whether such an approach is reasonable might be assessed by examining its application to the comparison of the sizes of finite sets.

Problem 7.3.4 Find a bijection from $\{2, 4, 6\}$ onto $\{a, b, c\}$. Then show that there can be no bijection from $\{2, 4, 6\}$ onto $\{a, b, c, d\}$. Finally, show that there can be no bijection from $\{2, 4, 6\}$ onto $\{a, b\}$.

Problem 7.3.5 This problem generalizes Problem 7.3.4. Let A be a finite set having m elements, and let B be a finite set having n elements. Prove that there is a bijection $f: A \to B$ if and only if $m = n$.

Based on the results of Problems 7.3.2, 7.3.3, and 7.3.5, we define two sets A and B to have the **same cardinality** if there is a bijection $f: A \to B$. When two sets A and B have the same cardinality, we also say the sets are **equinumerous** and write $A \approx B$.

Problem 7.3.6 Show that the given sets have the same cardinality.

(a) $\{1, 2, 3, ...\}$ and $\{2, 4, 6, ...\}$.

(b) $\{1, 3, 5, ...\}$ and $\{2, 4, 6, ...\}$.

Problem 7.3.7 Verify that the function $f: \mathbf{N} \to \mathbf{Z}$ defined so that

$$f(n) = \begin{cases} (1 - n)/2 & \text{if } n \text{ is odd} \\ n/2 & \text{if } n \text{ is even} \end{cases}$$

is a bijection. Hence, we may conclude that $\mathbf{N} \approx \mathbf{Z}$.

Problem 7.3.8 Consider the function $g: \mathbf{R} \to \mathbf{R}^+$ defined so that $g(x) = 2^x$.

(a) Sketch the graph of g and use the graph to argue informally that g is a bijection. (A rigorous proof that g is a bijection would require a formal definition of exponentiation valid for any real number exponent; this is better left to a course in real analysis.)

(b) What does the fact that g is a bijection allow us to conclude about $\mathbf{R}$ and $\mathbf{R}^+$?

Problem 7.3.9   Show that the relation ≈ of equinumerosity is an equivalence relation on any given collection of sets. HINT: See Theorems 5.56 and 5.67, along with Problem 5P.4.

Technically, we have not as yet defined what we mean by a *finite* set or an *infinite* set, instead appealing to prior informal experiences with these concepts. We are now in a position to create formal definitions. A set is **finite** if it is either empty or equinumerous with $\{1, 2, 3, ..., n\}$ for some positive integer n. The **cardinal number** of an empty set is 0, and the **cardinal number** of a finite set that is equinumerous with $\{1, 2, 3, ..., n\}$ is n. That is, the cardinal number of a finite set is its number of elements. A set that is not finite is called **infinite**.

Problem 7.3.10   Verify that the given set is finite and determine its cardinal number.

(a) $\{2, 4, 6\}$.

(b) $\{\sqrt{5}, \sqrt{6}, \sqrt{7}, ..., \sqrt{1001}\}$.

Problem 7.3.11   Here you will prove that the set $\mathbf{N} = \{1, 2, 3, ...\}$ is infinite. Assume to the contrary that $\mathbf{N}$ is finite. Because $\mathbf{N}$ is nonempty, we would then be able to conclude that there is a bijection $f: \{1, 2, 3, ..., n\} \to \mathbf{N}$ for some $n \in \mathbf{N}$. It follows that $\{f(1), f(2), f(3), ..., f(n)\} = \mathbf{N}$. Show that $\{f(1), f(2), f(3), ..., f(n)\}$ does not include the successor of one of its elements, contradicting one of the axioms for $\mathbf{N}$.

To compare the "sizes" of sets, it is not enough to know only when two sets should be viewed as having the same number of elements. We also need a mechanism for determining when one set has "fewer" or "more" elements than another set. At least in the case of finite sets, it would seem that one set has fewer elements than another if any one-to-one pairing of the elements of the sets always leaves at least one element of the "larger" set unpaired with any element of the "smaller" set.

Problem 7.3.12   Find all one-to-one functions from $\{2, 4, 6\}$ into $\{a, b, c, d\}$. Then verify that none of these functions is a bijection.

Recall that the term *injection* is just another name for a one-to-one function. When there is an injection from a set A into a set B, we say that A is **no more numerous** than B (equivalently, B is **no less numerous** than A), and we may write either $A \le B$ or $B \ge A$. When A is no more numerous than B and A is not equinumerous with B, we say that A is **less numerous** than B (equivalently, B is **more numerous** than A), and we may write either $A < B$ or $B < A$. When A is less numerous than B, we also say that A has **smaller cardinality** than B (equivalently, B has **greater cardinality** than A).

To illustrate, because your work in Problem 7.3.12 shows that there is an injection from $\{2, 4, 6\}$ into $\{a, b, c, d\}$ (in fact, you found several), we may conclude that $\{2, 4, 6\}$ is no more numerous than $\{a, b, c, d\}$ (i.e., $\{2, 4, 6\} \leq \{a, b, c, d\}$) and that $\{a, b, c, d\}$ is no less numerous than $\{2, 4, 6\}$ (i.e., $\{a, b, c, d\} \geq \{2, 4, 6\}$). Furthermore, as you also showed in Problem 7.3.4 that there is no bijection from $\{2, 4, 6\}$ onto $\{a, b, c, d\}$, we may conclude that $\{2, 4, 6\}$ is not equinumerous with $\{a, b, c, d\}$. Hence, $\{2, 4, 6\}$ is less numerous (has smaller cardinality) than $\{a, b, c, d\}$, and $\{a, b, c, d\}$ is more numerous (i.e., has larger cardinality) than $\{2, 4, 6\}$. Symbolically, $\{2, 4, 6\} < \{a, b, c, d\}$ and $\{a, b, c, d\} > \{2, 4, 6\}$.

Problem 7.3.13 Verify that $\{1, 2, 3, 4\} < \{-2, -1, 0, 1, 2, 3\}$.

Problem 7.3.14 Prove each of the following.

(a) For any set A, $A \leq A$.

(b) For any sets A and B, if $A \approx B$, then $A \leq B$.

(c) For any sets A and B, if $A \subseteq B$, then $A \leq B$.

(d) For any sets A, B, and C, if $A \leq B$ and $B \leq C$, then $A \leq C$.

(e) For any finite set A, $A < \mathbf{N}$.

Problem 7.3.15 Apply the appropriate results from Problem 7.3.14 to explain why each of the following is true.

(a) $\mathbf{Z} \leq \mathbf{R}$.

(b) $\{-2, 0, 2\} < \mathbf{N}$.

(c) $\varnothing \leq \varnothing$.

(d) $\mathbf{Z} \leq \mathbf{N}$.

(e) $\{a, b, c, d\} \leq \mathbf{R}$.

It would seem reasonable to believe that, for any sets A and B, if A is no more numerous than B and B is no more numerous than A, then A and B should be equinumerous (i.e., if $A \leq B$ and $B \leq A$, then $A \approx B$). This is a result known as the *Schröder-Bernstein Theorem*. Its proof is more challenging than you might suspect.

Problem 7.3.16 In this problem, we will help you to verify the Schröder-Bernstein Theorem. First, note that to establish the desired result, we need only show that if A and B are sets and there exist injections $f:A \rightarrow B$ and $g:B \rightarrow A$, then there exists a bijection from A onto B. So assume A and B are sets, and $f: A \rightarrow B$ and $g: B \rightarrow A$ are injections.

Consider any $a \in A$. There may or may not exist a member of B that g maps to a. If there is, because g is an injection, there is only one such member of B; call it b_1. Now there may or may not exist a member of A that f maps to b_1. If there is, because f is an injection, there is only one such member of A; call it a_1. If possible, obtain a (necessarily unique if it exists) member of B that g maps to a_1; call it b_2. Again, if possible, obtain a (necessarily unique if it exists) member of A that f maps to b_2; call it a_2. Continuing in this way, note that the procedure will either come to an end or continue forever. If it continues forever, we obtain the sequence $a, b_1, a_1, b_2, a_2, b_3, a_3, \ldots$; if it comes to an end, we obtain the initial portion of the sequence up to the stopping point (a or a, b_1 or a, b_1, a_1 or a, b_1, a_1, b_2, etc.).

Let A_1 be the members of A for which the procedure just outlined never comes to an end, let A_2 be the members of A for which the procedure ends with a member of A, and let A_3 be the members of A for which the procedure ends with a member of B. Note that every member of A is in exactly one of the sets A_1, A_2, or A_3.

The identical procedure described above can also be applied beginning with any member of B and results in a finite or infinite sequence that alternates between members of B and members of A, but which this time begins with a member of B. Let B_1 be the members of B for which the procedure never comes to an end, let B_2 be the members of B for which the procedure ends with a member of A, and let B_3 be the members of B for which the procedure ends with a member of B. Note that every member of B is in exactly one of the sets B_1, B_2, or B_3.

(a) Given any function $k\colon X \to Y$ and any subset S of the domain X of k, the **restriction** of k to S is the function $k|S\colon S \to Y$ defined so that $(k|S)(x) = k(x)$ for each $x \in S$. In other words, the restriction operates on inputs in exactly the same way as the original function but only allows members of a specified subset of the original function's domain to serve as inputs.

Prove that $f|A_1$ is a bijection from A_1 onto B_1, that $f|A_2$ is a bijection from A_2 onto B_2, and that $g|B_3$ is a bijection from B_3 onto A_3.

(b) Prove that the function $h\colon A \to B$ defined so that

$$h(a) = \begin{cases} f(a) \text{ if } a \in A_1 \cup A_2 \\ (g|B_3)^{-1}(a) \text{ if } a \in A_3 \end{cases}$$

is a bijection (the desired bijection from A onto B).

Problem 7.3.17 Recall that an interval in **R** is a subset of **R** that includes all real numbers between any two of its members.

(a) Let a and b be real numbers with $a < b$. Sketch the graph of a bijection from the interval (a, b) onto the interval $(-\pi/2, \pi/2)$, thus demonstrating that (a, b) and $(-\pi/2, \pi/2)$ have the same cardinality.

(b) Use the graph of the tangent function, restricted to the interval $(-\pi/2, \pi/2)$, to argue that $(-\pi/2, \pi/2)$ has the same cardinality as $\mathbf{R}$.

(c) Let I be any interval in $\mathbf{R}$ having more than one element. Show that I has the same cardinality as $\mathbf{R}$. HINT: Use the result of (c) in Problem 7.3.14 to get $I \leq \mathbf{R}$. How can you use the same result, along with (a) and (b) of the current problem, to show that $\mathbf{R} \leq I$? What does the Schröder-Bernstein Theorem now allow you to conclude?

Intuitively, *counting* and *listing* go hand-in-hand. When the members of a nonempty finite set have been listed, we can count them off, "one," "two," "three," and so forth, to determine the cardinality of the set. For example, to count the members of the set $\{a, e, i, o, u\}$, we point to a and say "one," to e and say "two," and so forth, thus creating the bijection

$$
\begin{array}{ccccc}
a & e & i & o & u \\
\updownarrow & \updownarrow & \updownarrow & \updownarrow & \updownarrow \\
1 & 2 & 3 & 4 & 5
\end{array}
$$

between the set $\{a, e, i, o, u\}$ that we want to count and the set $\{1, 2, 3, 4, 5\}$; we then conclude that $\{a, e, i, o, u\}$ has five members.

It would seem possible to use a similar "counting" procedure in the context of an infinite set, provided the elements of the set can be put into a (never-ending) list

$$a_1, a_2, a_3, \ldots .$$

We could then point to a_1 and say "one," to a_2 and say "two," and so on, but we have to imagine continuing on without ever stopping. Of course, the listing $a_1, a_2, a_3, \ldots$ is actually a sequence, and if we assume all elements of the set in question appear exactly one time in the list, the sequence represents the bijection

$$
\begin{array}{cccc}
a_1 & a_2 & a_3 & \cdots \\
\updownarrow & \updownarrow & \updownarrow & \\
1 & 2 & 3 & \cdots
\end{array}
$$

between the set $\{a_1, a_2, a_3, \ldots\}$ and the set $\mathbf{N}$ of all counting numbers. An infinite set for which there is a bijection from $\mathbf{N}$ onto the set is called **countably infinite**. That is, a countably infinite set is one that is equinumerous with $\mathbf{N}$.

Note that anytime we are able to put all the members of an infinite set into an unending list (sequence), we have demonstrated that the set is countably infinite. For example, the bijection in Problem 7.3.7 effectively puts the members of $\mathbf{Z}$ into the infinite list

$$0, 1, -1, 2, -2, ...,$$

thus showing that $\mathbf{Z}$ is countably infinite.

Problem 7.3.18 Show that the given set is countably infinite by finding a way to put all of its members into an infinite list (sequence):

(a) The set of all rational numbers that can be expressed as $a/3$ for some positive integer a;

(b) $\{a + bi \mid a, b \in \mathbf{Z}\}$.

Problem 7.3.19 Show that every infinite subset of a countably infinite set is countably infinite.

Problem 7.3.20 Show that the union of any two finite sets is finite. Conclude, then, that if A is an infinite set and B is a finite subset of A, it must follow that $A - B$ is infinite.

Problem 7.3.21 Use the result of Problem 7.3.20 to show that every infinite set contains a countably infinite subset (thus, we may view countably infinite sets as the "smallest" among infinite sets).

Problem 7.3.22 In this problem, you will show that a countably infinite union of countably infinite sets is countably infinite. To do this, consider countably many sets A_1, A_2, A_3, ..., each of which is countably infinite, with $A_1 = \{a_{11}, a_{12}, a_{13}, ...\}$, $A_2 = \{a_{21}, a_{22}, a_{23}, ...\}$, $A_3 = \{a_{31}, a_{32}, a_{33}, ...\}$, and so forth. Now consider the following array displaying all of the elements of all of these sets.

$$a_{11}, \ a_{12}, \ a_{13}, \ a_{14}, \ a_{15}, \ a_{16}, \ \cdots$$

$$a_{21}, \ a_{22}, \ a_{23}, \ a_{24}, \ a_{25}, \ a_{26}, \ \cdots$$

$$a_{31}, \ a_{32}, \ a_{33}, \ a_{34}, \ a_{35}, \ a_{36}, \ \cdots$$

$$a_{41}, \ a_{42}, \ a_{43}, \ a_{44}, \ a_{45}, \ a_{46}, \ \cdots$$

$$a_{51}, \ a_{52}, \ a_{53}, \ a_{54}, \ a_{55}, \ a_{56}, \ \cdots$$

$$a_{61}, \ a_{62}, \ a_{63}, \ a_{64}, \ a_{65}, \ a_{66}, \ \cdots$$

$$\vdots$$

Use the array to identify the pattern being employed to obtain the infinite sequence

$$a_{11}, a_{12}, a_{21}, a_{13}, a_{22}, a_{31}, a_{14}, a_{23}, a_{32}, a_{41}, \ldots$$

which lists all of the objects in the array, hence all of the elements in all of the sets $A_1, A_2, A_3, \ldots$ (we may then conclude that $A_1 \cup A_2 \cup A_3 \cup \cdots$ is countably infinite). NOTE: It is possible that some of the sets $A_1, A_2, A_3, \ldots$ may have elements in common. If this is the case, once an element has been listed in the above sequence, we do not need to list it again. In this way, the sequence represents a bijection between $A_1 \cup A_2 \cup A_3 \cup \cdots$ and $\mathbf{N}$.

Problem 7.3.23 Use the result of Problem 7.3.22 to demonstrate that $\mathbf{Q}$ is countably infinite.

Problem 7.3.24 Adapt the technique used in Problem 7.3.22 to prove that if $A = \{a_1, a_2, a_3, \ldots\}$ and $B = \{b_1, b_2, b_3, \ldots\}$ are countably infinite sets, so is $A \times B$.

Problem 7.3.25 Use induction to extend the result of Problem 7.3.24, establishing that for any natural number n, if $A_1, A_2, \ldots, A_n$ are all countably infinite sets, so is $A_1 \times A_2 \times \cdots \times A_n$.

A set is **countable** if it is either finite or countably infinite. Thus, among the countable sets, we have identified all finite sets, $\mathbf{N}$, $\mathbf{Z}$, and $\mathbf{Q}$.

Problem 7.3.26 Show that any subset of a countable set is countable.

Problem 7.3.27 Prove that a set A is countable if and only if there exists an injection $f : A \to \mathbf{N}$.

We began our study of cardinality in this section with a suggestion that, just as it is useful to distinguish among the sizes of finite sets, it might be convenient and enlightening to explore whether similar distinctions among sizes of infinite sets could be formulated. After all, though all finite sets are "small" compared with infinite sets, some finite sets are "larger" than other finite sets. Why might it not then be reasonable to compare sizes of infinite sets, rather than simply view them as "large" compared with finite sets?

Note, though, that to this point we have not demonstrated the existence of two infinite sets having different cardinalities. Could it be that every infinite set is countably infinite? The next problem reveals the answer to be "no."

Problem 7.3.28 Here, you will verify that the infinite set $[0, 1)$ of all real numbers greater than or equal to 0 and less than 1 is *not* countably infinite. The approach is due to Georg Cantor, the mathematician to whom most of what we have explored in this section can be credited. The proof proceeds by contradiction.

So suppose there is a way to put all of the numbers in $[0, 1)$ into an infinite sequence, say

$$a_1, a_2, a_3, a_4, \dots .$$

Each of these numbers can be uniquely represented by an infinite decimal expansion that is allowed to be 0 from some point on, but that is not allowed to be 9 from some point on. (For example, $1/4$ would be represented by $0.250\overline{0}$ rather than $0.249\overline{9}$.) For each $n \in \mathbf{N}$ and each $k \in \mathbf{N}$, let $a_n(k)$ be the decimal digit (0, 1, 2, 3, 4, 5, 6, 7, 8, or 9) in the kth place to the right of the decimal point in this representation of a_n.

We will now construct a number b that is in $[0, 1)$, but which is not equal to any of $a_1, a_2, a_3, a_4, \dots$. For each $n \in \mathbf{N}$, let $b(n)$ be defined so that

$$b(n) = \begin{cases} 1, & \text{if } a_n(n) \neq 1; \\ 2, & \text{if } a_n(n) = 1. \end{cases}$$

Let b be the number between 0 and 1 whose nth decimal digit is $b(n)$.

(a) Explain why b is not equal to any of the numbers $a_1, a_2, a_3, a_4, \dots$.

(b) How are we now able to conclude that $[0, 1)$ is not countably infinite?

A set is called **uncountable** if it is not countable. Thus, Problem 7.3.28 demonstrates that $[0, 1)$ is uncountable. Then, because Problem 7.3.17 showed that every nonempty interval in $\mathbf{R}$ having more than one element is equinumerous with $\mathbf{R}$, it follows that $\mathbf{R}$ itself is uncountable.

Problem 7.3.29 Show that if A is uncountable and $A \subseteq B$, then B is also uncountable.

Problem 7.3.30 Consider any set A.

(a) Show that there is no bijection from A onto $\mathcal{P}(A)$, the collection of all subsets of A. HINT: Assume to the contrary that $f : A \to \mathcal{P}(A)$ is a bijection and consider the subset $\{a \in A \mid a \notin f(a)\}$ of A.

(b) Argue that $A \leq \mathcal{P}(A)$. Then use this result, along with the result from (a), to conclude that $A < \mathcal{P}(A)$.

Problem 7.3.30 establishes that two uncountable sets need not have the same cardinality. In particular, the power set of a given uncountable set will have larger cardinality than the given set itself. For example, because $\mathbf{R} < \mathcal{P}(\mathbf{R})$, it follows that $\mathcal{P}(\mathbf{R})$ is an uncountable set having larger cardinality than the uncountable set $\mathbf{R}$.

Taking this process a step further, we may observe that, for instance,

$$\mathbf{N} < \mathcal{P}(\mathbf{N}) < \mathcal{P}(\mathcal{P}(\mathbf{N})) < \mathcal{P}(\mathcal{P}(\mathcal{P}(\mathbf{N}))) < \ldots,$$

which indicates how a sequence of infinite sets of ever larger cardinalities can be constructed. Thus, there are infinitely many different "sizes" an infinite set may have.

Problem 7.3.31   Here, you will verify that $[0, 1)$ and $\mathcal{P}(\mathbf{N})$ are equinumerous. Keep in mind that each number in $[0, 1)$ has a unique decimal expansion that is allowed to be 0 from some point on but that is not allowed to be 9 from some point on. In what follows, a decimal expansion is assumed to be of this type.

(a)   Define $f : \mathcal{P}(\mathbf{N}) \rightarrow [0,1)$ so that, for a given subset A of $\mathbf{N}$, $f(A)$ is the number in $[0, 1)$ that can be represented by the decimal expansion having 1 in the nth decimal place if $n \in A$ and 0 in the nth decimal place if $n \notin A$. Show that f is an injection.

(b)   Define $g:[0, 1) \rightarrow \mathcal{P}(\mathbf{N})$ so that, for a given $a \in [0, 1)$ having decimal expansion $0.a_1 a_2 a_3 \ldots$ of the required type (see above), $g(a) = \{a_n \cdot 10^n \mid n \in \mathbf{N}\}$. Show that g is an injection.

(c)   Use the results from (a) and (b) to explain why it now follows that $[0, 1) \approx \mathcal{P}(\mathbf{N})$.

Problem 7.3.32   Use the results from Problems 7.3.31, 7.3.17, and 7.3.9 to explain why $\mathbf{R}$ and $\mathcal{P}(\mathbf{N})$ have the same cardinality.

Knowing now that $\mathbf{N} < \mathcal{P}(\mathbf{N}) \approx \mathbf{R}$, it is reasonable to ask whether there is a set whose cardinality is greater than that of $\mathbf{N}$ but less than that of $\mathbf{R}$ (i.e., strictly between the cardinalities of $\mathbf{N}$ and $\mathbf{R}$). In 1936, Kurt Gödel proved that assuming such a set does not exist (an assumption known as the *Continuum Hypothesis*) will not contradict any of the standard axioms of set theory (some of these standard axioms appeared back in Chapter 2). Then in 1963, Paul Cohen proved that assuming such a set does exist also will not produce any contradictions among the standard set-theoretic axioms. The work of Gödel and Cohen reveals that the usual assumptions mathematicians make about sets are not enough to resolve the question of whether there is a set A for which $\mathbf{N} < A < \mathbf{R}$. From the point of view of logic, the question is *undecidable* within the framework of standard set theory. There are many other interesting foundational questions, often involving either logic or set theory (or both), that some mathematicians attempt to study.

Index